# A Theory of
# Human and Primate
# Evolution

# A Theory of Human and Primate Evolution

Colin P. Groves

*Senior Lecturer*
*Department of Prehistory and Anthropology*
*Australian National University*
*Canberra*

CLARENDON PRESS · OXFORD

1989

Oxford University Press, Walton Street, Oxford OX2 6DP

Oxford New York Toronto
Delhi Bombay Calcutta Madras Karachi
Petaling Jaya Singapore Hong Kong Tokyo
Nairobi Dar es Salaam Cape Town
Melbourne Auckland

and associated companies in
Berlin Ibadan

Oxford is a trade mark of Oxford University Press

Published in the United States
by Oxford University Press, New York

British Library Cataloguing in Publication Data
Groves, Colin P.
A theory of human and primate evolution.
1. Primates. Evolution
I. Title
599.8'0438
ISBN 0–19–857629–3

Library of Congress Cataloging in Publication Data
Groves, Colin P.
A theory of human and primate evolution/Colin P. Groves.
p. cm. Includes indexes.
1. Human evolution. 2. Primates—Evolution. I. Title.
GN281.G77 1989 573.2—dc19 88–18990
ISBN 0–19–857629–3

Typeset by Colset Private Ltd., Singapore
Printed in Great Britain by
Butler and Tanner Ltd, Frome, Somerset

To the memory of
John Napier, teacher, colleague, and friend
and of
Vratja Mazák, colleague and friend
this book is dedicated. I hope they
would approve of it.

# Preface

The ferment that has been going on in evolutionary theory for some fifteen years has largely bypassed the anthropological community. Textbooks on human evolution still present Natural Selection, acting on Mutation, as The Way in which evolution works; anagenesis, the story of human evolution. Punctuated Equilibria may get a paragraph; but neutral evolution, macromutations, speciation theory, internal processes, and so on, seem generally not to rate even, a mention in most works on human evolution. There are, of course, honourable exceptions, but they are few, often timid, always restricted in scope.

Similarly with what may be called the New Taxonomy. Few textbooks, and (surprisingly) few technical publications either, have tried to ask such essential questions as, 'In the light of modern taxonomic thinking, what (if anything) is *Australopithecus*'? Cladistic analysis has, at least, crept into the specialist literature, and there are some very well-handled examples of its use, although as far as human evolution is concerned too many of the cladistic analyses start by assuming the reality of too many taxa, which surely such a study should be trying to test? Again, all too many textbooks ignore this field, and the assumption of anagenesis pre-empts further critical discussion.

I set out to write a book which began by looking at evolutionary and taxonomic theory as they stand today, sifting through the theories and trying to shape them into some consistency, and then applying them to palaeoanthropology to see if they made sense when faced with this new source of data, and whether they in turn shed any new light on the human evolutionary story. I think they do. Cladistic analysis, when used with some caution, sheds considerable light on the interrelationships of fossil protohominids; when we stop racking our brains to think of the selective advantage of a character—from an incisor lingual pillar up to brain size—we open our minds to possibilities that we had never thought of before.

Of all the ideas mulled over in this book, the one that I personally found most enlightening was that of Centrifugal Speciation. It was with some trepidation that I presented my version of it at a symposium in Brisbane in May 1987; to my astonishment it was not damned or thrown out of court; instead one participant after another stood up and said that this exactly corresponded to their findings in such diverse species as egrets and small dasyurid marsupials. Experiences like this have given me confidence that by bringing these ideas out into the open and giving them an airing I am stimulating my colleagues to look at their own data afresh and find that some forgotten theory has something to say to them.

What I have done in this book is trot out old theories, dust them off, examine them in a new context, fit them together, and apply them to humans and other primates. The innovators were Brown, Eldredge, Gould, White, Kimura, Løvtrup, Dutrillaux, Wilson, Sarich, Hershkovitz, Hennig . . . . Most of these people are alive; I hope they approve of what I have made of their theories.

Willi Hennig lived to know he had irrevocably changed the way taxonomy works; Michael White certainly had a considerable impact on speciation theory in his lifetime, and was a profound influence on me during our too-short acquaintance, and I can only hope that he would have looked kindly on the way I have used, or misused, his ideas in this book.

Finally: have I written a textbook? I started out to do so, but it has grown enormously. Introductory it is not; but I believe that advanced undergraduates, and certainly graduate students, may find some value in it. Many textbooks tell the student what to think; here, on the other hand, you will find the evidence laid out (or references given in detail), so that students can examine it themselves. Personally, I think that is what a textbook should do; so perhaps after all I have written one. It is for the reader to judge.

*Canberra, Australia*                                                         C.P.G.
*May* 1988

# Acknowledgements

Nobody but myself is responsible for the theories put forward in this book. I do, none the less, owe an enormous debt of gratitude to many people who, whether they know it or not, have contributed in discussion to my thinking. Some of them may be very surprised to read their names here, but if they read the whole book they will see what I mean. In alphabetical order: Roland Albignac, Peter Andrews, Mike Archer, Peter van Bree, John Campbell, Ken Campbell, Eric Delson, Bernard Dutrillaux, Tim Flannery, Val Geist, Peter Grubb, Vladimir Hanak, Helmut Hemmer, Philip Hershkovitz, John Edwards Hill, Teuku Jacob, Richard and Maeve Leakey, Jerry Lowenstein, Gerard Lucotte, the late Vratja Mazák, George Miklos, Guy Musser, the late John Napier, Prue Napier, Francis Petter, Jean-Jacques Petter, Sartono, Ted Steele, Chris Stringer, Alan Thorne, Elizabeth Vrba, Sherry Washburn, Allen Wilson, the late Michael White, Bernard Wood, Wu Rukang, Wu Xinzhi, Adrienne Zihlmann. Peter Grubb, as well as Pierre Dandelot and Roland Wirth, have at times made delightful travelling companions, very much on the same wavelength, and full of interesting ideas; but above all I owe a debt to my wife, Phyll, who has been with me all the time, measuring, giving opinions and advice, and sharing my thoughts and encouraging them to grow.

# Contents

# 1
# The taxonomy of animals

## 1.1 WHAT IS A SPECIES

### 1.1.1 The meaning of species

Species are defined by one authority (Mayr 1963) as 'actually or potentially interbreeding populations which are reproductively isolated from other such populations': they correspond, crudely, to 'kinds of animals' in the popular parlance.

The meaning and significance of 'reproductively isolated' has been very commonly misunderstood. The requirement is that there should be no gene-flow between two species under natural conditions; i.e. over their ranges in the wild. It is not a requirement that they must not interbreed when brought together in captivity.

Species are kept separate in the wild by 'reproductive isolating mechanisms' (RIMs for short). These may be either pre-mating or post-mating. Pre-mating RIMs include ecological mechanisms (the two species exploit different habitats, and so never, or rarely, happen to meet) and etho-logical mechanisms (they respond to different behavioural cues, especially colour or shape recognition signals or courtship rituals, so do not regard each other as potential mates). Post-mating RIMs come into play if there are no pre-mating ones, or if the latter have failed: some genetic incompatibility between the parent species affects the development of the hybrid offspring, so that it dies *in utero* or fails to reproduce in its turn. Gross differences in karyotype are the most obious sign that a post-mating RIM exists, since in a hybrid with incompatible chromosome sets correct meiotic pairing is nearly or quite impossible.

There has been some argument about how RIMs arise: whether they are positively selected for, or are merely by-products of speciation. Paterson (1982) has argued strongly that they are by-products: post-mating RIMs must be, as there is no way hybrid sterility could actually be conducive to fitness, whereas pre-mating RIMs are part of the specific mate recognition systems without which (in the absence of post-mating RIMs) two species would merge into one very rapidly.

It is the existence of pre-mating RIMs which often misleads even biologists in captive breeding experiments. Commonly, these mechanisms prevent members of different species coming within a certain critical distance for mating; or they direct mate preference so that individuals choose conspecifics in preference to theoretically available mates of different species. In the unnatural restrictions of captivity, critical distances are commonly violated, and the caging together of opposite-sexed pairs of two different species renders mate preferences inoperative. It is surely significant that in cases of occasional interbreeding between 'good' species in the wild just such unnatural conditions can be shown to be operative. Struhsaker (1970) noted a few cases of interbreeding between two related but widely sympatric *Cercopithecus* spp., *C. mona* and *C. pogonias*, in Cameroun: these were in areas where there had been drastic human interference with the natural environment, bringing members of the two species into proximity. Bernstein (1966) recorded two hybrid macaques in a troop of *Macaca fascicularis* in an isolated forest region of Peninsular Malaysia; *M. nemestrina*, the other parent species, had been differentially eliminated from the region, the last male or males having evidently been denied conspecific mates. Aldrich-Blake (1968) recorded a hybrid between two not-very-closely-related species of *Ceropithecus*, *C. mitis* and *C. ascanius*, in a troop of the former in Uganda: the second of these species is here rare, being on the margin of its distribution, making mate preference difficult to operate.

Consequently, if two putative species prove capable of interbreeding in captivity, no conclusions can be drawn about the absence of pre-mating RIMs in the wild. If the resulting hybrids are inviable or sterile, post-mating RIMs are of course demonstrated, and it can be concluded that the two are indeed different species. On the other hand, the survival and fertility of hybrids merely illustrates the absence of post-mating RIMs, and still says nothing about whether pre-mating RIMs operate under natural conditions.

Fooden (1967*b*) described a case of apparent morphological incompatability in the genitalia between two species of *Macaca*, *M. mulatta* and *M. arctoides*. Conceivably this could act as a kind of 'intramating' RIM. The same could be the case for the differences in phallus (especially baculum) morphology between some sibling species.

Operationally, the existence of RIMs is generally easy to infer from museum series, the stock in trade of the traditional taxonomist. If two samples can be distinguished by a number of different characters, such that all specimens can at once be referred to one sample or the other, then the existence of two reproductively isolated species can confidently be inferred (provided, of course, that age and sex differences have been taken into account). The larger the samples, the more securely can chance differentiation be ruled out; the longer the list of differentiating features, the more securely can simple polymorphism be ruled out. It is even, and quite frequently, valid for a specialist to pick out a single specimen from a larger sample and affirm that it is so different, in so many characters, that it cannot possibly be part of the same species. In a most remarkable paper Musser (1969) describes how he was able to recognize the misassignment of a skin in a series of the Murid Rodent *Rattus penitus*, and the similar misallocation of a skull; by examining the collector's field notes he discovered that the skin and the skull belonged together and represented a highly distinctive new taxon, not only a new species but a new genus as well, *Tateomys rhinogradoides*, whose existence he was subsequently (1982) able to demonstrate in its natural habitat.

### 1.1.2 The nomenclature of species

Carl Linnaeus (1707–78) proposed the system of classification and nomenclature that is still in use today. More will be said of this below, but in the meantime it is necessary to describe the rules of species naming. A species has two names. The first is the name of the genus, the grouping to which the species belongs; the second, the name of the species itself. It is a common practice, at least when first referring to the species, to put after the specific name the name of the describer (and name-giver) and the date of the first description. The generic name is always spelled with a capital initial letter, the specific name always with a lower case initial even if the species was named after a person or a country. An example will make this clear.

There are six species of wild cat living in the East African savannah and bush country as follows:

> *Felis silvestris* Schreber, 1777 ('Wild Cat', or Caffre Cat);
> *Felis serval* Schreber, 1776 (Serval);
> *Felis caracal* Schreber, 1776 (Caracal, or Desert Lynx);
> *Panthera leo* (Linnaeus, 1758) (Lion);
> *Panthera pardus* (Linnaeus, 1758) (Leopard, or Panther);
> *Acinonyx jubatus* (Schreber, 1775) (Cheetah).

It so happens that Schreber described four of these species, though in different years, while Linnaeus described the two others. The six species are not all equally related: the lion and leopard are different enough from the smaller cats to be placed in a separate genus, *Panthera* [apart from size, there are differences in the skull, teeth, hyoid bone, structure of the feet, and karyotype; also in the voice (Hemmer 1967)]. The cheetah is so different from all other cats that it is placed in a genus by itself; a genus with only one species is called monotypic.

Note that in some cases the describer's name and date are in parentheses, in others not. Parentheses are used if the describer referred the species in question to a genus different from the one now adopted. Linnaeus and Schreber put all cats in one genus, *Felis* (thus *Felis pardus*, etc.); in 1816 Oken described a new genus, *Panthera*, for the leopard, and most subsequent commentators have regarded this step as warranted, and that the lion belongs here too; and in 1828 Brookes described a further new genus, *Acinonyx*.

Note that the names are treated as if they were Latin, and those specific names which are adjectives (like *jubatus*, meaning 'bearded') have to agree in gender with the generic name, which is always a noun. Schreber originally called the Cheetah *Felis jubata*, as the Latin word *Felis* (a cat) is feminine; but *Acinonyx* (which means 'not-moving-claw', referring to the fact that alone among the cats the claws of the cheetah are not retractile) is masculine, so the -*a* feminine termination becomes the -*us* masculine one when the Brookes genus is recognized.

Note too that a scientific name does not have to be appropriate or descriptively accurate in any way. Cheetahs are often hardly bearded at all and certainly much less so than some other cats, such as male tigers: but *jubatus* cannot be changed for that reason. The Wild Cat is not the only cat to live in forests (*silvestris*), and in many parts of its range does not live in forests at all, but the name has to stand. (*Mustela africana* Cuvier, 1820 is the correct name for a species of weasel thought by its describer to be from Africa but now known to be from South America!)

The objective rules for deciding the correct name of an animal species (or genus) are (1) priority, (2) availability, (3) the type system. The subjective rule is subjective synonymy. The International Code of Zoological Nomenclature (3rd edn, 1985) is the source which lays down these rules:

(a) The earliest name awarded to a species (or genus) is, other things being equal, the one which has to be used for it. (The earliest date a description can bear is 1758, the date of the tenth edition of Linnaeus' *Systema naturae*, which formally encoded the binomial system of nomenclature.) Hermann (1804) gave the name *Felis guttata* to the Cheetah, unaware that Schreber had already described and named the species. Schreber's name has priority and so is the senior synonym; Hermann's is a junior objective synonym.

(b) What may make other things unequal is unavailability. To be available, in the meaning of the Code, a name must have been accompanied by a description, or bibliographic reference to one (if not, it is called a *nomen nudum*); it must not be a homonym (the same as one previously awarded to a species in the same genus); and it must fulfil a number of additional criteria from various dates in the present century (e.g. a name awarded after 1950 must not have been given 'provisionally'). All these factors can overturn a name's priority. Homonymy is a very complex rule.

Primary homonymy exists where the same name was given twice to different species in the same genus. Kerr (1972) gave the name *Viverra maculata* to a newly discovered Australian marsupial, the Tiger-cat (now called *Dasyurus maculatus*): in those days, the marsupials were poorly known and different species were allocated to different mammalian orders along with placental mammals. In 1830, Gray described a new species of genet, a small placental carnivore, also as *Viverra maculata*. Today, neither of these two species remains in the genus *Viverra* (which is restricted to the civets), and one of them is even placed in a different group of mammals altogether, yet the two names '*maculata*' are primary homonyms, and only the earlier of them can be used. So although *Genetta maculata* (Gray, 1830) is the prior name for the species of genet in question, it is a primary homonym so unavailable, and the next earliest name, *Genetta poensis* Waterhouse, 1838, is the correct name.

Secondary homonymy is yet more difficult, because it is a fluctuating concept. The bandicoot rats, common agricultural pests of India, belong to two closely related genera, *Nesokia* Gray, 1842 and *Bandicota* Gray, 1873; some authorities consider them to be congeneric, in which case the name *Nesokia* of course has priority. Gray and Hardwicke (1830) described the Short-tailed Bandicoot Rat as *Arvicola indica*; it was later made the type species of *Nesokia*. Earlier, Bechstein (1800) had given the name *Mus indicus* to the Large Bandicoot Rat, later allotted its own genus *Bandicota*. For those specialists who do not today recognize the genus *Bandicota* but place all bandicoot rats in the genus *Nesokia*, the name *indica* cannot be used for the short-tailed species, and Gray (1837) indeed created a replacement name, *hardwickei*, which must be used by all who believe them both to belong to *Nesokia*. Secondary homonymy here exists because there cannot be two names '*indica*' in the same genus; i.e. *Nesokia*. So for those who believe the two genera to be distinct, the two bandicoot rats will be *Nesokia indica* (Gray and Hardwicke, 1830) and *Bandicota indica* (Bechstein, 1800); for those who combine them in one genus, they will be *Nesokia hardwickei* (Gray, 1837) and *Nesokia indica* (Bechstein, 1800).

Secondary homonymy thus exists where two species, described in different genera under the same specific name, are brought together into the same genus. The junior secondary homonym must now change its name, and the next earliest available name must be used. If, however, the senior homonym is removed from the genus into a separate genus, the homonymy lapses and the junior ex-homonym can be used again.

(c) Every species name has, or should have, a type specimen. This is not, repeat not, a 'typical' specimen. It is simply the specimen to which the name in question is irrevocably bound (NB except by appeal to the rule-makers themselves, i.e. the International Commission on Zoological Nomenclature). If the type specimen is found to have been wrongly assigned to a particular species, the name(s) attached to it are transferred along with it to the correct species, and if necessary a new name must be found for the species it was formerly thought to belong to.

A number of deer specialists suspected, throughout the present century, that there were two different species of muntjac deer (genus *Muntiacus*) in Borneo, one being *M. muntjak* (Zimmermann, 1780), the Red Muntjac, a species widespread in Southeast Asia. Those who believed in the second species called it *Muntiacus pleiharicus* (Kohlbrugge, 1890); most authors, however, did not accept that there were two species. Recently Groves and Grubb (1982) were able to confirm that there really are two species in Borneo;

but the type specimen of *pleiharicus* (which consists of a frontlet and antlers in the Leiden Museum) is actually an example of *M. muntjak*. Therefore *pleiharicus* became a synonym of *M. muntjak*, and the second Bornean species (now properly defined, for the first time) needed a new name, which was consequently supplied: *Muntiacus atherodes* Groves and Grubb, 1982.

There are various categories of type specimen:

(i)   Holotype: designated by the actual person who awarded the species name; or inferentially if only one specimen was available to the describer.

(ii)  Syntypes (or cotypes): specimens made joint type specimens by the describer. This is not really sufficient fixing of a name, as the syntypes may turn out to be a heterogeneous lot, so it is the duty of a future reviser to nominate one of them as the

(iii) Lectotype: in which case the rest of the syntypes become Paralectotypes.

(iv)  Neotype: a specimen chosen to be the type specimen where the holotype (or other category of type) is lost or was never preserved.

A specific name must have a type specimen, even if only an illustration. It cannot be based just on hearsay.

If two names were based, inadvertently one presumes, on the same type specimen they are objective synonyms: while both available, only the senior synonym has a chance of being the correct name of a species. On occasion, when type specimens are untraceable, the same neotype may be designated for more than one name, so making them all objective synonyms.

Consideration of subjective synonymy is outside the purview of nomenclature as such, and part of the realm of taxonomy. The Code of Nomenclature still holds, of course, but decisions as to what names are subjective synonyms are taxonomic ones. The rules of nomenclature are purportedly designed to preserve the 'freedom of taxonomic thought'. Based on comparisons of series of specimens, a taxonomist must decide how many species exist in the group under study; then, by comparing type specimens with the resultant species, the names which are available for each species will become clear. Very often two or more type specimens will fall into the range of a single species; that is when they become subjective synonyms, and according to the Code of Nomenclature the senior available synonym becomes the correct name for the species. The term for the correct name of a taxon is its valid name.

## 1.2 WHAT IS A SUBSPECIES

### 1.2.1 The meaning of subspecies

Subspecies are geographic segments of a species, which differ morphologically to some degree from other such segments.

It is important to grasp the geographic nature of subspecies: they are never, by definition, sympatric. They correspond to some degree to the term 'race'—indeed, this word is often informally used to mean subspecies—although there is dispute whether human 'races' do actually qualify as subspecies.

While it is possible to envisage subspecies that differ only in some aspect of ethology (vocalizations, perhaps), or in karyotype or differential fixation of some enzymic allele, the basis of the concept is morphological differentiation. Traditionally the 75 per cent rule is used: 75 per cent of the individuals classified in one subspecies are distinguishable from 100 per cent of the individuals belonging to the other subspecies of the same species, which is statistically equivalent to 90 per cent joint non-overlap (Mayr *et al.* 1953). Lesser cut-off points are sometimes used, but clearly it is meaningless to award a formal taxonomic designation to a population, or series of populations, of which only half or less of the members are distinctive.

During the first half of the century most species, at least of mammals and birds, were divided up into numerous subspecies by specialists. Specimens from a given locality were found to differ from those from other sampled localities in characters *x*, *y*, and *z*. The White-tailed Deer (*Odocoileus virginianus*) was split into 38 subspecies, within a total range from southern Canada south to Peru; smaller sized mammals were split even more finely. But in 1954 Brown and Wilson proposed to discard the subspecies concept altogether, noting that at any rate in the animals with which they were familiar (insects) geographic variation was neither concordant (i.e. two or more different characters varied independently over a geographic range) nor discrete (i.e. no boundaries could be drawn; geographic variation was clinal). The discussion that followed in the pages of the same journal reached no definite conclusion; in general authors tended to agree that much non-concordant, clinal variation is prevalent but that discrete categorization is possible especially in land vertebrates.

Corbet (1970) is certainly speaking for a majority of mammalogists when he suggests that most described subspecies are really samples taken at points along clines. That these samples may differ from one another quite strongly is less significant than that they are clinally arranged, the characters in which they differ changing smoothly according to geography, either concordantly or not. He proposes to reject all subspecies which are known to be just points on a cline, and to recognize as 'provisional subspecies' those which are suspected but not yet demonstrated to be so. There are on the other hand also 'definitive subspecies'. In some cases they represent isolated populations; in others they are in genetic contact with one or more other subspecies, but along a narrow zone in which are found intermediate individuals assignable to neither but spanning the breadth of variation between the two. The same author (Corbet 1978) considers that in such cases the contact is almost certainly secondary; but, as will be discussed below, it is possible that the pattern of two discrete subspecies with a so-called hybrid swarm between them could arise *in situ* (Endler 1977).

The Lion (*Panthera leo*) is an example of a species which can be genuinely divided into subspecies. One was confined to the Mediterranean slopes of the Atlas in Morocco, Algeria, and Tunisia, but today survives only in captivity. A second subspecies was distributed from the Palestine coast east into northern India; it is extinct except in a national park in Gujarat, North-west India. A third is still fairly common throughout the savanna zones of subsaharan Africa, the southernmost portion excepted. A fourth was confined to the Cape region, and has been totally extinct since about 1850. All these four can be distinguished by characters of size, colouration, skull, and teeth, and by the different shapes of the mane and other hair-tracts of the males; a specialist could probably assign every specimen to one of these subspecies. The Asian and North African subspecies have been, in recent times at least, isolated from one another and from the subsaharan lions by tracts of unsuitable habitat: they fall into Corbet's 'definitive subspecies' category, isolated subtype. The Cape and 'common' African lions probably interbred where their ranges met, although there is no possibility of checking this now; but in that the range of each was discrete, every known specimen from whatever part of the range being equally typical of its respective subspecies, they too were certainly also 'definitive' subspecies, but probably of the intergrading subtype.

(Note that these four are absolutely allopatric; they are not 'actually' but are assumed to be 'potentially' interbreeding and so conspecific. Here is a case where objectivity breaks down: see Section 1.3 below.)

A number of subspecies have been described within the subsaharan region as a whole: one from the Kruger National Park, one from the Kalahari, several from East Africa, two from West Africa, one from Nubia. In modern revisions (Hemmer 1974) some of these have sometimes been recognized as valid subspecies, but it seems likely that they are 'provisional' only, and the evidence tends to suggest that they are points on a cline, and that a full-scale analysis would find that specimens from intermediate localities would show gradients from one to the other.

A species with recognized subspecies is said to be polytypic; if it has none, it is monotypic. It is nonsense to speak of a species having only 'one subspecies': it has none, or it has at least two.

### 1.2.2 The nomenclature of subspecies

A subspecies name is a trinomial: a third word (with a lower case initial) is added on after the species name.

If a species at some point in its taxonomic history comes to be divided into subspecies, then the subspecies occupying the region from which the species was first described, and so occupying the species' type locality, is called the nominotypic subspecies; it has the same type specimen as does the species, and its name is a doubling of the specific name. The other subspecies have their own names, their own type specimens, and their own type localities.

The four lion subspecies mentioned above are as follows:

*Panthera leo leo* (Linnaeus, 1758) (North African Lion);
*Panthera leo persica* (von Meyer, 1826) (Asian Lion);
*Panthera leo senegalensis* (von Meyer, 1826) (Common African Lion);
*Panthera leo melanochaitus* (Hamilton Smith, 1842) (Cape Lion).

The specimens to which Linnaeus had access are said to have come from Constantine, Algeria; this is therefore the type locality of *Felis leo* Linnaeus, 1758 [now *Panthera leo* (Linnaeus, 1758)], and so the subspecies occurring at that locality is the nominotypic subspecies and simply repeats the specific name. It is interesting to note that von Meyer's two names were described by him as 'varieties', a category not recognized now but translated into modern terms as subspecies, while the Cape Lion was originally described as a distinct species of lion—*Leo melanochaitus*. This is in fact a very common occurrence; taxa described as species, especially in the last century, have often been discovered to be no more than subspecies, and their names are retained but as trinomials instead of binomials. Names of species and subspecies are said to be 'names of the species group' and inter-changeable—i.e. the name is kept if downgraded from species to subspecies, or indeed if upgraded from subspecies to species—but they are completely separate from 'names of the genus group' and cannot be interchanged with them. So Hamilton Smith's *Leo* is to be regarded as a generic name in its own right, and not as some kind of an upgrading of the specific name in the Linnaean combination *Felis leo*. (The genus *Leo* is now generally sunk; it is a subjective junior synonym of *Panthera* Oken, 1816.)

## 1.3 THE RELATIONSHIP BETWEEN SPECIES AND SUBSPECIES

The line between species and subspecies is not always clear-cut. There are objective criteria for regarding a taxon as a valid species if it is either (1) sympatric with its relatives or (2) parapatric with no indication of inter-breeding. But there are other cases where the reproductive isolation criterion is less satisfactory:

1. *Cases of total allopatry*. Two taxa that are totally allopatric (i.e. their ranges do not come into contact anywhere) may objectively be classed as different species if they are different in karyotype, or if experimentally attempted interbreeding reveals the presence of a post-mating RIM. The karyotype may not however be known; or captive breeding groups may not be available; or the data may be available but negative; and yet very marked, absolute morphological differences may incline the taxonomist to separate

them at the species level notwithstanding. Conversely, but again if there are not reproductive data or if such data are negative, the degree of difference between two allopatric taxa may be of the same order as geographic differences (whether subspecific or clinal) within one or both of them; and the taxonomist may prefer therefore to class them in one species (see the discussion of the lion data, above). Here, simple 'degree of difference' is being employed as a taxonomic indicator.

2. *Cases of parapatry with restricted interbreeding.* It has been stated above that where the ranges of two discrete, homogeneous taxa meet with the formation of a hybrid swarm between them, there are good grounds for recognizing definitive subspecies. Similar cases may also occur where inter-breeding occurs which is less than total: hybrids are formed, even beyond the $F_1$ generation, but the parental types predominate or at least are commonly encountered. In such cases, the taxonomist often prefers to recognize the parental forms as distinct species. Thaeler (1968) contrasted two such cases with a case which he interpreted as intergrading between two subspecies of a single species. Two cases of hybridization between the pocket gophers *Thomomys bottae* and *T. townsendii* were studied; in both cases the hybrid zones were about 1 mile across, the hybrids accounting for 40 per cent of the population within the hybrid zone in one case, and for 60 per cent in the other. This contrasts with a hybrid zone between two subspecies of *T. bottae* in which the zone was 3–4 miles across: the hybrids formed 45 per cent of the population towards one side of the zone, where one of the parental types was completely absent, and 76 per cent of the population in the middle of the zone, where both parental types could be found, but in very low frequencies. The parental types were frequent in the interspecific example presumably because hybridization was not sufficient to swamp them; their low frequency of occurrence in the second could even, perhaps, be accounted for by simple recombination. The point is none the less established that we have a continuum between specific and subspecific differentiation.

The status of two baboon taxa that form a hybrid zone in the Awash Valley, Ethiopia, has been much discussed. Nagel (1973) regards the parental forms as distinct species, *Papio hamadryas* and *P. anubis*, while Shotake (1981) demotes them to subspecific rank [the prior name, if the two are combined, is *Papio hamadryas* (Linnaeus, 1758)]. The same population is being discussed, admittedly from an ethological and a genetic standpoint respectively; the question, difficult as it is to resolve taxonomically, illustrates the continuum that may occur between the two levels. The term semispecies (a term of evolutionary biology rather than of taxonomy) applies to such cases.

The question must be asked: Is this invariably the case? Do species always start life as subspecies; and, conversely, are (definitive) subspecies destined, if they survive long enough, to achieve species status?

The answer is surely 'no' to both questions. In the second case, Endler (1977) has found mathematically that a hybrid zone can remain indefinitely stable, while Nagel (1973) concludes from old records that the Awash baboon hybrid zone has remained stable for at least a century. The first case is more complex; much of the argument will have to await the discussion on speciation below, but in the meantime a mention of sibling species seems appropriate.

Sibling species are species which differ little, or (in theory) not at all, morphologically. Their existence emphasizes that reproductive isolation, not morphological differentiation, is the essence of the species category, despite the apparent violation of this precept in the discussion on the taxonomy of allopatric forms above. (Even here, however, it is morphological discontinuity, rather than degree of difference as such, that is in question.) Sibling species are found in every group of sexually reproducing animals which are fairly well known taxonomically: first identified in insects such as *Drosophila* and *Anopheles*, they have since been uncovered in abundance among mammals. One of the earliest cases was the disentangling of two remarkably similar species of bats, *Plecotus auritus* and *P. austriacus* (Hanak 1966). While their ranges are still not entirely clear, it is evident that they are sympatric over huge areas of Europe and western Asia, though so far only *P. austriacus* has been identified in North Africa and only *P. auritus* is known east of Lake Baikal. A still more intense similarity exists between *Myotis mystacinus* and *M. brandti* (Gauckler and Kraus 1970), whose distributions once again seem to be nearly coterminous. Both these species pairs were demonstrated to me by Dr Hanak, and I found the *Myotis* differentiation far more difficult (see Fig. 1.1): the two *Plecotus* species became relatively easy to recognize with experience, which perhaps indicates that the more familiar one becomes with sibling species the less 'sibling' they often become.

The literature of the small mammals of Eurasia has recently become replete with sibling species, especially through the work of Soviet karyologists on such genera as *Sorex* (the common shrew). There are, as one might expect, fewer cases as one goes up the scale of size. *Mustela putorius* (Common Polecat) and *M. eversmanni* (Steppe Polecat) are very difficult to distinguish, as are *Martes martes* (Pine Marten) and *M. foina* (Beech Marten), though whether they really rank as sibling species pairs is a matter of definition. Again, however, one notes that the species pairs show extensive sympatry. What all these cases imply is that sympatric speciation has occurred among mammals (a claim which will be dealt with at greater length below, but if it is true then there are abundant instances where species have not arisen as subspecies).

Ayala (1975) examined a number of *Drosophila* samples (of the *D. willistoni* group) electrophoretically, and compared the degree of genetic differentiation to be found between (1) different local populations, (2)

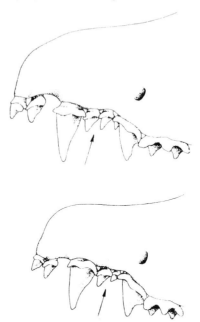

**Fig. 1.1.** Maxillary dentition of *Myotis brandti* (above) and *M. mystacinus* (below). The arrow points to the most conspicuous craniodental distinction between the two species: the comparative sizes of the premolars. From V. Hanak, 1971, *Vestník Československé společnesti zoologické*, **35**, 181.

subspecies, (3) semispecies, (4) sibling species, and (5) morphologically well-differentiated species. Nei's I (genetic identity) averaged 0.970 between local populations of the same species, 0.795 between subspecies, 0.748 between semispecies, 0.563 between sibling species, and 0.352 between morphologically different species. Fish and mammal samples showed comparable values. The ranges of I between categories overlapped very widely, showing that in electrophoretically detectable proteins as in morphological characters species may be less differentiated than subspecies, implying that species do not have to pass through a subspecies stage.

The largest mammal to be differentiated into species that could rank as sibling species is the mouse-deer, *Tragulus*. Two species, *T. javanicus* and *T. napu*, are sympatric throughout Peninsular Malaysia, Sumatra, and Borneo, with one or other occurring in additional outlying areas such as Thailand, West Java, and Palawan. Where they occur together they are easily differentiated by size and colour. But on numerous small islands, notably the Riau Archipelago of Indonesia and the offshore islands of the Malay Peninsula, one or the other occurs, and it is nearly impossible in many cases to tell which of the two is present (Chasen 1940). They are thus siblings where allopatric, but not where sympatric. They may be demonstrating the disputable principle of character displacement (Brown and Wilson 1956), whereby two similar species diverge more strongly where they are sympatric, presumably for adaptive reasons (they have to split the habitat between them rather than each inhabiting the same habitat range); but it is more likely that two initially dissimilar forms are converging where they do not have to share the habitat and so can undergo ecological release.

It is a not uncommon principle, even among large mammals, that different species are more alike where allopatric than where sympatric; in most cases, however, it is less total than in *Tragulus* and restricted in scope (e.g. body size, dental proportions, or pattern), and so more likely to be due to convergence than to character displacement. The instances where sibling species, totally similar but widely sympatric, represent convergence rather than initial divergence are therefore likely to be severely limited.

To sum up, then, there is no qualitative difference between species and subspecies. The sorts of differences one finds between species are the same as those between subspecies, or indeed between morphs: indeed, the very same characters may be involved (Lewontin 1974). Reproductive isolation is the criterion; even here, intermediate degrees are possible.

## 1.4 SPECIATION

How, then, do new species arise? This is really the same as asking how do RIMs arise: because, as we have seen, there is no qualitative morphological distinction between species and subspecies, so it is potentially at least a different question from the origin of new morphological categories.

### 1.4.1 Modes of speciation

By far the most painstaking analysis of speciation (White 1978) lists seven possible modes of new species formation:

1. *Strict allopatry without a population bottleneck.* Two or more large segments of a species simply become isolated by geographic barriers and diverge gradually. It has been argued (below) that gradual change in this manner does not generally accomplish much in evolution, although it is perfectly conceivable that species divergence could eventually thus occur. At any rate, it would seem to correspond to 'form-making in immobilism', the mode of diversification favoured by Croizat (1962).

2. *Strict allopatry with a narrow bottleneck of one component.* This is the 'founder principle' mode, favoured by Eldredge and Gould (1972), Stanley (1977), and others, as discussed in Chapter 2, Section 2.3; unlike the previous mode, it results in only one, not two, new species.

3. *Extinction of intermediate populations in a chain of races.* This is like

the first mode above, except that the populations destined to become separate species are already to some extent differentiated.

4. *Clinal speciation*. In this a local steepening ('step') in a cline might develop as a result of selection against individuals tending to migrate; Endler (1977) calculated mathematically that strong selection pressures on either side of such a step could cause a break and so *in situ* speciation.

5. *Area-effect speciation*. 'Area effects' happen where there are large areas over which strong differences in gene frequencies or gene complexes occur; if these gene complexes are incompatible, there can be strong selection against hybrid formation, so again *in situ* speciation will occur. Paterson (1982) argues that in fact such 'speciation by reinforcement' is a logical impossibility, but specific mate recognition systems (SMRSs) are positively selected for.

6. *Stasipatric speciation*. In this a chromosomal rearrangement occurs within the range of a given species, and spreads because of the reduced fecundity of heterozygotes. This will occur where there are high degrees of inbreeding, so that despite low heterozygote fecundity there are soon sufficient homozygotes to predominate locally. About one individual in 500 (judging by findings in such diverse groups as lilies, grasshoppers, and humans) carries a gross chromosomal mutation; White (1978) suggests that perhaps one in $10^4$ or $10^5$ individuals survive, and that a reduction in heterozygote fecundity of only 5–10 per cent would initiate divergence. Chromosomal rearrangements might even spread by species selection (see Chapter 2, section 2.3) if the new rearrangement protects an adaptive linkage group.

7. *Sympatric speciation*. This is based on the arguments of Maynard Smith (1966): individuals of different genetic constitutions may actively select different habitats; or assortative mating could be a by-product of some other polymorphism, especially if the morphs respond to different habitats.

White (1978) concentrates particularly on the stasipatric model and discusses possible examples of it and the conditions which would promote it. Among mammals, certain types of social structure, especially the harem system, would lead to the inbreeding likely to permit the initial spread of chromosomal rearrangements and their appearance in homozygous form; it could be argued that such a social system would transform the homozygous occurrence of a rearrangement (provided it is at all viable) from a nearly impossible into a probable event, so overcoming the 'Fleeming Jenkin effect' (see Chapter 2, subsection 2.1.2) at least in as far as these chromosomal mutations are concerned. Ripley (1979) has speculated on the ecological conditions likely to lead to the predominance of stasipatric modes of speciation in primates: essentially, coarse-grained environments, which are indeed best exploited by demic clusters of close-knit social groups without surplus males.

Arnason (1977) observed that among the mammals it is the marine groups,

Cetacea and pinniped Carnivora, which would least correspond to the requirements for speciation by chromosomal rearrangement; and these do in fact show the lowest karyotypic diversity among mammals, indicating that non-stasipatric modes alone have operated. Bush *et al.* (1977), taking the other side of the argument, point to equids and some primate groups as being prime candidates for stasipatric speciation, and such appears to have been the case to judge by the extreme karyotypic diversity within the genera *Equus* and *Cercopithecus*. They add that chromosomal rearrangement may, in addition to providing the requisite sterility barrier, produce morphological divergence as well by creating new linkage groups to give supergenes and by producing an altered pattern of gene expression.

In a later survey, White (1979) argues that speciation by chromosomal change, whether sympatric (stasipatric) or parapatric, does not follow the rectangular model (below), in that morphological divergence follows the chromosomal rearrangement, rather than accompanying it. Paterson (1981), however, suggests that any such subsequent divergence must be strictly limited, as there can only be positive selection of SMRSs, not negative selection against hybridization. It must be supposed that most phenotypic change will be a sheer consequence of the chromosomal reorganization, probably simply by giving old genes new neighbours, and so new contexts in which to express themselves (Grant 1971). But it must be admitted that a vast amount of karyotypic change seems to be feasible without any phenotypic consequences whatever.

Only Bickham and Baker (1979) refuse to associate karyotypic change with speciation as such; rather they see repeated chromosomal change occurring phyletically until an 'adaptive plateau' is reached. While such a concept tends to be subjective, and probably undefinable, the value of increasing protection of newly linked gene complexes has been mentioned above. The accumulating evidence for karyotypic diversity at the species level simply will not be denied.

At any rate, sympatric or parapatric speciation, of some kind or another, is no longer to be considered an unusual occurrence. Maynard Smith (1966) argued for its occurrence between ecological races (White's type 7, above), and Bush (1969) recorded an actual example in the parasitoid dipteran *Rhagoletis*.

### 1.4.2 Testing the geography of speciation

Many karyotypically differentiated species are not sympatric, but parapatric, a fact of which White (1978, and especially 1979) is well aware. One of the most remarkable cases is that of the mole-rats (*Spalax*) of Israel, which have four parapatric, chromosomally different sibling species which replace one another from north to south in the Lebanon/Israel region (Wahrman *et al.* 1969). In that they live in small, isolated inbred populations they would, as noted by Bush *et al.* (1977), fall into the category liable to undergo speciation by chromosomal reorganization.

Anderson and Evensen (1978) propose a number of tests of geographic speciation models, using living North American mammals. They compare 130 geminate species pairs (pairs which seem to have separated very recently) with 50 non-geminate but still congeneric pairs, and classify them as being sympatric (A, having considerable overlap; B, some overlap), parapatric (C), or allopatric (D, ranges nearly meeting; E, widely separated). Pairs of species that have more recently separated should have changed their distributions less than those that became separate longer ago. They find that parapatry predominates, accounting for over 40 per cent of cases, between geminate species pairs, with type A, the 'considerable overlap' category, next largest (about a quarter). On the other hand, between non-geminate species pairs, type E, widely separate ranges, predominates, with type-D allopatry the next largest category.

These findings are not entirely expected. Less closely related species seem to move apart geographically, instead of coming into sympatry as most speciation models assume. Still less predictable was the finding that if geminate species are sympatric at all they tend to be very widely so; Anderson and Evensen (1978) do not draw attention to the deficiency in type B, but it is a notable feature of their results.

In an attempt to extend these results I performed the same calculations on (1) African and (2) Palaearctic species, using the data of Meester and Setzer (1971) and Corbet (1978) respectively. In the first instance, the data are probably poorer than the North American data, being a compilation of the opinions of a large number of different authorities, of widely differing taxonomic philosophies; but better in the second instance, being based on the well-considered conclusions of a single author, with the benefit of recent high-quality taxonomic (especially cytotaxonomic) work by Soviet and other European authors. My results, compared with those of Anderson and Evensen (1978), are presented in Table 1.1. [In this table, the North American percentages have been read off from Anderson and Evensen's (1978) figure and so are approximate only; and subtotals of allopatry and sympatry are given because it was not always easy to detect the precise degree of allopatry without maps (in the African case) or with only approximate ones (in the Palaearctic case).]

Initial parapatry seems evident for the African sample as for the North American. In the Palaearctic sample, however, sympatry is the overwhelming condition for geminate species; this could not have been predicted by traditional models of speciation, but supports a sympatric mode: especially, stasipatric speciation seems relevant, a great number of the cases being of near-sibling pairs which are karyotypically distinguishable. Frankly, allopatric geminate pairs are in a distinct minority in all three samples, though there are more of them in the African sample—a finding that may

**Table 1.1**   Geographical relationships between congeneric species pairs in mammals.

| Species origin | % of pairs in each classification | | | | | | | |
|---|---|---|---|---|---|---|---|---|
| | A | B | A + B | C | D | E | D + E | Total |
| *Geminate species pairs* | | | | | | | | |
| N. America[1] | 26 | 15 | 41 | 41 | 12 | 6 | 18 | 130 |
| Paleoarctic | 22 | 30 | 52 | 24 | 6 | 19 | 25 | 81 |
| Africa | 6 | 10 | 16 | 50 | 18 | 17 | 35 | 97 |
| *Non-geminate species pairs* | | | | | | | | |
| N. America[1] | 8 | 18 | 26 | 10 | 20 | 44 | 64 | 50 |
| Palaearctic | 36 | 28 | 64 | 0 | 8 | 28 | 36 | 36 |
| Africa | 31 | 23 | 54 | 5 | 12 | 29 | 41 | 65 |

[1] Read off from Anderson and Evensen's (1978) Fig. 2.

simply reflect the poorer state of taxonomic knowledge, with many supposed geminate pairs being better interpretable as subspecies.

As they diverge, it would seem that North American species tend to drift apart geographically, while in the other two continents they become either allopatric or sympatric. Consistent in the three samples, at any rate, is the decline of parapatry.

Until the appearance of White's (1978) book, the orthodox view of speciation was the peripheral model—variant 2 in White's list, with variants 1 and 3, the other strictly allopatric modes, being also countenanced. This orthodoxy dates largely from the influential work of Mayr (1942, see also 1963), who argued very strongly, and (given the evidence then available) plausibly, that geographic isolation was essential for diversification. Maynard Smith's (1966) arguments for sympatric speciation received little critical attention in this climate of opinion, and another model that failed to make an impact was the centrifugal speciation of Brown (1957).

Brown (1957) pointed out that there is a recurring pattern in the distribution of genera: a more evolved ('derived', in cladistic terminology) taxon is often central, surrounded or flanked by more primitive relatives. He suggested that fluctuations in population density would break up the range of a species, permitting diversification to occur, such changes being greater in the centre of a species' range where genetic diversity is greatest, followed by re-expansion of the (now altered) central population which would tend to overwhelm the old peripheral type. White (1978) criticizes aspects of this model, but agrees that, in modified form, it is the essence of his own stasipatric speciation.

That this state of affairs (more derived forms in the centre, more primitive ones peripherally) is common in taxonomic analysis is one which has taken a long while to be acknowledged since Brown's (1957) analysis. I myself was initially surprised at such a pattern in my own work on the genus *Dendrolagus* (tree-kangaroos), until I realized that it was staring me in the face in countless cases in my own previous work and that of others. Here are just a few instances:

1. The species of *Dendrolagus* exhibit three grades of morphological derivation according to foot structure and other characters (Groves 1982):

(a) The most derived species, *D. dorianus*, has quite un-kangaroo-like short feet and long, powerful arms, a high-crowned skull, and complex secators (the cutting premolars). Its range extends through the highlands of Papua New Guinea west to the Wondiwoi peninsula of Irian Jaya.

(b) The intermediate level is represented by two allopatric species, *D. ursinus* and *D. matschiei*, which together cover much of the range of *D. dorianus*, plus the Huon peninsula (*D. matschiei*) and the Bird's Head of the far west (*D. ursinus*): that is, a central-plus-peripheral distribution.

(c) The three primitive species of the genus have an entirely peripheral range. *D. inustus* is sympatric with the intermediate group in the Bird's Head and along the north coast of Irian Jaya, and is found, in addition, down part of the west coast and on at least one of the offshore islands; *D. bennettianus* and *D. lumholtzi* live in Queensland. It is of interest that *D. bennettianus*, found further north than *D. lumholtzi* and so nearer to the centre of the genus, is somewhat more derived if not in the direction of groups (a) and (b).

The result in the genus is a Christmas-tree-like effect: the most derived species with the smallest, most central range, the intermediate and least-derived groups having successively more peripheral ranges with, in each case, some overlap with the level above (Fig. 1.2). It is as if new species had been generated in the centre of the range and expanded outwards, partly replacing earlier species. This is, in effect, the sympatric geminate species pattern just as described in subsection 1.4.2. (above), with an added element of evolutionary polarity.

2. The lions described in subsection 1.2.1 (above) fit precisely this model (Fig. 1.3). We have a central subspecies (*Panthera leo senegalensis*), surrounded by three peripheral ones (Fig. 1.4). The peripheral ones share primitive features; in the central form the mane is more concentrated around the head pole (so potentially a more effective signal), the cranial capacity is relatively greater, and the social organization is, perhaps predictably, more complex (Hemmer 1974). So we have again derived central and primitive peripheral taxa: not, in this case, distinct species, but subspecies of one species, as if to emphasize that there is no difference in principle between these two taxonomic levels.

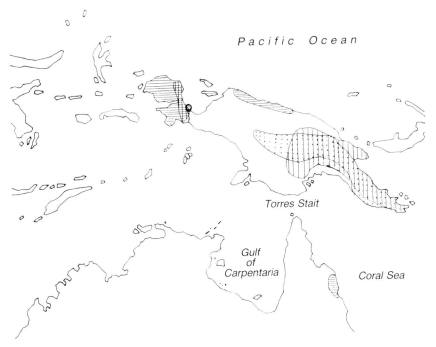

**Fig. 1.2.** Distribution of the species-groups of *Dendrolagus*, to show centrifugal patterns. Horizontal hatching: primitive species *D.inustus* (New Guinea) and *D.lumholtzi* and *D.bennettianus* (Australia). Vertical hatching: 'middle' species *D.ursinus, D.spadix* and *D.matschiei* ( = *goodfellowi*). Stippling: the highly derived species *D.dorianus*. The blank area on the map is more likely to represent a gap in knowledge rather than a gap in distribution.

3. The Gabon–Cameroun region is the centre of the African rain-forest block, today as in the past (Grubb 1982), and here is found both the greatest diversity of mammals and the most derived members of their respective species groups. I will cite primate examples in a later chapter, but will restrict myself to a non-primate case here. The *Cephalophus ogilbyi* group of duikers is distributed over the whole rain-forest belt, from Sierra Leone via Zaire to the Rift Valley in Kenya. The central (Gabon–Cameroun) species, *C. callipygus*, has the most derived colour pattern, hair-tracts, and skull form; other highly derived states are found on the margins of the region, in colour pattern (*C. ogilbyi crusalbum*, of the Gabon coast) and skull form (*C. o. ogilbyi*, of the Cameroun–Nigeria border); while the primitive species of the group are peripherally distributed (*C. brookei*, West Africa; and *C. weynsi*, Zaire to Kenya).

So, why such patterns? Thorne (1981) has drawn attention to the diversity gradient within a continuously distributed species, of high diversity at the

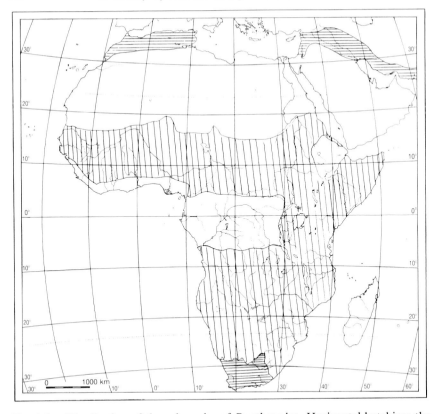

**Fig. 1.3.** Distribution of the subspecies of *Panthera leo*. Horizontal hatching: the primitive small-brained subspecies *P.l.leo, persica* and *melanochaita*. Vertical hatching: the large-brained *P.l.senegalensis*. The distribution represents, as far as can be reconstructed, that of the early nineteenth century.


centre and much reduced diversity at the periphery (the centre-and-edge hypothesis). While care must be taken to distinguish between geographical and ecological marginality (Lewontin 1974; White 1978), there is undeniable validity in this generalization: the centre of a species' range is a generator of genetic diversity. Moreover, as Lewontin (1974) and others have emphasized, the sorts of characters that differentiate species are the same as those subject to polymorphism or polytypism within a species; often enough, the very same characters are involved. This gives rise to the strong presumption that the same processes are involved in each case: all that is needed is a mechanism capable of causing reproductive isolation. Stasipatric speciation would be just such a mechanism.

For the die-hard peripheral speciationist, of course, all this will cut no ice. Ranges will be presumed to have changed, so that species that arose allopatrically will have moved into the centre of the ranges of their less-derived

**Fig. 1.4.** The four subspecies of *Panthera leo*. The distribution of the mane of the adult male differs characteristically. (a), (b), and (c) are the primitive (peripherally-distributed: see Fig. 3) subspecies, in which the mane extends back along the underside; (d) is the derived subspecies in which the mane is concentrated around the head-pole.

(a) *P.l.leo* (North Africa)      (c) *P.l.melanochaita* (South Africa)
(b) *P.l.persica* (Asia)      (d) *P.l.senegalensis* (most of Africa)

relatives. This would not only be far less parsimonious than the centrifugal/stasipatric model, but would be difficult to bring about on the basis of known past climatic changes. For example, the rain-forest blocks of Africa expanded and contracted under different climatic regimes (Grubb 1982): they did not go walkabout and end up interpenetrating each other.

I would see the following arguments as favouring some sympatric, or at least centrifugal, model of speciation, as being at least as common in mammals as the peripheral model:

1. it is geographically parsimonious;

2. karyotype differences between species are the usual state of affairs, offering at least the chance for stasipatric effects to occur;

3. breeding systems are regularly such as to promote wide dissemination of mutations, even to the extent of involving some inbreeding (Wilson *et al.* 1977);

4. only a 'minute fraction' (White 1978) of all individuals in a species, and so a similarly minute fraction of its diversity, is peripherally located;

5. central environments are likely to be more diverse than those at the edge of a species' range, and a new species generated there is more likely to be successful, having been subjected from the moment of its origin to a full range of environmental challenges.

### 1.4.3 Species and subspecies in evolution

It must again be insisted that what can lead to speciation may stop at a lower level of differentiation. Stasipatric quasi-speciation events, i.e. a chromosomal rearrangement, may produce only a new case of chromosomal polymorphism, such as are widespread in mammals, at least: *Sorex araneus* and *Mus musculus* are examples (White 1978). Allopatric, parapatric, and centrifugal differentiation may produce only a new subspecies, with a stable hybrid zone between it and the parent form; or a less distinct form, not deemed worthy of taxonomic differentiation either because absolute differences have not resulted or because of the smoothness of the cline arising between the locus of differentiation and the parent heartland. It is worth reiterating that there is no difference in principle between speciation and such lesser degrees of differentiation.

This being so, it is theoretically possible that an event resulting in full speciation may occur within a species which is already divided into subspecies, or otherwise polytypic or polymorphic. The consequences of this have been analysed by Endrödy-Younga (1980), who refers to it as heteromorph speciation. If the speciation occurs locally, within the parent species' range, the characters predominating in that region will be incorporated into the new species. If speciation is parapatric, for example by intensification of a hybrid barrier, the characters of a former subspecies will become those of the new species. If the parental species is polymorphic, more than one of the coexisting morphs may be inherited by the daughter species. Thus a species may actually be younger than the subspecies of its parent species, or even than the latter's polymorphic variants, and may be more closely related (even by evolutionarily derived character states) to one of its parent's subspecies than to others. Endrödy-Younga illustrates these points by reference to the Tenebrionid Beetle *Anomalipus expansicollis* of the Transvaal/Zimbabwe/Mozambique border region; a second (unnamed) species, recently derived from it but in characters as little differentiated as the latter's subspecies, is sympatric with it in some areas.

A good example from mammals is seen among pocket gophers (*Thomomys*) of western North America. *Thomomys bottae* is a widespread species with a number of subspecies, divided into two informal sections by karyotype differences. Within one of these sections are three subspecies (*T. b. bottae, jacinteus,* and *amitae*), and, in a morphological sense, the distinct

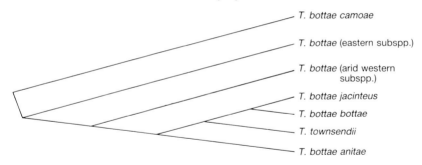

*T. bottae camoae*

*T. bottae* (eastern subspp.)

*T. bottae* (arid western subspp.)

*T. bottae jacinteus*

*T. bottae bottae*

*T. townsendii*

*T. bottae anitae*

**Fig. 1.5.** Phylogeny of *Thomomys bottae* and *T. townsendii* (species of Pocket Gophers), based on electrophoresis of 21 loci. After James L. Patton and Margaret F. Smith, 1981, *Journal of Mammology* **62**, 493–500.

species *T. townsendii*. That the latter is a good species is evident (see section 1.3 above), yet morphologically it is much less differentiated from the *T. b. bottae* group than members of this group are from other subspecies (Fig. 1.5). Thus a new species has apparently arisen from within a cluster of taxa which remain subspecies, and in cladistic terms *T. bottae* is paraphyletic with respect to *T. townsendii*: nor can there be any alternative conclusion in taxonomic terms.

## 1.5 CLASSIFYING THE HIGHER CATEGORIES

The purpose of classifying is to indicate degrees of relationships. The question of how to classify above the species level therefore resolves itself into the question: What do we mean by being 'interrelated'?

### 1.5.1 Cladistic analysis

Until some 20 years ago, this seemed to be no problem, simply because few had asked themselves or others exactly what they thought they were doing in taxonomy. Crowson (1970) has characterized this phase of taxonomic practice the 'Aristotelian essence' phase: practitioners seemed to be aiming at discovering the essence of organisms, classifying them into one group or another because they seemed to be 'essentially' similar or different. That these taxonomists were in reality working without a guiding philosophy at all is in fact a tribute to their competence: somehow, working by feel and experience, they managed to come up with results which often need little or no revision when re-examined today by the elaborate methodologies that have been dreamed up in the meantime. The overwhelming bulk of revisions have been necessitated by just one factor which they considered important, and we do not: evolutionary rate. Of the need to base taxonomy on phylogeny there is no disagreement; and this in turn takes us back to an appreciation of the pre-evolution taxonomists, who did not even have this philosophical prop for their endeavours.

It was Hennig (1966) who first insisted that the only objective meaning that biological 'relationship' could possibly have was recency of common ancestry. Two taxa whose common ancestor lived a million years ago are more closely related to one another than are two taxa whose common ancestor lived one-and-a-quarter million years ago; and that is all there is to it. There may be a feeling among some biologists that one taxon has evolved faster and changed more than its sister group (i.e. taxon descended from the same ancestor), but there is no objective method of quantifying this, and it just cannot be taken into account in erecting a taxonomic structure. There are, it is true, ways of estimating (NOT measuring) amounts of change (see, for example, Cherry *et al.*, 1978; Groves 1986), and these methods tend to indicate that differences in evolutionary rate are usual, not exceptional, rendering it likely that, even were such things as 'amounts of change' eventually able to be measured accurately, and even were it thought permissible to use them in taxonomy, it would be impossible to take them into account in practice.

In theory, the phylogeny of a given group of organisms must be known before a taxonomic scheme can be made out. In practice, there are ways in which the phylogeny can be clearly enough reconstructed to permit taxonomic work to proceed. Out-group comparison (see below), in particular, may allow the character states of the members of the group to be sorted into plesimorph (primitive) and apomorph (derived), so that synapomorph (shared derived) character states emerge. It is these shared derived conditions which alone constitute evidence for relationships. This method of analysis, called cladistic analysis, has achieved such a wide following among taxonomists that it is probably true to say that it is now the usual way of reconstructing phylogeny, and so of beginning a taxonomic revision.

The most logical way to sort out character states into primitive and derived is to examine their condition in one or more taxa which are definitely not members of the group being studied, i.e. an out-group. The out-group should obviously bear some relationship to the group under study, if the condition of its character states is to be of any relevance: thus if we are interested in finding out the polarity (primitive vs. derived condition) of character states among the Hominoidea, in order to work out a phylogeny and make a taxonomic scheme, birds would not seem to be a very suitable out-group; rather, other 'higher primates', the Cercopithecoidea (preferably with the Ceboidea in reserve, as a referee to resolve difficult cases), would be useful.

But how do we know that the out-group is indeed an out-group? If we are to be sure of this, we should first perform another cladistic analysis, using a group that is an out-group to them both; and, to make sure that this latter is indeed an out-group, a further cladistic analysis. . . . We will end up with a very satisfactory classification of living organisms; or perhaps we will have to go even further than that, as it is not certain that all living organisms have a common ancestor ('Rocks as out-groups'—Humphries 1983). At some stage we will have to make an assumption, that we really have identified a genuine out-group; or else use a different method of analysis.

The other method of cladistic analysis commonly used is the ontogenetic method (Nelson 1978). This assumes that the earlier in ontogeny a character appears, the earlier in phylogeny it appeared. Two organisms that diverge only late in their ontogeny are thus deduced to be more closely related than two which begin to diverge at an earlier stage. The assumptions here are, first that nothing has been inserted into an early stage of ontogeny, or withdrawn from it; and second that there have been no phase shifts or truncations. This scheme has been appropriately criticized by de Queiros (1985); a few words will be added here. The first assumption is very frequently flouted: thus Berrill (1955*b*) makes a good case that Tunicate larvae (Phylum Chordata, Subphylum Urochordata) are not pre-existing but have been inserted into the tunicates' ontogeny as an aid to the dispersal of these sessile organisms. Similarly, coelom formation by enterocoeles characterizes the echinoderms and the non-vertebrate chordates, yet the vertebrates themselves—on other grounds obviously related to these other groups—form the coelom as a schizocoele. The vertebrates may be supposed to have altered their embryology here in the interests of developmental economy; but the validity of the ontogenetic method in cladistic analysis is severely challenged. As for phase shifts and so on, these have been so extensively documented by Lovtrup (1974), Gould (1977), and Alberch *et al.* (1979) that the whole idea of relying on ontogeny for cladistic analysis is thrown into disrepute, and it becomes at best a fall-back or check method to out-group analysis.

All this is not to imply that all is well with out-group analysis. Parallelism (see especially Chapter 3, section 3.3) is a problem that simply will not go away, because it is one of the facts of evolution itself. The practising cladist's answer to this is parsimony: William of Occam maintained that it is illogical to multiply one's basic postulates unnecessarily, and this idea, the parsimony principle, has been widely used to choose between competing hypotheses of phylogenetic relatedness. But to William of Occam we have a counterbalance in William of Baskerville, who warned that coincidence is rampant: the most similar appearances may have entirely different origins (Eco 1983). So it is that on occasion the results of applying the parsimony principle are clear-cut; but at times one hypothesis appears almost as good as another. It may be that there are groups (of organisms) which have been characterized as a whole by high rates of parsimony, whereas others have not; if this is so, then even the parsimony principle becomes dangerous and could lead to false conclusions. An example of the latter case is Eaglen's (1983) application of cladistic analysis to lemurs: here even the most parsimonious scheme required no fewer than 10 parallelisms in 30 characters, and was only slightly preferable to the next most parsimonious, which required 10 parallelisms and one reversal. Under such circumstances Eaglen refused to choose any one hypothesis as being 'most probable'.

Why? The concept of evolutionary directionality will be discussed below. Heads (1985) has also raised the question of the validity of the very concept of 'uniform ancestor', or, rather, he has drawn attention to Croizat's (1962) discussions on it.

At this point it would seem most useful to turn to an analysis of the characters themselves; their functional meaning, their expression, their detail, their correlations with other characters. Szalay (1981) condemned many proponents of cladistic analysis for utterly ignoring function: some character states will virtually not function unless they are associated in the same organism with certain others. In a sense, therefore, they will be the same character, which is being counted twice. At other times, a character state may take on a different aspect when analysed in more detail: short-facedness in rhinos, when analysed not in terms of face length relative to some other skull measurement but of positions of various key features (infraorbital foramen, nasal notch), becomes a different character and can be seen to have evolved more than once because facial shortening is differently expressed in different taxa (Groves 1983). All this brings up the old question of 'What is a character?'; a question on which simplistic character-counting must inevitably founder, in cladistic analysis as in the old numerical taxonomy of Sokal and Sneath (1963).

The final refuge, of course, is the fossil record. In theory cladistic analysis enables phylogenetic reconstruction without any reference to fossils at all (if the analysis has as its aim the elucidation of relationships among extant organisms); but the fossils are there, and while it must be agreed that they cannot offer the same scope for comparisons as extant organisms—only hard parts fossilize in the main, and even these are unlikely to be preserved in their entirety—some key parts may be preserved, and these may demonstrate an unpredicted derivation or an unexpected parallelism. Gingerich (1978a, etc.) recommends a method called stratophenic analysis, whereby the fossils are arranged in stratigraphic sequence and attempts are made to link them up in an ancestor descendant relationship, as superior to cladistic analysis; and at times this could be the only way of going about things, especially where there are no convenient living referents, and for all we know a set of *Hyopsodus* molars at one stratigraphic level might indeed have given birth to those at the next. When it comes to extant organisms, and even those fossils well known by a good representation of body parts, stratophenic analysis can be very misleading. For Simons (1972) the middle Miocene primate *Pliopithecus* made a good ancestral gibbon: it had gibbon-like teeth, and it was, as expected in such a stratigraphically early form, more primitive in its cranial and post-cranial anatomy; but Groves (1974b) argued that cladistically the gibbon is a hominoid, sharing with others certain character states which are obviously derived, while the dental similarities with *Pliopithecus* might be primitive retentions: in other words that there was no indication of parallelism involved. Here was a case where cladistic analysis seemed to work, and flatly contradicted stratophenic analysis; subsequent work (Corruccini and Ciochon 1977) has entirely supported the cladistic conclusions.

I conclude, therefore, that cladistic analysis is potentially a very useful tool for working out the relationships of organisms (in fact an essential first step); but it must be used very carefully, with due attention to the meaning of the characters involved, to the problem of parallelism, and to the information which may (or may not!) be yielded by the fossil record. Eaglen (1983, p. 271) put it well: 'Those who feel that [cladistic methods] will unveil all of the obscure events buried in the unrecorded course of evolutionary history are likely to be disappointed. Those who expect such methods to clarify assumptions, to point out gaps in our knowledge . . . will find cladistic parsimony methods well suited to their needs'.

### 1.5.2 Can cladists have ancestors?

It often happens that cladistic analysis of a group, especially if fossils are included, yields one or more branches which appear to have no uniquely derived (autapomorphic) character states at all. In cladograms—the branching trees which are the first stage of cladistic analysis—such a taxon is given a twig to itself, even though this twig has no derived states to support its existence. Those willing to go to the next stage of analysis, the phyletic tree (which strict cladists refuse to do), may well suppress these apomorphless twigs and represent the taxa concerned as ancestors. To the strict cladist this is shocking, and there is (and can be) no evidence for such a placement.

Clearly, there were such creatures as ancestors, and whatever the inherent unlikelihood of such an event, they may sometimes be discovered. In theory, at least, they will be lacking autapomorphic states, though I suppose evolutionary reversals may also occur and deprive an ancestor of a previously acquired autapomorphy. True, we can never actually prove a given taxon to be an ancestor: it is a working hypothesis, but surely a reasonable one, and—like all hypotheses—susceptible to falsification and so scientifically tenable.

### 1.5.3 Are dichotomies inevitable?

The principle that, when a lineage splits, it splits into two, seems to have slipped unnoticed and unchallenged into cladistic theory. There appears little justification for this. Let us imagine a continuously distributed, but clinally varying, widespread species. If allopatric speciation is to occur, the range will have to be broken up—either by some local event (a river changing course, perhaps) or by some more widespread occurrence such as climatic change or a rise in sea level. The local event will, it is true, split the species' range into two; but an increase in aridity will as likely split a forest belt into three or more segments as into two, and a rise in sea level will cut off numerous islands, not just one. A polychotomy will result.

If speciation is centrifugal, the initial split will, of course, be dichotomous. If the new derived species is successful, however, it will soon spread out in all directions, either co-existing with or replacing its primitive parent. If it replaces it, the range of the latter will eventually be split into two—or more. Given that no change has occurred in the meantime in the parent species, the clinal variation that previously existed across it will *ipso facto* result in the remnants of it being differentiated.

A final point needs to be made. Polymorphic variants (allelomorphic variants within a population) are primitive or derived, just as are fixed character states; and the essence of clinal variation is geographic gradients in the frequencies of these primitive and derived states. If a widespread species does become split into three, by a process such as the one envisaged above, then each of the three daughter populations will have its own mix of these morphs. At locus A (in the simplest possible model) the derived allele may increase in frequency from east to west; at locus B the derived allele may increase form west to east. After a split into three, the central isolate will find itself sharing a high frequency of the derived allele at locus A with the western isolate, and a high frequency of the derived allele at locus B with the eastern isolate. There is now a fair chance that, in the vicissitudes of gene frequency fluctuation, the central population may end up sharing one derived state with one of its relatives, and a different one with the other, resulting in a trifurcation that resists breakdown along cladistic lines. If this sounds familiar, it is meant to be (see Chapter 4).

### 1.5.4 Are there primitive species?

In the Hennigian convention, when a species splits (assumed to be into two!) two new species are considered to be formed, the parent species immediately ceasing to exist. This is a taxonomic convention; but is it soundly based in evolutionary theory?

We have seen above that there are many models of speciation in which a small founder population only is the source of a new species, the parent species remaining unchanged by the removal of such a tiny fragment of its gene pool. There is no sense in maintaining, in such cases, a fiction that two daughter species have been formed: only one has. While the aim of a cladistic taxonomy is to record relationships in the form of lineage splitting, it must at the same time keep its roots in the real world and be prepared to admit of lineages budding, rather than splitting. There may be expected, thus, to be surviving primitive species even in a highly speciose lineage; they are the unchanged stem species, and there is no logic in classifying them separately from the actual stem forms if such are known from the fossil record. Indeed, whether one of the daughter species differs significantly from its parent—i.e. has many, or any, autapomorphic features—might be one way of deducing what sort of speciation event occurred.

# 2
# The progress of evolution

Ten or even five years ago a chapter with this title would have explained how slow and gradual evolution is and how it works by natural selection and random mutations and recombination. There would then have been brief descriptions of industrial melanism and the evolution of the horse, before turning to an attempt to apply the same model to the human fossil record. Such a simplistic viewpoint is no longer possible: the ferment in evolutionary biology that has been taking place in the last decade, in one of the most exciting and fast-developing fields in biology, demands consideration. That no real analysis of human evolution, taking these new ideas into account, has been made is surely an anthropological scandal.

## 2.1 NATURAL SELECTION

### 2.1.1 The adaptionist model: the role of natural selection

According to the neo-Darwinian model, the evolutionary synthesis, natural selection acts upon mutations to alter the gene frequencies within a given population, finally bringing the new mutation to fixation. The process is slow and gradual, but it is considered that this is the most important, if not the only, way in which evolution happens. The many small changes due to selection accumulate to produce the major changes that demonstrably occurred during evolution, ultimately producing the present diversity of genotypes, within and between the one-million-plus species of animals and the quarter-million or so species of plants, from a single primaeval organism.

Natural selection is the differential survival and/or differential reproduction of particular genotypes to produce a population or species that is 'fitter', that is to say, better adapted to its environment. In that adaptedness, or fitness, is the quality of having survived and/or reproduced, the concept of natural selection involves a tautology: it is 'survival of the fittest', but what is fittest can be defined only by it having it survived. If we limit the definition of natural selection to 'differential survival', then we avoid tautology but automatically include processes like sexual selection and genetic drift in our definition. There is probably no clear-cut way out of this dilemma except by reference to intuitive concepts that genotypes change to make populations more 'adapted'. Dunbar (1982) also argues the necessity for differentiating adaptation from fitness, and criticizes operational definitions of the former (i.e. in terms of fitness) which would only lead to reinstatement of the tautology; the alternative he offers, defining it in terms of 'some problem set by nature', would seem, however, to place the matter firmly in the realm of subjectivity. While it certainly does make intuitive sense that in polluted English woods, with their soot-blackened trees, the black morph of *Biston betularia* (the Peppered Moth) is better adapted than the normal-coloured morph and that this is the reason why it has progressively replaced the latter in such environments since 1850 (Kettlewell 1955), there is no objective test which will allocate this well-studied case to a category natural selection while refusing admission into this same category to a so-called neutral mutation that in precisely similar fashion progressively replaces its precursor allele by differential survival (see section 2.2. below).

Dice (1947) demonstrated differential survival experimentally, by exposing two colour morphs of the Deer-mouse *Peromyscus maniculatus* to predators (Barn Owls and Long-eared Owls) against different coloured soils. Contrastingly coloured mice were taken in preference to concealingly coloured ones, in an approximately 2 : 1 ratio. Like the Peppered Moth case, this is intuitively a case of natural selection, and has been quantified with an 'index of selection', but contributes no more satisfactorily to a non-tautologous definition of natural selection.

## 2.1.2 The adaptationist model: critique

Darwin's contemporary Fleeming Jenkin pointed out that if one individual in a panmictic population showed a favourable variation while its neighbours did not, the variation would soon be lost by crossing; Darwin considered this to be the best criticism ever made of his work (quoted by Willis 1940). It remains as true today under the hypothesis of particulate inheritance as in Darwin's time (when inheritance was thought to blend) such that a single favourable mutation will have a very poor chance indeed of spreading in the population at large.

Haldane (1932) was well aware of this; and he was painfully aware of the slowness of evolution in general if it had to depend only on natural selection acting upon new mutations, as in the neo-Darwinian theory. 'Haldane's dilemma', that the process of gene substitution must be agonizingly slow otherwise an unbearable gene load would be imposed on the population, has been argued to be a statistical artefact (Weiss and Goodman 1972). The slowness of the process remains true, however: selective values at different loci tend to cancel each other out so that individuals with very high fitnesses are extremely rare (Weiss and Goodman 1972). Again, Haldane appreciated the point: selection involving several genes simultaneously is always slower than for single genes—whether dominant or recessive—and 'evolution must have involved the simultaneous change of many genes, which doubtless accounts for its slowness' (1932, p. 103). No palaeontologist, Haldane was not in a position to investigate whether selection has really been quite as slow as to rely entirely on his maximum calculated rate of adaptive gene substitution of 1 per 300 per generation, let alone simultaneous substitution at several loci, with selection intensity decreasing linearly with the number of loci (Lewontin 1974, p. 309). In addition, Stearns (1977) has noted that the tenets of population genetics, hence of natural selection itself, are limited to the single locus case.

Johnson (1976), well aware of such constraints, proposes that in effect genes are selected not individually but rather in linkage groups whose very existence is, in selective terms, no coincidence; certainly such a model goes a long way to restoring to selection its evolutionary role.

Pleiotropy may also reduce selection pressures; or may cause unexpected phenotypes to be fixed as a consequence of selection for some gene's most powerful effect. Deol (1970) describes some of the quite unanticipated influences on size, behaviour, and so on that are pleiotropic effects of colour genes in mice. Gould and Lewontin (1979) list a variety of other constraints on a purely adaptationist scenario. Organisms, they point out, are the raw material of evolution: not genes. Thus an organism cannot be, as they put it, 'atomized into traits' but operates as a whole, imposing limits to such improvements in fitness as may be promised by new genes; and the genes in any case function amid the strong constraints of epigenesis and phyletic history. The multifactorial basis of many phenotypic characters reduces the intensity of selection possible on each locus. Gould and Lewontin make Fleeming Jenkin's point about the tremendous odds against the incorporation of an isolated new mutation. The strongest selection, they suggest, will in fact favour genes that increase fecundity, hence an individual's relative contribution to the next generation, irrespective of traditional criteria of adaptiveness: this claim seems more dubious, at least for K-selected groups such as large mammals.

Many supposed adaptations may in fact be related to environment or habit in a *post hoc* fashion, i.e. serve a useful function despite not having been selected for it. Such features have recently been characterized as 'exaptations' (Gould and Vrba 1982). Equally significant is Lovtrup's (1974) observation that an organism of its very nature is adaptable, and that the essence of natural selection is to maximize this adaptability.

A much emphasized role for natural selection is 'balanced polymorphism', in which opposing selection pressures maintain alternative alleles together in a single population. It seems likely that the role of balanced polymorphism has been overestimated: not only is there much too much polymorphism in most studied populations for every case to be accounted for in this way (Lewontin 1974), but the effect of such selection pressures will be to reduce still further the effect of selection pressures tending towards evolutionary change. Many of his and Ayala's (1975) arguments for the 'balance' theory of natural gene pools could be explained by appeal to Mayr's (1942) 'co-adapted gene complexes' or Johnson's (1976) theory of linkage groups: balance remains, but the number of independent selection pressures involved falls back to realistic levels.

A final point to make is that, as anyone with field experience knows, animals do not merely live in an environment and await mutations that may make them better adapted to it: to a surprising extent they select their own niches and their own habitats. It is surely not such an outrageous claim to make that animals, rather than being adapted to an environment because they live in it, may live in an environment because they are adapted to it. This feeling has been growing among biologists lately. One aspect of the problem is, do related sympatric species evolve mechanisms to enable them to avoid competition, or are they, rather, enabled to coexist because their niches are different? Lee and Cockburn (1985) argue very strongly for the latter view as far as marsupials are concerned, and Richard (1985) seems to opt, if not so firmly, for the same conclusion in primates.

## 2.2 NEUTRAL MUTATION THEORY

The final stumbling block to a purely adaptationist model is 'neutral mutation theory'. While relatively uncontroversial when applied, under the name genetic drift, to small populations in which random effects would, it was agreed, override all but the strongest selection pressures, its application to large populations as a normal part of their evolutionary change (Sarich 1967; Kimura 1968) was generally accepted with reservations if at all.

In the form of the molecular clock, neutral mutation theory has gradually worn down opposition and come into general acceptance. The question of whether lineage splitting times as calculated from the molecular clock do or do not agree with the fossil record cannot be resolved at the moment, the record itself being in many cases based on much shakier foundations than many molecular biologists are aware or many palaeontologists have cared to admit, at least until recently! Conclusive evidence that the molecular clock does keep reasonable time, at least approximately, is its internal consistency. If in a theoretical phylogeny, elucidated by some method independent of the fossil record or the molecular clock, threre are three higher taxa (genera, perhaps, or families) A, B, and C, such that B and C form a common stem separate from A, then members of groups B and C should, if substitution rates are constant, show approximately equal molecular distances from any member of group A. Again, any member of group B should be more or less equidistant from any member of group C. In well-studied examples such as the primates, this is overwhelmingly found to be the case (Sarich 1967). Cases where there appears to be a conflict between the 'clock' and the fossil record, as in the splitting time of *Ateles* from its nearest relatives (Setoguchi 1985), may well turn out on further evidence to be a consequence of unpredicted homoplasy, or some other form of misinterpretation, as was a more notorious previous case (Andrews and Cronin 1982).

That the molecular clock works in the primates is of considerable significance, because different primate taxa have very different generation times: some five years for most monkeys (less than this in small New World monkeys, such as marmosets), about 10 years for apes, 20 years for humans. Yet the clock is internally consistent: there has been no slowing down in the rate of substitution since the splitting off of the hominoid (ape plus human) line, nor even since the splitting off of the human line from the latter [see Goodman *et al.* (1982) for a contrary claim, that rates have varied at different times: this does not however affect the argument about generation lengths]. Selection pressures operate per generation: were natural selection responsible for the changes in the proteins studied by the molecular clock, there would be a noticeable slow-down as generations lengthened. That no slow-down occurs confirms that substitution rate reflects some purely time-dependent factor—probably, as its proponents claim, mutation rate. If new mutations have an equal probability of being incorporated, they are effectively neutral.

The characters on which the molecular clock depends are invariably blood proteins: albumin, transferrins, cytochromes, and so on. Different proteins do change at different rates (Wilson *et al.* 1977), but the rate of change of each has proved constant through as wide a range of taxa as may have been investigated, even if not constant over time (Goodman *et al.* 1982: but see Kimura 1981). The question arises whether the proteins of other systems might also change at statistically regular rates. A study by Romero-Herrera *et al.* (1973, 1976) suggests that this might not be so for myoglobin. Differential rates of protein and 'organismal' change (Cherry *et al.* 1978), indeed the very existence of so-called 'living fossils', show that not all evolutionary change is regular; but it is still not excluded that plenty of other proteins may none the less experience effectively neutral mutations and so change, at least inter-mittently, at regular rates.

## 2.3 MACROEVOLUTION

The idea that the major features of evolution may not, after all, be the consequence of a simple accumulation of small-scale changes has recently been revived. As I hope to show, the concept is acceptable but incomplete.

### 2.3.1 The peripheral population model

Under adaptationist and neutral mutation models alike, the pace of evolutionary change is slow, even, and gradual. That so few cases of one major taxon emerging from another could be demonstrated in the fossil record—predicted to be quite a common finding under any gradualist theory—was ascribed to the incompleteness of the record. Eldredge and Gould (1972) were the first in modern times to challenge this view. As palaeontologists, they were aware that in some instances at least the fossil record is after all not really that incomplete, and that there ought to be many more lineage splits and origins of major taxa documented in their entirety (see, however, Behrensmeyer 1982). Instead, sustained unidirectional change of any kind is rarely shown. They therefore proposed that major change, usually involving splitting of lineages, happens very rapidly and punctuates what are essentially long periods of stasis or equilibrium. They thus erected a new model, the 'punctuated equilibrium model', to replace the traditional gradualism model. As indeed they point out, their theory is really a synthesis of the genetic drift of Wright (1968), the founder principle of Mayr (1963), and especially the hopeful monster model of Goldschmidt (1940).

Contributions to the punctuationist model have been made by Eldredge and Cracraft (1980), Stanley (1979), Vrba (1980), and others. The basic event of evolution is seen as lineage splitting, designated speciation. Typically, a peripheral population becomes isolated, allowing the operation of founder effect (atypical individuals disproportionately represented in the new isolate) and release of constraints on variation under conditions of relaxed intra-specific competition, throwing up new character combinations, one of which may, by genetic drift, become the norm. The radically different morph, which has been generated relatively rapidly [Stanley's (1979) rectangular evolution], is by its very nature reproductively isolated from the parent species. By range extension, made possible perhaps by removal of the isolating barrier, it then competes with the parent species and will either (1) prove competitively inferior and become extinct (2) prove competitively superior and replace the parent species, (3) be able to coexist with the parent species, or (4) be unable to spread but persist where it was formed. The contribution of selection to this process is restricted to an interspecific role, when the newly generated species is 'tested' in competition with its parent, and perhaps with other congeners, a process Stanley calls species selection.

In this model of the evolutionary process, especially as conceived by Stanley, evolutionary change is principally dependent on speciation. There is gradual phyletic change, i.e. without lineage splitting, but it plays a very minor role: macroevolution (i.e. evolution above the species level) is essentially 'decoupled' (Eldredge and Cracraft 1980) from microevolution (what goes on within populations). The direction of speciation is random; hence extinction of newly formed species is the rule, and the probability of their long-term survival is very low. Only if a species lives under ecological conditions which favour the continual isolation of peripheral demes will the rate of generation of new species be high enough to have much chance that a successful one will arise.

The existence of long-term trends in evolution is hard to deny, but to explain them has been a challenge to successive generations of evolutionary biologists. The evolution of the Equidae from polydactyly to monodactyly, from brachyodonty to hypsodonty, and from bunodont to lophodont conditions was ascribed by Osborn (1936) to orthogenesis, which he saw as a kind of internal urge to progress always in the same direction (not what the originators of the term intended: Craw 1985). Simpson (1951) found that the process was not constant and uniform but proceeded at different rates over time, and proposed that what operated was simple orthoselection, that is to say, natural selection favouring change fairly consistently in one direction as long as local environments favoured it. For Stanley (1979), species selection is a sufficient explanation: from a given equid stock new species will be generated by rectangular evolution whenever conditions allow: some are more, some less lophodont that their parent stock (and, clearly, some will be equivalent in this character to the parent, differing in quite other characters), but only those which tended to greater lophodonty (and/or hypsodonty, oligodactyly, etc.) would have an enhanced prospect of survival in competition with the parent.

Vrba (1980) has a somewhat different explanation which she dubs the 'effect hypothesis'. The conditions which favour high speciation rates include stenotopy: the propensity to live in narrowly defined, hence patchily distributed, already semi-isolated environments. 'Specialists' tend therefore to speciate more (increaser clades) than 'generalists' (survivor clades); of those successful products of speciation from specialist lineages, those which are the more specialized will again tend to speciate more. Hence there is a kind of 'speciation pressure' which becomes more and more marked as more and more stenotopic types are generated. Thus hysodont equids speciate more than brachyodont, and a resulting pattern is one of evolutionary trend. Vrba's illustration of this is a bovid lineage: the generalist antelope *Aepyceros* (the impala) remained almost the same, producing a mere two to three new species, over six million years, while its sister group, the more narrowly specialized Alcelaphini (hartebeest/gnu group), in the same period of time generated 27 species. (One may object that surely stenotopic types will become extinct much more often, increaser clades notwithstanding, but the argument does appear to work in some cases).

Paradoxically, high mobility of individuals also favours speciation, in Vrba's view, because animals that move far more frequently encounter new environments, and so have greater potential to form isolates. Taxa which live in small demes, with tightly organized social structure, have the same high probability of forming small isolates and so of speciating.

It seems undeniable that stasis (if in this term one includes short-term excursions about a mean value) is the usual condition of populations as shown in the fossil record. A higher rate of change must have preceded the static phase in order to produce it (Vrba 1980).

A number of authors have attempted to test the punctuationist and gradualist models on the fossil record. Gingerich (1977*a*) defended the gradualist view with reference to the condylarth *Hyopsodus* and the primate *Pelycodus* (both Eocene), and suggested that a spurious pattern of punctuated change could emerge if stratigraphic sampling were too coarse, while gaps in the fossil record do undeniably exist; and the characters used to define a species could alter the perception of whether change occurred or not along series, a point also made by Levinton and Simon (1980). A rejoinder to the point about coarse sampling might be that it could have just the opposite effect: three successive samples of a species, sampled too far apart stratigraphically, might appear to show an intraspecific evolutionary trend when a speciation event had in fact intervened between each pair. A renewed attempt to analyse evolutionary modes from the same data (Bookstein *et al.* 1978) proposes a further source of error, namely that a sustained pattern of emigration and immigration can give a false impression of stasis: to which again an appropriate rejoinder would be that gradually changing proportions of two closely similar species, mistakenly analysed as a conspecific sample, could mimic gradualism, as could fluctuating geographic clines. Bookstein *et al.* (1978) go on to analyse the *Pelycodus* and *Hypsodus* data, finding that the interpretation which best fits the data admits of 12 cases of gradualism, four of punctuation, and one of equilibrium. A detailed critique of their findings would be out of place here; suffice it to say that their initial decisions as to how to cut up the data seem not always beyond criticism: there are big gaps in the scatters of measurements at many levels, not all of which do they pick out as anomalous; any conclusions would depend very much on how their lineages for analysis are built up, and on the 'species' being non-overlapping in their dental measurements. When all is said and done, both the quality of the material (only dental metrics for the most part) and the fineness of the stratigraphy leave much to be desired: that these are two of the best-represented terrestrial sequences from the Eocene says much about the quality of the fossil record and suggests that except in very exceptional cases adequate sampling is likely to come mainly from more recent sequences in which the fine stratigraphy is more readily accessible.

Such sequences have recently been analysed (see also subsection 2.3.2, below). Williamson (1981) sampled 19 mollusc species lineages from different levels of the Koobi Fora Formation, and reported detailed evidence for rectangular evolution. In the best-studied lineage, that of the species *Bellamya unicolor*, there was long-term stasis through the sequence, with three marked episodes of change, each resulting in the genesis of a new species. Just below the level of the Suregei Tuff, where one of these new species occurred, was an intermediate population with a remarkable release of variability: the total variance was five-times that of the mean for the species *B. unicolor*. In what is identified as the novel species, the variance was only twice as great as the mean for *B. unicolor*, though still well outside the range of variances for that species at different levels.

In fact, all 19 lineages were transformed at the Suregei Tuff level, generally with a noteworthy increase in variance, occurring however at different stages in different lineages. Above the Tuff, in each case there was a sudden return to the original morphology. Later in time, at a point in the Lower Member of the Koobi Fora Formation, several of the lineages showed a different kind of change: new species entered the record and coexisted for a while with their parent(?) species, but soon vanished again. Higher up still, in the Guomde Formation, a speciation event similar to the Suregei one appeared to occur to all lineages, but the unconformity between this and the preceding Koobi Fora Formation has destroyed all documentation of any intermediates that might have existed.

Williamson (1981) considers that these stratigraphically sudden changes correspond enxcellently to the predictions of the rectangular model of speciation, especially as they correspond to falls in the level of Lake Turkana which would have isolated the molluscs from conspecific populations elsewhere. [He notes also that the extinction of every one of the new species immediately following Lake level rises (with the consequent competition with immigrant conspecifics) conforms to the species selection model of Stanley (1979).] The increase in variance reflects developmental disruption, and the survival of the variants in the intermediate stages reflects reduced competition. But the necessary intervention of founder effect, a key stage in the rectangular model, is refuted by the Turkana mollusc data. The changes occur over a large area of the Lake, 'in thick faunal units of millions of individuals' (p. 442); there is also the curious circumstance that one of the species is asexual, yet shows exactly the same changes as the others.

Several other questions are raised by Williamson's (1981) study. If gene pool reduction and cessation of gene flow are not indicated, yet rectangular speciation occurred, what was the mechanism? Clearly large numbers of individuals changed at once, at first chaotically, then converging in each lineage on a single new morphology. The new conditions—lower Lake level, and hence higher salinity—might indeed have produced very strong selection pressures: hardly support for a gradualist model, but resurrecting natural selection as an agent of speciation and refuting the 'decoupling' felt to be necessary by Eldredge and Cracraft (1980). Yet the stage of the new morphology in the lowest level was demonstrably preceded by a marked increase in variability promoted by developmental instability: we do not know whether it was an increase in genetic variance or just an ontogenetic destabilization caused by a higher salinity, but either way a relaxation-of-competition model fails, as population sizes are not thought to have changed.

In fact, as Williamson (1981) notes, the changes were not excessively rapid. The best-documented change, at the Suregei Tuff level, took between 5000 and 50 000 years to occur: although the exact length of time is uncertain, because of the uncertainty of estimation of deposition rates, we at least have boundary rates, and because of the nature of the intermediate stages any question of immigration from outside seems to be ruled out. This estimate of the time involved, especially if the upper limit is the correct one, calls into question for some commentators (Maynard Smith 1981) whether there is really anything at all new and in need of a special explanation about the punctuationist model. As we shall see, however, gradualist explanations will certainly not suffice.

A further query is the precise nature of the transition in the lower member of the Koobi Fora Formation. Here, in several lineages, but not this time in all, novel forms suddenly enter the record and coexist for a while with the normal species. As the stratigraphic sampling, on the evidence of the Suregei documentation, seems ideal, the absence of documentation of change at the second level is extremely puzzling.

The final question raised by the study is the precise meaning of 'peripheral'. The point could have been anticipated by Vrba's (1980) comments on the nature of populations likely to undergo speciation; but the Turkana mollusc study emphasizes that in the middle of the range of a species a transition can still occur, and in extremely large populations, whether they do or do not reproduce by exchanging genes. Which is exactly what we concluded at the end of section 1.4 in Chapter 1.

## 2.3.2 Macroevolution without lineage splitting?

Stanley (1979) dismisses the possibility of a concerted change in a whole species as being theoretically possible but in practice likely to be rare: phyletic speciation, for him, is almost entirely gradualistic. For a sudden change to occur in a whole species, bottlenecking would be necessary, and such an event is likely to lead to extinction rather than to genetic reorganization and renewed expansion.

There are, none the less, species which do periodically undergo population contraction and re-expansion, at least at the demic level: it is true that this is not species bottlenecking, but it is far from the peripheral population model (see also previous subsection). Carson (1975) has founded a hypothesis of speciation on just such an eventuality: the founder–flush speciation theory, in which the genetically random survival effects of a population crash leave an atypical remnant which then undergoes a new expansion in which high survival rates, for several generations, permit disorganization of epistatic 'supergene' systems and the organization, by natural selection, of new such systems, protected by such mechanisms as chromosome inversions. This hypothesis closely resembles the 'punctuation of an equilibrium' as in Stanley's (1979) model but differs from it in a number of respects: it does not necessarily take place in a peripheral population, but generally in a central one; it hypothesizes certain sequelae to events which are known to occur quite ordinarily, but to certain types of species only; and it attempts to explain in cytogenetic terms the necessity for a 'punctuation' under such conditions, an aspect always left unsatisfyingly vague in the model as recounted by palaeontologists.

Populations subject to regular founder–flush cycles are best known among temperate and arctic rodents, especially of the genera *Lemmus* (Arctic lemming) and *Microtus* (short-tailed voles). If there is indeed a higher probability of genetic reorganization due to such vast demographic fluctuation, these genera should evolve very rapidly in major jumps. Even if a major reorganization occurs only once in every hundred cycles, a species such as *Lemmus lemmus*, the Norway Lemming, with a reputed cycle length of four years, should experience a spciation event every 400 years. The genus *Lemmus* contains only four species, and their allopatric distributional pattern would seem to argue against founder–flush speciation having occurred even once. That is not to say that genetic change of any kind might not be expected as a consequence of the population cycles, but the evidence suggests that the changes are minor only; e.g. Tamarin and Krebs (1969) were able to demonstrate changes at the transferrin locus in populations of *Microtus ochrogaster* and *M. pennsylvanicus*.

Population cycles occur also among insects, and Powell (1978) attempted to mimic them experimentally in *Drosophila pseudoobscura* and track the genetic changes by examination of the giant (polytene) salivary gland chromosomes. No such changes were detected after four founder–flush cycles with a variety of different manipulative programmes. What did occur was positive assortative mating when crossing between different lines was attempted; that is, an incipient pre-mating RIM. The experiment may thus be said to have produced an approach to speciation, but in quite a different fashion from that intended by the founder–flush theory.

Perhaps the best and most suggestive documentation of a natural founder–flush cycle was made over half a century ago (Ford and Ford 1930). Isolates of the strongly demic butterfly *Euphydryas* (formerly *Melitaea*) *aurinia* undergo unexplained catastrophic declines at unknown intervals of time, followed by relatively sudden reappearances. The fortunes of a colony near Carlisle, northern England, could be followed from the late-nineteenth century. Abundant during the early 1890s, the colony steadily declined from 1897 until 1912, by which time it had practically disappeared (due, the authors suggested, to parasitic infestation). It remained low until 1920, after which it increased again, approaching its former abundance by about 1924, after which the rate of increase slowed.

In March 1920, 16 larvae were taken and hatched by the authors: the imagines were all unusually dark in colour, and all were more or less deformed. Up to 1924, malformations persisted but progressively reduced; as the population expanded an extraordinary range of other, harmless variations flourished, in size, marking, colour tone, and intensity. Eventually, as maximum abundance was reached, stability re-emerged, but the new norm was markedly different from the old, being darker in colour with brighter straw-coloured blotches in both sexes, and larger-sized markings than previously. The interesting comment was made that the new morph in some respects resembled the Irish form of the species, ranked as a separate subspecies.

As predicted in both Carson's and Stanley's models (apparently formulated without knowledge of Ford and Ford's study), a population crash was succeeded by a rapid increase chracterized by extreme morphological variation, even to the extent of the survival of deformed individuals: relaxed competition seems the most obvious explanation of this phenomenon [see, however, the commentary on Williamson's (1981) findings, above]. The morph that finally became the norm was noticeably different from the previous norm, again as predicted in Carson's and Stanley's models. But there was no major genetic reorganization: the result was demonstrably the same species; indeed it had converged upon an already existing macro-geographic variant.

It is to my mind impossible to overstate the value of the Ford and Ford study: well-documented at all stages, clearly yet succinctly described, it should have been cited as a major prop of any punctuationist model of evolution. That it has not been can perhaps be ascribed to three implications which could be unwelcome to the strict punctuationist: there was no speciation; the population that changed was not geographically peripheral; and the result was not diversification but parallelism.

A recent experimental analysis of a founder–flush cycle, this time in a laboratory population of *Drosophila mercatorum*, is that of Templeton (1979). Here, following the resumption of bisexual breeding by a partheno-genetic strain, a minor genetic revolution did indeed occur: but in epistatically linked functionally adaptive loci, leaving electrophoretic variants untouched. Templeton dubs this process 'genetic transilience', and argues that it applies widely to any sufficiently polymorphic population subjected to a founder effect; and that it is to be distinguished from 'chromosomal transilience', which will occur only where the parent population was already strongly divided into demes [see also Templeton (1980a, 1980b), where the genetics of the process are explained in detail].

The evidence, then, indicates that the 'founder effect' of a population crash followed by the maximal survival of the flush phase do indeed, taken jointly, offer unusual opportunities for species genotype alteration. It is possible, even probable, that very occasionally these alterations may be so major as to be in the nature of *de facto* speciation; it is also possible that peripheral populations have a somewhat greater chance of undergoing such an experience than central ones; but neither postulate is a necessary let alone a sufficient description of such an event. It is also possible (Petry 1982) that a whole species may undergo such a bottlenecking; in such a context, it would be interesting to compare 'before' and 'after' specimens of taxa which have undergone such a process in recent times, i.e. have been threatened with extinction and have since recovered under the influence of careful conservation measures. [The Yellowstone population of the American Bison (*Bos bison bison*), the Southern White Rhinoceros (*Ceratotherium simum simum*), or the Javan Rhinoceros (*Rhinoceros sondaicus*) would seem to be excellent candidates for such a study; and material would certainly be available for it in each case.]

The fossil record does none the less seem to record at least one well-documented case of bottlenecking: indeed, double-bottlenecking. McDonald (1981) finds that American Bison suffered a crash in population numbers after about 11 000 BP, and the numbers were not restored until after 9000 BP (Fig. 2.1); during this period there were high frequencies of abnormalities, and in the later stages (between 9400 and 7850 BP) there was noticeable

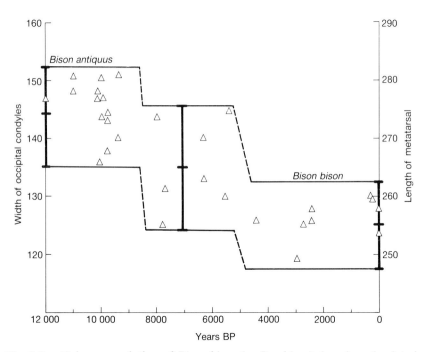

**Fig. 2.1.** Holocene evolution of *Bison bison* ( = *Bos bison*), based on the data in McDonald (1981). Cranial and Postcranial measurements are generally not available for specimens from the same subfossil sites, so a method was devised to plot them both on the same graph in order to increase the size of the data set: utilizing the evidence of living samples, and of those few fossil sites where both sets of measurements were available, an approximate match was made between the two. The resulting graph plotted against time shows two sharp size-reduction episodes; in between, a short-lived intervening form (*B.antiquus occidentalis* of McDonald) appears to have been subject to evolutionary stasis while it lasted, as has *B. bison bison* since c. 5000 B.P., and as was *B.a.antiquus* in the late Pleistocene.

dwarfing in some characters such as the width across the occipital condyles. The bison of the post-recovery phase are ranked by McDonald as a chrono-subspecies, *Bison antiquus occidentalis*, consistently different from its Pleistocene precursor *Bison antiquus antiquus* in several respects, though incompletely stabilized. At the end of the sixth millenium BP there was a further population decline accompanied by a drastic range reduction, resulting in the evolution of the still smaller (and, especially, shorter horned) modern American Bison, *Bison bison*. Although even in this case the data are still not as complete as they might be, the evidence implicates the northern Great Plains—the Alberta/Saskatchewan/North Dakota border area—as the locus of the last survivors and hence of the transition to the new taxon. The latest site containing *B. a. occidentalis* is Zap, North Dakota, dated at 5440 BP; the earliest containing *B. bison* is Head-Smashed-In, Alberta, dated at 5080 BP.

The taxonomic interpretation is, as McDonald (1981) admits, a matter of personal preference: I have earlier indicated (Groves 1981) that bison are best regarded as belonging to the genus *Bos*, and throughout their million or so years of separate evoluton there seems no need for more than a single morphospecies, and there seems not to have been sympatry between any two bison taxa at any time (McDonald 1981). Be that as it may, there is no doubt of the major morphological transitions undergone over very short periods of time, attended in at least one instance by a release of variability and in both instances by a founder–flush sequence.

Phyletic bottlenecking is quite likely to be rare, as Stanley (1979) suggests [and indeed McDonald (1981) ascribes both bison declines to overhunting by Palaeoindians]; and the evidence of the fossil record, at least of large mammals, does lead to the conclusion that a high rate of change in time is, in the great majority of cases where it can be studied, accompanied by high rates of diversification (as emphasized by Vrba 1980). The conclusion of the present section must be, however, that this diversification (1) need not be actual speciation and (2) need not be strictly allopatric.

## 2.4 DIVERSIFICATION AND ITS TAXONOMIC CONSEQUENCES

It has been seen (Chapter 1, section 1.3) that between specifically and sub-specifically distinct populations there is no sharp break, but a continuum, in terms of both genetic differentiation and reproductive isolation. The same, indeed, is true between populations that are subspecifically distinct and those that are not taxonomically recognizable in any sense. This being so, it is hard to understand why the rectangular model is invariably phrased in terms of speciation alone.

The evidence cited in section 2.3 (above) supports the implication of this, namely that the genotype changes that accompany population bottlenecking and subsequent expansion may yield a new species, a new subspecies, a somewhat differentiated consubspecific population, or no change at all. Any change that occurs is indeed sudden and rapid; presumably the amount of change expected will be strongly skewed towards the low end, so that only species with a high propensity to undergo demic bottlenecking would do so frequently enough to bud off an occasional species.

The literature abounds with cases of rapid evolutionary change in small populations, in the form of species introduced (whether deliberately or accidentally) into new environments. In some cases a special subspecific allocation for the newly differentiated population has been proposed, but never, to my knowledge, more than this.

The rabbits (*Oryctolagus cuniculus*) of mainland Australia are thought to have originated mainly or entirely from a small founder stock released near Geelong, Victoria, in 1859. Taylor *et al.* (1977) examined the degree to which Australian rabbits in general have diverged from British ones, and regional populations have diverged from those persisting near the release site. Within the southeast, populations diverged morphologically from the latter according to distance; those inhabiting semi-arid regions had not necessarily diverged further than those in well-watered areas, nor did altitude affect biological distance. All Australian rabbits differ in a consistent direction from British rabbits, but the populations near Canberra most closely approached the latter. From the major differences existing between even closely neighbouring populations it seems evident that deme differentiation is well-marked, so that a certain degree of bottlenecking might be expected to occur as the species spread from site to site; in this respect the apparent reconvergence of Canberra rabbits on the original parent stock might be taken as support for the concept that the direction of change is not entirely directional, although it seems not to be random either.

Western Australia's rabbits were also studied, but the results (surprisingly little change from the Geelong population) are difficult to interpret: it is uncertain whether they derive from a natural spread across the arid Nullarbor Plain, from introductions by human visitors, or from a separate release altogether.

Ashton (1960) studied the differentiation of the Green Monkey (*Cercopithecus aethiops sabaeus*) on the West Indian island of St Kitts, where they have been wild for about 300 years (i.e. at most, 70–100 generations), from the parent population in West Africa. The monkeys of the daughter population are an average 4 per cent larger than those of the parent stock, but this overall size difference is not uniform, being concentrated in the jaws and teeth which on average are more than 6 per cent larger. The variances for most measurements are considerably less than those of the West African stock, but bilateral variance is actually greater, and certain dental malpositions are commoner: paradoxical findings not readily explicable by the theoretical models described above but recalling McDonald's (1981) characterization of the intermediate (*occidentalis*) phase of American Bison evolution.

McCluskey *et al.* (1974) observe that the evolutionary rates indicated in both the rabbit and the Green Monkey cases are extraordinarily high: 116 darwins in the former case, 160 darwins in the latter (where a darwin is a change of 1/1000th part in 1000 years). These rates are 5000–10 000-times those indicated according to a gradualist interpretation of dental evolution in the Equidae over 5–16 million years.

Stanley (1979) cites a number of cases of the apparent origin of new species of fish in Holocene times. The most remarkable is probably that of five species of *Haplochromis* which appear to have evolved in Lake Nabugabo, Uganda, in the 4000 years since it was formed from a bay of Lake Victoria which was isolated by the extension of a sandbar across its entrance.

Carson (1976) discussed the endemic *Drosophila* species of the Hawaiian Islands, commenting on the rapid speciation that must have taken place to produce such a plethora on this young archipelago. In particular, the main island of Hawaii is only 700 000 years old, so that the two sympatric (non-sibling) species of the *D. planitibia* group living there are younger than this. Reconstructing the relationships of the endemic species by cladistic analysis, and placing maximum ages on the lineage splits by reference to the known ages of the islands, he found a good fit to the values of Nei's genetic difference. Apart from offering support to the time-related divergence postulate of neutral mutation theory, the calculations find excellent evidence of rapid and high speciation rates.

Perhaps the best-known studies on rapid evolution are those of Berry and his colleagues on populations of the house mouse, genus *Mus*, accidentally introduced outside the natural range of the genus. [It should be noted that some of these new populations are derived from *Mus musculus*, others from *Mus domesticus*, and that it was not always possible to distinguish which of the two was the parent as they have only recently been adequately differentiated and diagnosed (Marshall 1981), none the less, many valuable results still emerged from the studies.] The mice on Macquarie and Marion Islands, on the edge of the Antarctic Circle, are both very small compared to northerly mice (Berry *et al.* 1978*a*). Those of the Faeroes, Danish-owned islands at 62°N, derive from introductions by Vikings in the ninth century AD, but the two islands containing the most divergent mice were said to be still mouse-free as late as 1800; relationships among the islands are complex, and conclusions derived from genetic and different types of biometric data are inconsistent (Berry *et al.* 1978*b*). Mice were probably introduced to Skokholm, a small island off the coast of South Wales, UK, in 1894, and have diverged dramatically since then (Berry and Jakobson 1975).

It is uncertain how much of the biometric size differences between these new insular mice and the parent populations is purely genetic; but the existence of proportional differences, as well as the genetic frequency differences, indicates that major genetic reorganization has occurred. Even within the Faeroes, whose mice plausibly derive from a single introduction (at the oldest settlement, Thorshavn, on Streymoy Island), great divergence has occurred as they have spread from island to island; there also is a strong suggestion of *in situ* evolution, as the Esterase $2^c$ gene is fixed on two islands but is quite absent on Streymoy, the island from which all other islands were surely populated. Variances are little or no lower than those of mainland populations, and heterozygosity is reduced on five of the six mouse-inhabited Faeroes but in none of the other new island mouse populations.

The data from introduced populations, then, support the spirit, if not the letter, of the founder–flush model, and so of the rectangular model of divergence. Taken together with the comments in previous sections, we may restate the model of the progress of evolution as follows:

1. Change in genetically complex features is most rapid when a small isolated population (founder effect) undergoes a rapid increase (flush effect); but the first of these phases may not be necessary, i.e. large populations may change *en masse*.

2. The second phase of this process can often be seen to involve a remarkable release of variability, even to the extent of allowing the survival of deformed individuals.

3. The populations involved in the process may or may not be peripheral to the range of the species as a whole.

4. The result may be the formation of a new species, a new subspecies, or a morphology which while new is not deemed sufficient for taxonomic distinction.

5. Whole-species bottlenecking, if and when it occurs, presumably has the same consequences, but is probably less common than isolation of particular demes.

6. Gradual evolution does occur but has been documented only in single-gene characters, which arise, spread, and become fixed stochastically.

7. Gradual change and punctuated equilibria may be difficult to differentiate in the fossil record; it may depend simply on the fineness of sampling.

## 2.5 DIRECTIONALITY IN EVOLUTION

So far I have written as if evolutionary change were unpredictable; as if there were nothing in principle to prevent the genesis of a lineage of horned tigers, or elephant-trunked apes. It is evident to most biologists, of course, that constraints exist: tigers are more likely to evolve nastier teeth and claws than horns, while manipulative ability in apes would probably continue to be concentrated in the anatomy of the hand. But does it go further than this? Is there some direction in which a given evolutionary line is likely, even forced, to go?

Grehan and Ainsworth (1985) have recently raised this point again, calling it orthogenesis (and pointing out that this, the reappearance of the same characters again and again in related lineages, is the original and correct usage of this term, not the curious way in which Osborn (1936), for example, misused it).

The widely accepted theory of metachromism (Hershkovitz (1968, 1970b) is precisely orthogenesis. In this model, primitive colouration in any given mammal lineage is agouti; i.e. each hair having alternating eumelanic (black or brown) and phaeomelanic (red or yellow) bands; evolutionary changes involve gradual saturation (reduction and eventual elimination of one or other type of banding), then bleaching (paling of the remaining colour, until white is reached). Other evolutionary changes involve the diversification of the colour pattern: contrasts between tones on different parts of the body, forehead bands, throat collars, and so on. That such changes are invariably one-way has been hard for many people to swallow, and it may be that they are not in every case: but the concept is in full agreement with what is known of coat colour genetics, and independant evidence, where available, tends to support evolutionary scenarios constructed on the basis of metachromatic pathways.

### 2.5.1 Non-random change in karyotype

If minimum evolutionary pathways are deduced for primate karyotypes, as represented by different modes of banding, it emerges that some of the major primate groups have diversified predominately by one particular mode (Dutrillaux *et al.* 1981). By counting up the changes in Dutrillaux *et al.*'s Fig. 4, p. 184, the modes of change documented in Table 2.1 can be seen to occur within three of the groups (superfamilies). That is, within each of the three superfamilies there is a particular type of mutation which has occurred most often: in lemurs, Robertsonian translocation; in Old World monkeys,

**Table 2.1** Modes of chromosomal change within three primate super-familiar (From Dutrillaux *et al.* 1981).

| Superfamily | Robertsonian translocation | Other inter chromosomal | Pericentric inversion | Other intra chromosomal | Fission |
|---|---|---|---|---|---|
| Lemuroidea | 44 | 1 | 10 | — | — |
| Cercopithecoidea | 2 | 5 | 21 | 3 | 24 |
| Hominoidea | — | 4 | 21 | 5 | — |

fission; in hominoids, pericentric inversion. In the first and third cases the 'preferred modes' account for 70 per cent or more of the changes. In the cercopithecoid case the predominance is not so marked; but of non-fission events eight have occurred in combination with a fission.

If chromosomal mutations occurring naturally in new-born human babies are counted, translocations are commonest, followed by pericentric inversions; but, remarkably, many of these abnormal human chromosomes resemble their (normal) non-human homologues, mostly by 'reverting' to a deduced ancestral state still occurring in some other hominoid, but occasionally actually converging on the more derived state of some other species! Considering that random inversions would have virtually zero probability of resembling other species' normal chromosomes, it is evident that the possibilities of rearrangement are actually very limited indeed, and that those which occur as mutations are very often those which have actually occurred during the evolution of related species, or back-mutations to ancestral forms. *In vitro* irradiation of human lymphocytes leads to the same conclusions (Dutrillaux *et al.* 1981).

The postulate of predetermined mutation is consistent with findings on other species too. Irradiation *in vitro* of cells from the Old World monkey *Cercopithecus cephus* produces mainly fissions, both daughter products having active centromeres; in one case, the two daughters so produced resembled their equivalents in *Cercopithecus aethiops*.

It could be, of course, that many more mutations occur than are evident in these studies, some of which base their results on survivors of the mutation process. In part, we may therefore be dealing with directionality of selection rather than of mutation, although the observations of natural selection and indeed mutations imply that this is not all there is to it. The orthodox hypothesis that the direction of mutations is random is in any case violated. Similar studies, after phylogenetic reconstruction of chromosome changes, in other animal groups, should be very revealing.

### 2.5.2 Directional change within the chromosome

Brace (1967) noted that a simple mutation, involving a change in a single base-pair, tends to result in the failure to function of the enzyme that is the product of the gene in which it occurs; and so the partial reduction of the structure which depends on it. If not particularly disadvantageous, the mutation will persist; and such mutations, leading to structural reduction, will tend to accumulate in the population's gene pool. It is as if there is a pressure of mutations tending to reduce organs in the organism; thus structures of no vital function will be reduced, even eliminated, by the accumulation of mutations alone. This model he called the 'probable mutation effect'. The general tenor of evolutionary thought in the 1960s was opposed to the concept of neutral mutations and functionless organs, but in the present climate the probable mutation effect is overdue for revival, and if non-selectionist models are in countenance (as is argued here) the logic of the model becomes compelling.

More recently, attention has turned to non-genic DNA, especially the highly repeated sequences. Singer (1982) has reviewed knowledge of this still somewhat enigmatic portion of the genome. Some highly repeated sequences, known as satellites, are tandemly repeated; others are interspersed among gene-coding DNA sequences. The interspersed type is divided into a *Drosophila*-type, with families of units of several thousand base-pairs in length, separated by tens of thousands of base-pairs of single-copy DNA; and a *Xenopus*-type, with families of units of several hundred base-pairs in length, separated by at most a few thousand base-pairs of single-copy DNA. Most mammals have *Xenopus*-type, divided in turn into two classes, short and long. The satellite type at least is always associated with heterochromatin. The repeats, which are not always absolutely identical, seem to be derived by amplification of a sequence, followed in many cases by transposition to a different site, which is why some sequences can be interspersed in different parts of the genome. Sequences of related species are alike, but not identical, and their differences are characteristic and consistent, which is unexpected, because theoretically duplicates once formed would seem to be free to vary within a species and would be inherited by new species.

The function of repeated sequences would seem to be involved in gene expression, i.e. in timing of transcription and probably also of replication. This in itself would have interesting implications, stemming from the late replication of heterochromatin and its tendency to underreplicate. From this, Krassilov (1980) has speculated that environmental stimuli—perhaps climate, in some cases season, and certainly the age of the individual organism—may induce underreplication to the extent that some sequences may become extinct, carrying with them any associated neighbouring genes, or joining genes formerly separated by repeated sequences; and the amounts of different heterochromatic sequences could affect the timing of transcription of their associated genes, resulting in heterochronies (see Chapter 4, section 4.1). There is in fact some evidence that such processes do take place, at least within a single organism: Krassilov instances the seasonality of chiasma frequency in geckos (highly heterochromatic regions rarely have chiasmata), the absence in the lymphocytes themselves of the intercalary sequences found in embryonic cells which separate the genes for the constant and variable regions of the immunoglobulin molecule, and a documented tendency to underreplication of repeated sequences with increasing generations of mitosis.

Krassilov's (1980) theory is intriguing, and if supported by future investigations would have powerful implications for evolutionary progress. The idea that parental age might be implicated in the production of heterochronies is especially thought-provoking, implying as it does a greater tendency in this direction in long-lived organisms such as large mammals.

The concept of 'molecular drive' (Dover 1982) has already achieved considerable notoriety. This term designates processes leading towards the fixation of repeated DNA sequences (or, potentially, of single genes); but Jeffreys, quoted by Lewin (1982, p. 553), describes it as 'a catch-all for anything that changes in the genome', and indeed in some aspects molecular drive does appear to be used to cover processes that are essentially a form of mutation. There are two stochastic and two directional processes involved, as follows:

(a) *Stochastic processes*

(1) Unequal chromatid exchange at meiosis, creating a duplicate on one chromatid and a corresponding deletion on the other.

(2) Gene conversion, or non-reciprocal recombination, where a sequence is duplicated at the recipient site but not lost from the donor site.

(b) *Directional processes*

(1) Duplicate transposition. Some DNA sequences have an especially high propensity to duplicate and their copies to move, so that such sequences will spread rapidly throughout the chromosome complement, and hence through the population.

(2) Biased gene conversion. Mutants may occur as much as a hundred-times more commonly in one direction than in the other. This will enormously speed up the rate of fixation of a mutant under selection.

Of the two stochastic processes, gene conversion escapes being a synonym for 'point mutation' only by its asymmetry, while unequal chromatid exchange is a similar principle; both lead essentially to the duplication side by side of a sequence, whether repeated sequence or gene. Ohno (1970) had previously drawn attention to the significance of such an event, in that one of the duplicates is presumably now free to vary leaving the other, in theory, to perform the original function: the primate (including human) haemoglobin $\alpha$- and $\beta$-chain loci are thought to have separated in this way. Of the two directional processes, the second (biased gene conversion) is again nothing new (geneticists are well aware of the tendency of mutations at homologous loci in different species to occur in the same direction, and of the rarity of back-mutations); what is new is the analysis of the effect of this on evolution. Duplicative transposition is, similarly, an attempt to place a (much less familiar) cytogenetic process in an evolutionary context: transposable DNA has become a major topic of speculation and bemusement in recent years (Doolittle and Sapienza 1980; Orgel and Crick 1980), and Dover's (1982) insight has begun to make some sense of it.

Taken together, the different elements of molecular drive (especially the non-random processes) help to explain the previously rather puzzling consistency of species differences in repeated sequences. Dover (1982) also points out the manner in which, in a sexual population, chromosomes can be thought of as belonging to a single 'chromosome pool': they are randomly reassorted both within the genome (at meiosis) and within the population (at zygote formation), and that in species with rapid generation turnover, so that the second process is rapid with respect to the first, a new variant or duplicated sequence will be 'driven' through the population in a manner that involves no increase in variance of differences in fitness. While this effect is speculative, a mechanism for a concerted change in large numbers of individuals has been sought by evolutionary biologists ever since the evolutionary speculations of Berg (1924: see English edition, 1969), and seems to be needed again in the Williamson (1981) model of speciation in the Lake Turkana molluscs (subsection 2.3.1 above).

### 2.5.3 Genetic hitchhiking

Hedrick (1982) draws attention to another condition under which an allele can be pushed in a non-random fashion towards fixation. If there is association between alleles at two loci (gametic disequilibrium), selection acting upon one will result automatically in an increase in the frequency of the other as well. In a sexually reproducing population, such gametic disequilibrium can be sustained—if not by some functional association—by linkage or by inbreeding. The former is the mechanism emphasized by Hedrick, but in any theory of evolution in which a founder population is envisaged as having especial importance the role of inbreeding cannot be ignored; the prevalence of selection under such conditions can of course be queried, as it can under a stationary population regime where a gene is envisaged to hitchhike to high frequency by linkage with an advantageous partner.

## 2.6 THE INHERITANCE OF ACQUIRED CHARACTERS?

On the pedestal of Lamarck's statue, in the Jardin des Plantes in Paris, are written the words of his daughter: 'On va vous venger, mon père; la posterité vous admirera' ('You will be avenged, my father; posterity will admire you'). And indeed in the present century there have been a number of findings which do tend, at least in muted or attenuated form, to vindicate him. The suicide in 1926 of Paul Kammerer laid Lamarck's ghost to rest for a while, but as argued by Koestler (1971) something does remain to be explained by Kammerer's findings amid the fakery by person or persons unknown; and the 'Dauermodifikations' (changes induced in an individual which persist for generations) of Jollos (1913) and the canalization/genetic assimilation model of Waddington (1957) have tended to remain an embarrassment. But because the idea of the inheritance of acquired characters remains the ultimate heresy, the most obvious explanation has been ignored.

Lately, however, Steele (1979) drew attention to experiments performed in the early 1970s in which somatically selected idiotypes in male mice were inherited, at least in the $F_1$ generation. The same worker reported with a colleague (Gorczynski and Steele 1981) a more elaborate version of the same experiment, in which this time two different tolerances were transmitted at least as far as the $F_2$ generation, being inherited independently. While Brent *et al.* (1981) failed to obtain positive results when trying to duplicate the experiment, the original findings are not dismissed and add to the growing unease with what is known as 'Weismann's doctrine', that the germ line is insulated from what happens in the soma, and further data, including a review of Brent *et al.*'s experiments (Steele *et al.* 1983), seems to support this challenge to orthodoxy. E.J. Steele (personal communication) continues to work on the problem, but because his work is still in progress and his interim results are not yet published, no further details will be given here, except to note that the findings are remarkable and positive.

As suggested by Steele (1979), the possibility of exceptions to Weismann's doctrine is strongly implied by the 'provirus hypothesis' of Temin (1976), in which the 'central dogma of molecular biology'—that the DNA to RNA to protein transcription sequence is irreversible—is also flouted. A provirus is integrated into the host genome after certain virus infections, notably infection of chicken cells by (RNA-based) rous sarcoma virus, and acts as the blueprint for the manufacture of progeny virus. The means of integration is the enzyme reverse transcriptase which transcribes the virus's RNA code into DNA for incorporation as part of the host's DNA. (A complement to this model is the 'protovirus hypothesis', that RNA viruses evolved in the first place from normal cellular components.) Proviruses, especially in connection with transposable genetic elements (both being, effectively, mobile DNA), have recently been reviewed by Temin and Engels (1984).

The provirus hypothesis has been taken further by Beneviste and Todaro (1982) who propose that such virogenes can actually transfer between different host species, in the form of infective viral particles (retroviruses). The close relationship of certain virogenes of Old World monkeys to those of the genus *Felis*, but of no other Felidae, suggests that such a retrovirus was pased from an Old World monkey to the ancestral *Felis* and has persisted in the genomes of living species (where it is found at 10 to 50 copies per genome). The authors cite further cases, and suggest in addition that retroviruses may carry other host genes with them and so 'maintain a species in contact with its ecologic neighbours' (p. 1202): an idea with quite shockingly revolutionary implications if it is valid.

If an RNA virus can infect a different species and incorporate its instructions into the latter's genome, carrying host genes with it, then it seems inescapable that it can do the same not only between individuals of the same species but within a single individual. Steele's (1979) model proposes that the somatic mutants (possibly non-stochastic), which in Macfarlane Burnet's theory (Burnet 1940) are the source of specific antibodies, are 'captured' (probably in the form of mRNA) by RNA viruses, which then infect ova or sperm and synthesize DNA copies to be inserted into the genome. Such a process would add to the directional push of mutants (discussed in section 2.5 above) and could account for the apparent adaptedness of newly arisen genetic types; Steele instances the very high rates that have been calculated for HbA to HbS mutations in highly malarial environments. Could even behavioural repertoires, he asks, become incorporated into the genome in this way?

If all this is revolutionary, a still more recent proposal at the same time 'tames' and explicates the cytogenetic advances, and extends them yet further. Campbell (1982) categorizes gene activities into five 'levels', exclusive of level 0 which would be the classical Mendelian/neo-Darwinist gene, whose replication and transcription are automatic and uninfluenced by external agencies, and whose mutations are random. At level I are those genes which are susceptible to modification by gene-processing enzymes; like all higher levels they are transposable, and susceptible of influence in their expression by their neighbours. At level II are self-referent genes which actually code for the enzymes which process them. At level III are those genes that code for self-processing enzymes, and which are actually sensitive to external stimuli so that their basic sequence transcribes only under suitable environmental conditions. At level IV the code for the 'detector' apparatus is itself modified by environmental stimuli, altering the gene's future expressivity. For all these four levels Campbell (1982) cites examples; genes for antibodies are the best-known example of level IV. In level V are those genes whose structural alterations in somatic cells can be incorporated into the germ line, as in the Gorczynski/Steele experiment: he believes that the number of such genes will be found to be 'susbstantial'. Finally, Campbell postulates a level VI, an admittedly speculative category of genes which might perhaps change in anticipation: while this sounds outrageous, he draws attention to various modes of sophisticated information-processing which might render this category plausible.

## 2.7 THE NECESSITY FOR DIRECTIONAL MECHANISMS

Any taxonomist, any specialist on a given group of organisms, who has tried to list characters, sort them into primitive and derived states, and conduct a cladistic analysis of the group in question will be aware of the prevalence of parallelism. My own work on the Bovini (Groves 1981) and Rhinocerotidae (Groves 1983) and, especially, on the Hominoidea (Groves 1986) can be instanced. Competing hypotheses of the sequence of lineage splitting are sometimes almost equally good; and whichever hypothesis is ultimately selected as being most probable because most parsimonious has to acknowledge that parallel evolution in different lineages renders the solution more uncertain than would have been hoped. Thus in the hominoid case the conclusion that the gorilla split first from a common human/chimpanzee stem is based on the common possession of 25 presumably derived character states by human and chimpanzee compared to 'only' 12 in common to human and gorilla, seven to gorilla and chimpanzee. Considering that a trifurcation is preferred by the molecular clock school, a very large number of parallelisms must have been accumulated in a short enough space of time to render an apparent trichotomy irresolvable; while if the trifurcation is real *all three* sets of common character states must either have evolved in parallel—or be a relic of polymorphism in the common ancestor (Heads 1985).

Eldredge and Cracraft (1980) recommend that the whole concept of parallelism be rejected as indistinguishable from convergence, and believe its inclusion would render cladistic analysis unworkable without bringing in a weighting scheme. In this they have been appropriately criticized by Hull (1970). The usefulness of the cladistic method is not threatened by the acknowledgement that its results sometimes require interpretation of a nature different from that envisaged at the start; indeed one of the same authors (see Eldredge and Tattersall 1975 p. 231) had previously labelled parallelism 'perhaps the most profound generalisation yet to emerge' from cladistic analysis. While agreeing that there are instances in which the applicability of parallelism as opposed to convergence is a difficult question to resolve, unequivocal cases of parallel evolution are characterized by a duplication of characters that is remarkable in its exactitude. The concept of homologous mutations has proved consistent, to say the least, with observed facts of polymorphism in mammals, notably in the case of coat colour where what appear to be the same loci, with the same range of alleles, can be identified in the most diverse orders of mammals (Searle 1968). Such a finding, so surprising at first sight, becomes less so when the extensive homology of the karyotype among diverse eutherians is shown by different banding techniques (Dutrillaux *et al.* 1980).

Recently, too, the term orthogenesis has been revived (Grehan and Ainsworth 1985): not in its widely misunderstood, mystical sense, but in its original sense of evolutionary constraints, what Darwin called the 'laws of growth'. Under Stanley's (1979) model of species selection, a limitless array of mutations (and recombinations?) contributes to the differentiation of a very large number of new proto-species, and only those with an adaptively superior combination survive in competition with their parent species. If such events are truly random, the chance of the same one recurring even twice must be vanishingly small, while the probability of such a recurrence in a combination favourable enough to survive as a successful new species must be essentially zero. Directional control of mutation and recombination is a necessary concept if successful speciation (or quasi-speciation, as the rectangular process should be known) is to occur at all; as are the epigenetic constraints under the heading of orthogenesis.

Whether part of the secret of directionality should be ascribed to consistent tendencies to underreplication of repetitive DNA with consequent heterochrony, or to the genetic assimilation of somatic innovations, or even to the transmission to one species of adaptations preciously acquired by another in the same area—these are still in the realm of speculation as to their importance, if no longer as to their occurrence itself, and are perhaps the most important questions remaining to be asked by evolutionary biologists.

# 3
# Epigenesis and evolution

Ever since Haeckel the processes of ontogeny and phylogeny have been associated in biology, either causally in one direction or another, or more simply by ontogeny setting limits to what can happen in phylogeny. The present revival of interest in developmental shifts as evolutionary mechanisms is due largely to the insights of Gould (1977), although a beginning had been made previously by Lovtrup (1974).

## 3.1 HETEROCHRONY

Gould (1977) pointed out that growth, maturation, and development frequently become dissociated, resulting in what he calls 'recapitulation' and 'paedomorphosis'. In the former case, for which Alberch *et al.* (1979) have recently coined the more satisfactory term 'peramorphosis', a pre-adult stage of a descendant resembles the adult of an ancestor, a concept arrived at a century ago but given the status of a (spurious) universal law by Haeckel. In the case of paedomorphosis the opposite applies: the adult descendant resembles a pre-adult stage of its ancestor.

Gould points out, however, that the processes by which such end results are potentially achieved can vary radically. Thus peramorphosis is achieved either by acceleration of somatic development, or by retardation of the reproductive system allowing prolongation of growth (hypermorphosis). Similarly paedomorphosis is achieved either by acceleration of reproductive development so that growth is truncated (progenesis), or by retardation of somatic development (neoteny).

Peramorphosis by acceleration, Haeckel's orginal model of recapitulation, would involve the addition of new stages at the end of an ontogeny into which ancestral morphs are concertinaed. The process may occur, but I know of no clear-cut examples of it.

Peramorphosis by prolongation, or hypermorphosis, certainly does exist. At a stage when the ancestor would have been maturing sexually, the descendant is still immature, and a simple prolongation of the same growth trajectory until maturity is finally achieved produces a previously non-existent morphology. Hypermorphosis need not in fact involve a delay in sexual maturity at all: the essence is prolongation of growth. In many mammals sexual dimorphism is established by prolongation of the growth of the male, with its allometric consequences, beyond sexual maturity; among mammals, many large ruminants are cases in point (Georgiadis 1985); in males among many of these species social maturity awaits the final cessation of growth and the achievement of spectacular if 'predictable' new morphologies, even though sexual maturity has already occurred, whereas in females social and sexual maturity are coincident with each other and with the achievement of full growth.

Paedomorphosis by progenesis results in what are essentially breeding juveniles. The Pygmy Marmoset, *Callithrix pygmaea*, resembles the early juvenile of the Common Marmoset, *Callithrix jacchus*. At six months of age the young *C. jacchus* begins to change its coat from an undifferentiated agouti to an incipient marbled pattern, and to acquire white ear-tufts; over the next six months it gradually achieves the adult pattern, and at about 18 months of age it become sexually mature and ceases to grow or develop. *C. pygmaea* grows at the same rate as *C. jacchus* and closely resembles the latter at the same size; but at around 12 months of age, that is to say just before the stage at which its relative starts to change colour, it suddenly becomes sexually mature and ceases growth, so that throughout life it resembles *C. jacchus* of that age. An important consequence of progenesis is that the generation turnover becomes much more rapid, and this form of hetero-chrony could be successful in the irregular type of environment which is associated with r-selection.

Perhaps Gould's major achievement in his 1977 book is to demonstrate the totally distinct natures of progenesis and neoteny, even though both result in a paedomorphic condition. In neoteny aspects of shape are retarded with respect to the rest of development. It is a far more complex concept than progenesis, involving potentially the dissociation of many somatic and reproductive characters during ontogeny. The adult Bonobo or Pygmy Chimpanzee, *Pan paniscus* or *Pan troglodytes paniscus*, has the skull shape and body proportions of juvenile stages of other chimpanzees (*Pan troglodytes* subspp.), but is little or no smaller in size (Zihlman and Cramer 1977), achieves sexual maturity at the same age as far as is known, goes bald like other chimpanzees though perhaps later in life, and right from birth has the blackish face which other chimpanzees acquire only with maturity [see Shea (1983), who shows that the same can be said of body proportions as well]. This is perhaps the most obvious case of neoteny in primates; but Gould notes that taken further it can involve increased size and delayed maturity (and consequent association with K-selection regimes), and in eliminating the achievement of ancestral adult morphologies has an enormous potential for the superimposition of evolutionary novelties.

The vast potential for evolutionary change opened up by heterochrony has perhaps not been appreciated by most biologists; in the last analysis, it seems possible to ascribe a major proportion of evolutionary change to changes in rates of development. Extreme dissociation of developmental rates of different organ systems is what the Krassilov Effect (see subsection 2.5.2 above) is all about. I have no intention of insisting that 'heterochrony' is a synonym for 'evolution', but I would like to stress the uninvestigated explanatory potential embedded in the concept.

It is neoteny which, Gould (1977) argues (reviving an old theory of Bolk from the 1920s), especially applies in the case of human evolution. Neotenous chracters of the human condition, with reference to 'ape' ancestors, are:

1. Skull characters: rounded thin-walled braincase lacking superstructures (sagittal and nuchal crests, supraorbital torus), brachycephaly, small flat face located below the frontal lobes rather than in front of the braincase, and ventrally directed foramen magnum.

2. External characters: persistence of labia majora, ventral direction of vagina, retention of lanugo (primary or downy hair covering), failure of hallux to rotate. In some human 'races' characters such as loss of skin pigment, presence of epicanthic fold, and thick outrolled 'suckling' lips can be added to this list.

3. Prolonged growth: delayed closure of cranial suture and dental eruption, prolonged infant dependency and growth phase, and long life.

4. Large body size and lengthened generation time.

Superimposed on these purely neotenous features are others which are their consequences (development of a chin, a consequence of facial recession) or which are made possible by them (projection of the nose, made possible by facial flatness but not a simple consequence of it), and some which in heterochronic dissociation have become hypermorphic (elongation of legs, broadening of chest).

The second category of features will not be detectable in the fossil record, but the others should be, in whole or in part. If such hypothetical models as the Krassilov effect are valid, then heterochronic changes might be expected to appear suddenly in the fossil record, singly or in groups characterizing each time a new form (species or subspecies or infrasubspecific type). The story of human evolution should therefore be traceable as one of a series of punctuations involving for the most part neotenous events or changes that follow from these.

## 3.2 THE CONSEQUENCES OF DEVELOPMENTAL INSTABILITY

The case for epigenetic disruption as a major force in evolution has been put by Lovtrup (1974) and by Ho and Saunders (1979). The latter authors, while basing themselves in part on assumptions which are either incorrect ('intraspecific differences are not related to interspecific differences' p. 573) or illogical ('the environment is almost infinitely divisible into niches such that competition among morphs of a species is unlikely to occur' p. 573), by drawing attention to the way in which organisms adjust to a norm in ontogeny, and to their potentiality for being deflected from this norm by an event affecting an early ontogenetic stage, illustrate ways in which new evolutionary types may arise. In effect such events would be a form of heterochrony, allied to neoteny: Ho and Saunders dub it 'heterorhesis'. A mutation—or external event?—which deflects a developing organism from its normal developmental pathway will enormously affect the final adult phenotype: not precisely by slowing down or speeding up development, but simply by altering it. Presumably the earlier in ontogeny this deflection occurs the more different will be the result: a true punctuation, a Goldschmidtian 'hopeful monster' (see subsection 3.3.3 below), though again the earlier it occurs the less chance the resultant monster would have, one supposes, of surviving and establishing itself.

The other evolutionary model based on epigenetic deflection (Lovtrup 1974) is much more elaborate, based as it is on an extensive consideration of actual epigenesis in a variety of animals. Each stage in development is necessary for the normal production of the one that follows: including the features of the fertilized egg itself (a stage of ontogeny which the author unfortunately calls 'progenesis'). Extranuclear factors—whether of cytoplasmic inheritance or of the environment—are important at each stage, especially in the very earliest when the body plan is laid down. Each organism is by its very nature adaptable, and a major function of selection would be to increase the relative adaptability of some organisms. Interspecific selection will always be directional; but intraspecific selection will often be normalizing. Isolation, which exempts a deme from competition, is neces-sary for the 'constructive but undirected' phase of evolution; inter specific competition for the 'destructive but progressive' phase. The longer ontogeny lasts, the more chance for disruptive events to intervene; hence animals with long ontogenies, notably mammals, will have had more oppor-tunity to have experienced large jumps in their evolution, and the docu-mented rapidity of their evolution, despite their long generation times and low total individual numbers per species, demonstrates that they have indeed done so.

In this remarkable chain of argument and observation Lovtrup (1974) would appear to have anticipated many, perhaps most, of the advances of the new evolutionary school. That his evolutionary model is arrived at by quite a different route from the strikingly similar (yet less complete) models of palaeontologists (Eldredge and Gould 1972; Gould 1976; Stanley 1979) and an immunologist (Steele 1979) adds tremendous authority to the model itself.

## 3.3 ALTERNATIVE THEORIES OF EVOLUTION

### 3.3.1 Nomogenesis

The first of the 'complete theories' of evolution developed in modern times was the 'nomogenesis' of Berg (1969; translation of work published in 1922). While badly hampered by an ignorance of genetics, slow to reach the Russia of the early twentieth century, Berg's consideration of the Darwinian model led him to the conclusion that it was incompatible, at least as the major mechanism, with what he knew of the pattern of living organisms and their evolution. His conclusions are as follows (1969, pp. 406–7):

1. Evolution is based not on chance variations but on laws (hence the title 'nomogenesis'). The successive stages of character evolution are based on laws, and the laws governing ontogeny and phylogeny are the same.

2. Convergence (and parallelism) are as much a feature of evolution as divergence.

3. Evolutionary variations affect not single individuals but a vast number over a wide area.

4. Evolution proceeds not by slow continuous changes but by jumps.

5. Hereditary variations (mutations) are restricted in number and are directional, not numerous and random.

6. Natural selection is primarily an agency of conservatism, not of change.

7. New species arise by 'mutations' (meaning sharply distinct changes).

8. Much of evolution is the unfolding of pre-existing rudiments, rather than the formation of new characters as such.

The arguments here are in effect the arguments from palaeontology (punctuated equilibria), epigenesis (commonality of ontogeny and phylogeny), cytogenetics (directional mutation; molecular drive), taxonomy (rectangular speciation), and general evolutionary biology (role of selection; ubiquity of parallelism; concept of exaptation), as expounded above. The whole of Berg's book reads in fact like a remarkably modern piece of work; that one can find alternative explanations for many of his examples, especially his evidently non-genetic ones like the evolution of languages, in no way diminishes the extraordinary breadth of the whole. In the end one almost concludes that his freedom from the constraints of early genetic theory could be not a drawback but an advantage, especially in the light of the newest advances of modern genetics (see section 3.2 above).

Berg also believed that extinction, like evolution itself, is subject to internal laws; and that life arose polyphyletically, the different lines evolving very much in parallel. Neither of these two postulates seems to flow from the main thrust of the theory itself, and neither seems particularly plausible even today; but if we are to revive nomogenesis we must obviously examine the whole theory and not merely accept the parts which delight us, and remain open-minded. I suspect that, despite all the arguments of the last few years, we may still be unduly influenced by selectionist outlooks on the importance of competition in producing evolutionary change, and so perhaps in putting a stop to it.

### 3.3.2 Age and area hypothesis

Willis (1924, 1940) noted further drawbacks to the theory of evolution by natural selection alone: the Fleeming Jenkin criticism (Chapter 2, subsection 2.1.2), the slowness of the process especially where several characters are being selected at the same time, and the incredible destruction of inter-mediates that must have taken place. Like Berg and like many modern commentators, he concluded that natural selection is mainly an agent for weeding out unsuitable variations.

His substitute for the Darwinian catch-all was the 'age and area hypothesis', involving a form of nomogenesis affecting biogeography. A plant taxonomist, he had noted how families of plants throw out again and again the same unusual (family-restricted) variants such that the more specialized types occupy the most restricted areas. The Podostemaceae are water plants, mainly tropical in distribution, consisting of 160 species in 40 genera. During the course of their evolution numerous forms with varying degrees of dorsiventral orientation have arisen in independent lineages: a distinctly unusual morphology in plants. The most dorsiventral species are the most local in distribution; unless there have been wide evolutionary reversals, this means that they are new rather than relict species and that the older, simpler forms are those which are widespread. Willis (1940) asks the questions: why, without any differences in habitat requirement, have the most specialized forms of each genus acquired the same distinguishing features; and why has just this family evolved, time and again, this thoroughly unusual morphology? The example of the Podostemaceae is only one of very many which Willis (1940) describes, and taken together they point to a down-playing of natural selection in favour of a directional response: a nomogenetic process, perhaps.

Less far-reaching than Berg's model, the age and area hypothesis evidently enjoyed some popularity among the botanists of the day, but was eclipsed without being satisfactorily disposed of by the neo-Darwinian synthesis. Examples among animals come to mind: why have true (frontal) horns evolved, apparently independently, in the majority of families of ruminant artiodactyls (Janis 1982), but in only one other family in the whole of the Mammalia, including the macropodid and diprotodontid marsupials, Proboscidea, Perissodactyla, non-ruminant Artiodactyla, and giant rodents, lagomorphs, and hyraxes, all of which closely approach the ruminants in their habitat and dietary requirements and ways of life? The reader is referred back to the discussion above (section 3.2) of directionality in evolution.

### 3.3.3 The hopeful monster

The most famous protest against the emerging neo-Darwinian synthesis was that registered by Goldschmidt (1940). As reviewed by Stanley (1979) and Gould (1980), this author did raise some valid queries despite being nowadays mostly relegated to footnotes in evolutionary texts, his monsters having been deemed hopeless in probability terms by the compilers of the synthesis. Unlike Berg and Willis, Goldschmidt has exerted a notable influence on today's model-builders, and Gould (3.3.4 below) refers to the 'decoupling' of intraspecific and interspecific evolution (Eldredge and Cracraft 1980) as the 'Goldschmidt break'.

Goldschmidt could not believe that natural selection, with all its slowness and the slightness of its effects, could produce anything but microevolutionary changes. In particular, he found it difficult to accept that macroevolutionary changes are simply an accumulation of microevolutionary changes as required by the synthesis. From time to time, he proposed, a major mutation occurs producing a radically different form, which must then be tested against pre-existing forms.

In many respects the punctuated equilibrium model is a simple resurrection and elaboration of the hopeful monster theory. As I have indicated above, more than this is needed for a complete theory of evolution; but the accumulating evidence does suggest that some version of the Goldschmidt theory is by now inescapable (see also Macbeth 1974).

### 3.3.4 Gould's 'new and general theory'

In a progress report on the new thinking in evolutionary biology Gould (1980) described the 'new and general theory of evolution' which in his opinion seemed to be emerging. There are, he suggested, three major levels at which evolution occurs: variation within populations, speciation, and macroevolution (which he equates with trends of differential success among species). Between the first two levels is the Goldschmidt break; between the second and third levels, the Wright break. Natural selection is the driving force at the first level, species selection at the third level. Speciation remains the most puzzling of the three levels.

In a major advance over his own and Stanley's (1979) previous thinking, Gould accepts that, many species being divided into demes with little or no gene-flow between them, allopatry is not a necessary precondition for speciation. He also accepts the Carson founder–flush model, although he is not entirely correct in stating that it has been experimentally verified by Powell (see Chapter 2, subsection 2.3.2).

Gould's interim model, which is really all it is intended to be, is largely descriptive, with little attempt to elucidate mechanisms. As such, it needs to be elaborated by some attempt to explain processes, and certainly it badly needs the concept of directionality. This is why the most satisfactory evolutionary model is, I believe, best described by Berg's term, nomogenesis.

## 3.4 IMPLICATIONS FOR HUMAN EVOLUTION

If the theory propounded here, a modified and elaborated nomogenesis, is correct, certain expectations emerge about human evolution. If fulfilled, the theory will be shown consistent with the evidence; if not, the theory will be falsified or will need modification.

1. *Homo sapiens* as a species is highly autapomorphic (Groves 1986), and so should be (a) the product of many sequential quasi-speciation events [human evolution best represented as a bush rather than a ladder, in the words of Gould (1976)], and (b) itself geologically young.

2. The lineage leading to *H. sapiens* was stenotopic and demic right from the start.

3. New species will appear suddenly, usually as small founder populations expand; there may be high variability if they are examined at an early enough stage.

4. That the new species would have developed right in the centre of the ranges of their parent species (centrifugal speciation) is more probable than that they developed on the edge; many sympatric or parapatric quasi-species are to be expected, as there are several chromosomal changes that must have occurred (Yunis and Prakash 1982) in the human lineage since its individualization (stasipatric speciation). Such a species may arise from just one particular subspecies of its parent species (heteromorph speciation).

5. Extensive parallelism would have occurred between different hominine species; if gene transfer occurs, this will be greater between sympatric pairs.

6. The most marked evolutionary jumps will be heterochronic in nature, especially involving neoteny, and if the Krassilov effect is valid will occur with increasing frequency as generation length increases.

7. New character states will not necessary be adaptive for some new environmental condition when they appear, but may conversely, as exaptations, open up new niche possibilities for their possessors.

8. Cases of gradualism and punctuation will be better distinguished in the most recent geological period, that which is accessible to $C^{14}$ dating.

Before these propositions are examined on the evidence of the fossil record, I will survey the living primates to see whether the order to which we belong offers any insights into the evolutionary process.

# 4
# The taxonomy of the primates

In preceding sections I have, by and large, avoided using primates as examples. This is because there is in the anthropological community as widespread an ignorance today about what goes outside the primates as there used to be about what goes on outside our own species. Other primates are germane to human biology, and other mammals are germane to primate biology, or so I believe; and, if this is so, we can examine the taxonomy and, inferentially, the phylogeny of primates in the light of the discussions of the previous three chapters and see if any new light can be thereby shed on the subject.

**Table 4.1**   Taxonomy of the order Primates.

**Suborder Paromomyiformes**
   **Plesion *Purgatorius***
      Superfamily Plesiadapoidea
        Family Plesiadapidae
             Saxonellidae
             Carpolestidae
      Superfamily Paromomyoidea
        Family Paromomyidae
          Subfamily Paromomyinae
            Genus  *Palaechthon*
                   *Plesiolestes*
                   *Paromomys*
                   *Phenacolemur*
                   *Ignacius*
          Subfamily Micromomyinae
            Genus  *Palenochtha*
                   *Micromomys*
                   *Timimomys*
          Subfamily Picrodontinae
                *Picrodus*
                *Zanycteris*
       *Incertae sedis:*  *Navajovius*
                *Berruvius*

**Suborder Strepsirhini**
   **Infraorder Adapiformes**
      Family Notharctidae
        Genus  *Cantius*

           *Notharctus*
           *Pelycodus*
           *Smilodectes*
           *Microadapis*
           *Laurasia*
           *Copelemur*
           *Mahgarita*

Family Adapidae
  Subfamily Protoadapinae
    Genus *Protoadapis*
          *Agerinia*
          *Pronycticebus*
          *Europolemur*
  Subfamily Anchomomyinae
    Genus *Anchomomys*
          *Fendantia*
          *Huerzeleris*
          *Periconodon*
  Subfamily Adapinae
    Genus *Adapis*
          *Leptadapis*
          *Paradapis*
          *Alsatia*
          *Simonsia*
Family Petrolemuridae
    Genus *Petrolemur*
          *Hoanghonius*
          *Lushius*
Family Sivaladapidae
    Genus *Indraloris*
          *Sivaladapis*
          *Sinoadapis*
*Incertae sedis: Azibius*


**Infraorder Chiromyiformes**
  Family Daubentoniidae
    Genus *Daubentonia*


**Infraorder Lemuriformes**
  Family Lorisidae
    Subfamily Lorisinae
   Tribe Lorisini
    Genus *Loris*
          *Nycticebus*
   Tribe Perodictini
    Genus *Perodicticus*
          *Arctocebus*
          *Mioeuoticus*

**Table 4.1**   *contd*

Subfamily Galaginae
Genus  *Galago*
       *Euoticus*
       *Galagoides*
       *Otolemur*
       *Progalago*
       *Komba*
Family Lemuridae
Genus  *Lemur*
       *Hapalemur*
       unnamed
       *Varecia*
Family Megaladapidae
Genus  *Lepilemur*
       *Megaladapis*
Family Indriidae
Subfamily Indriinae
Genus  *Indri*
       *Propithecus*
       *Avahi*
       *Mesopropithecus*
Subfamily Palaeopropithecinae
Genus  *Palaeopropithecus*
       *Archaeoindris*
Subfamily Archaeolemurinae
Genus  *Archaeolemur*
       *Hadropithecus*
Family Cheirogaleidae
Genus  *Microcebus*
       *Mirza*
       *Cheirogaleus*
       *Allocebus*
       *Phaner*


**Suborder Haplorhini**
  **Plesion *Decoredon***
  **Infraorder Omomyiformes**
Family Microchoeridae
Genus  *Nannopithex*
       *Necrolemur*
       *Microchoerus*
       *Pseudoloris*
       *?Chasselasia*
Family Anaptomorphidae
Subfamily Teilhardininae

Genus  *Teilhardina*
   *Chlororhysis*
Subfamily Anaptomorphinae
Genus  *Anaptomorphus*
Subfamily Tetoniinae
Genus  *Tetonius*
   *Anemorphysis*
   *Altanius*
   *Trogolemur*
   *Absarokius*
   *Mckennamorphus*
Family Omomyidae
Subfamily Omomyinae
Genus  *Uintanius*
   *Omomys*
   *Chumashius*
   *Ourayia*
   *Macrotarsius*
Subfamily Washakiinae
Genus  *Hemiacodon*
   *Loveina*
   *Shoshonius*
   *Dyseolemur*
   *Washakius*
   *Rooneyia*
   *Arapahovius*
   *Ekgomowechashala*
Subfamily Utahiinae
Genus  *Utahia*
   *Stokia*
*Incertae sedis:  Donrussellia*
    *Copelemur*
    *Kohatius*

**Infraorder Tarsiiformes**
Family Tarsiidae
Genus  *Tarsius*
*Incertae sedis:  Afrotarsius*

**Infraorder Simiiformes**
**Section Platyrrhini**
Plesion *Branisella*
Family Homunculidae
Genus  *Homunculus*
Family Callitrichidae
Subfamily Callimiconinae
Genus  *Callimico*
Subfamily Callitrichinae
Genus  *Callithrix*

**Table 4.1**   *contd*

---

                        *Leontopithecus*
                        *Saguinus*
        Family Cebidae
            Genus  *Cebus*
                        *Saimiri*
                        *Dolichocebus*
        Family Aotidae
            Genus  *Aotus*
                        *?Tremacebus*
        Family Callicebidae
            Genus  *Callicebus*
        Family Pitheciidae
            Genus  *Pithecia*
                        *Chiropotes*
                        *Cacajao*
                        *Cebupithecia*
        Family Atelidae
            Subfamily Atelinae
            Genus  *Lagothrix*
                        *Ateles*
                        *Brachyteles*
            Subfamily Alouattinae
            Genus  *Alouatta*
                        *Stirtonia*

**Section Catarrhini**
    Plesion *Amphipithecus*
    Plesion *Pondaungia*
    Superfamily Parapithecoidea
        Family Parapithecidae
            Genus  *Qatrania*
                        *Parapithecus*
                        *Apidium*
    Superfamily Cercopithecoidea
        Plesion *Prohylobates*
        Plesion *Victoriapithecus*
        Family Colobidae
            Subfamily Nasalinae
            Genus  *Nasalis*
            Subfamily Colobinae
            Genus  *Procolobus*
                        *Colobus*
                        *Microcolobus*
                        *Presbytis*
                        *Semnopithecus*
                        *Trachypithecus*
                        *Pygathrix*

        *Mesopithecus*
        *Dolichopithecus*
  Family Cercopithecidae
    Subfamily Cercopithecinae
      Genus  *Cercopithecus*
            *Chlorocebus*
            *Erythrocebus*
    Subfamily Papioninae
      Genus  *Papio*
            *Theropithecus*
            *Parapapio*
            *Gorgopithecus*
            *Lophocebus*
            *Mandrillus*
            *Cercocebus*
            *Macaca*
            *Procynocephalus*
            *Paradolichopithecus*
            *Allenopithecus*
            *Miopithecus*
Superfamily Propliopithecoidea
  Family Propliopithecidae
      Genus  *Oligopithecus*
            *Propliopithecus*
Superfamily Pliopithecoidea
  *Incertae sedis: Turkanapithecus*
  Family Pliopithecidae
    Subfamily Pliopithecinae
      Genus  *Pliopithecus*
            *Laccopithecus*
    Subfamily Crouzeliinae
      Genus  *Crouzelia*
            *Plesiopliopithecus*
            *Anapithecus*
Superfamily Hominoidea
  Family Proconsulidae
      Genus  *Proconsul*
            *Rangwapithecus*
            *Limnopithecus*
            *Dendropithecus*
            *Micropithecus*
  Family Dryopithecidae
      Genus  *Dryopithecus*
  Family Oreopithecidae
      Genus  *Oreopithecus*
            *Nyanzapithecus*
  Family Hylobatidae
      Genus  *Hylobates*

**Table 4.1**  *contd*

*Krishnapithecus*
*Dionysopithecus*
Family Hominidae
*Incertae sedis: Afropithecus*
   Plesion *Kenyapithecus*
   Subfamily Ponginae
      Genus *Sivapithecus*
            *Lufengpithecus*
            *?Platodontopithecus*
            *Gigantopithecus*
            *Pongo*
   Subfamily Homininae
      Plesion *Ouranopithecus*
      Tribe Gorillini
      Genus *Gorilla*
      Tribe Panini
      Genus *Pan*
      Tribe Hominini
      Genus  unnamed
            *Paranthropus*
            *Australopithecus*
            *Homo*

## 4.1 THE DIVISIONS OF PRIMATES

Simpson (1945) divided the primates into two suborders, Prosimii and Anthropoidea. One still from time to time finds acceptance of this scheme, especially in textbooks written by physical anthropologists, on grounds such as that it is time-honoured. While antiquity is, to be sure, no guarantee of irrelevance (viz. nomogenesis), it is at the same time hardly a reason for over-looking decades of new work and rethinking. Simpson constructed his classification at a time when there was no articulated philosophy of systematics; one would have hoped that Hennig (1966), not to mention his elaborators and exegeticists, had changed all that.

The only thing about the higher level relationships of the living primates on which, I think it can safely be claimed, all specialists do agree, is that the lemurs (in the broad sense, i.e. including Lorisoidea) form a monophyletic group, so do the anthropoids (monkeys, apes, cladists, and pheneticists), and so do the three or more living species of tarsiers. The question about which there is much less agreement is the interrelationships of these three groups. Three views are possible, and each has been espoused in modern times:

1. Prosimii (lemurs and tarsiers) vs. Anthropoidea: Simpson (1945), recently resurrected by Schwartz (1986).

2. Strepsirhini (lemurs) vs. Haplorhini (tarsiers and anthropoids): Pocock (1918) and Hill (1953), perhaps the majority view among specialists at the present time.

3. Plesitarsiiformes (tarsiers) vs. Simiilemuriformes (lemurs and anthropoids): Gingerich (1976) and Schwartz *et al.* (1978).

Note that the subordinal classification of primates must depend on the extant forms, which by the very fact of their being extant, and so available for examination *in toto*, offer far more sources of evidence than do fossil forms. Luckett (1976) in particular has made this point very forcibly: study of the several, functionally independant, taxonomically varying characters of the foetal membranes shows clearly the correctness, if a cladistic approach is taken, of a Strepsirhini/Haplorhini scheme, and this is backed up by a study of other aspects of soft anatomy. Faced with the absence of evidence as to the placentation, eye structure, and so on of fossil forms, the most logical response is not to abandon the strepsirhine/haplorhine division, but to seek associated characters in the hard parts which *can* be detected in fossils; and failing this, to rank equivocal fossil taxa *incertae sedis* (pending, one hopes, future discoveries which will clarify matters). If this means that restudy necessitates the transfer of some fossil groups from one suborder to the other [as in the case of the Microchoeridae, Eocene 'tarsiids' recently proposed by Schmid (1982) to be in all likelihood strepsirhines], then so be it. If a classification is to have any hope of being (1) compatible with phylogeny and (2) stable, it must be based on *all* available sources of evidence, which means inevitably that it must be based first and foremost on extant forms.

What if new discoveries reveal the existence of previously unsuspected fossil taxa which appear to form the sister group of all living forms? Do we make the primary division that between the new fossils and the rest, relegating the division based on living forms to a lower taxonomic rank? The case is not hypothetical, for it exists in primates. The group variously called Plesiadapoidea, Plesiadapiformes, or Paromomyiformes (see, for example, Szalay 1973) is certainly more divergent, cladistically, from the Strepsirhini and Haplorhini than they are from each other, i.e. is the sister group of Strepsirhini plus Haplorhini. So, should the suborders become Paromomyiformes vs. Euprimates, with the Strepsirhini and Haplorhini being relegated to the status of infra-orders of the latter? And if, later on, another family, or even a single genus, were to be discovered which is a sister group of Paromomyiformes plus Euprimates, do these two groups now take the rank of suborders, the Paromomyiformes/Euprimates split being downgraded to infra-ordinal, and some intermediate rank invented for the strepsirhine/haplorhine split?

Understandably appalled at such a prospect, Patterson and Rosen (1977) proposed the concept of a *plesion*, an unranked quasi-taxonomic category, to be inserted at the appropriate level so as not to disturb a pre-existing classification. The concept of plesion is still not a well-defined one; as I would see it, it would be of most use to denote a group of limited diversity, whereas Archer and his colleagues (in Archer and Clayton 1984) use it to cover any extinct group, however, diverse, whose cladistic ranking would disturb the ranks of their extant relative; thus the Paromomyiformes would be a plesion, as would each and all of its subgroups all the way down.

The problem is essentially that of what Jefferies (1981, p. 490) has named 'stem groups'. If all members of a sister group are the total group, the crown group (Hennig's *group) is 'the latest common ancestral species of the living representatives of the group, plus all the descendants of that ancestor whether living or dead', and the stem group is the total group minus the crown group. In this rather helpful nomenclature, the paromomyiformes are a stem group: none of them, with the probable exception of the earliest known genus, *Purgatorius*, could be ancestral to any living primate. Yet the Paromomyiformes are highly diverse: divisible into a number of families, each with two or more (up to two dozen) genera.

I propose the following course of action for the assignment of stem-group taxa:

(1) an individual genus that is cladistically coordinate with or superordinate to a pair or set of higher category sister groups should be given the title plesion and listed, in a sequential classification, immediately antecedent to the sister groups to which it is itself the sister group;

(2) a more diverse group should be assigned the same taxonomic rank as the components of its sister group.

As intimated above, it appears plausible that *Purgatorius* could be ancestral to all subsequent primates, or at least is cladistically their sister group. It should therefore be removed from the Paromomyiformes, which thereby becomes an evolutionary clade, sister group to the Euprimates. We classify as follows:

*Order Primates*
Plesion *Purgatorius*
Suborder Paromomyiformes
Suborder Strepsirhini
Suborder Haplorhini

The Strepsirhini and Haplorhini, though together the sister group (Euprimates) of the Paromomyiformes, are by this scheme ranked coordinate with the latter. If future discoveries should show that *Purgatorius* does not stand alone but is a member of a clade with one or more other genera, then a fourth suborder would have to be erected for them. Such a scheme preserves (1) the Holocene-centredness of a taxonomy and (2) its general conservatism—both of which aspects I hold to be valuable in that they promote the scheme's stability (and so its maximum information-retrieval potential) through allowing the widest diversity of sources of evidence—while permitting its essential flexibility and dynamism.

### 4.1.1 Characters of the primate suborders

The suborder Paromomyiformes preserve the following primative states:

— small brain; marked post-orbital constriction;
— no post-orbital bar;
— laterally facing orbits;
— claws on all digits.

At the same time they have the derived states of:

— hypertrophy of central incisors in both jaws;
— reduction of lateral incisors, canines, and premolars, and development of a diastema.

The number of families is disagreed, largely because whether the Microsyopidae are primates at all—if not, they are remarkably convergent of the Paromomyiformes—is in dispute. Though mainly Palaeocene in time, *Ignacius* (Paromomyidae) and *Plesiadapis* (Plesiadapidae) survived until the late Eocene.

The euprimate stem developed an enlarged brain, a post-orbital bar, more convergent (Cartmill 1972) orbits, and nails on some, at least, of the digits: precisely how many digits were nailed in the euprimate ancestor is debatable (see below), but at least the hallux was. The suborder Strepsirhini is characterized by at least one derived state:

— tarsal rotation mechanism (Szalay and Decker 1974).

If the curious genus *Daubentonia* is related to the Indriidae, as held by Tattersall and Schwartz (1974) and Schwartz and Tattersall (1985), then the following derived states can be added to the list:

— dental comb in lower jaw;
— nails on all digits except the second pedal digit, which has on the contrary an enlarged claw, the toilet claw.

However, Groves (1974*a*) and Oxnard (1983) have given reasons for thinking that *Daubentonia* is the sister group of all the other strepsirhines; reasons which will be repeated below (section 4.2). If this latter view is correct, then the dental comb and toilet claw are shared derived characters of the other strepsirhines, not of the suborder as a whole.

The suborder Haplorhini is characterized by a considerable list of derived character states, of which the most obvious are:

— nose hairy, lacking rhinarium;

— upper lip free from gum;
— upper central incisors medially apposed, with no gap between them;
— placenta haemochorial;
— retina with fovea centralis and macula lutea;
— no tapetum lucidum behind retina;
— post-orbital plate developed;
— failure to synthesize Vitamin C (Pollock and Mullin 1987).

There seems, consequently, little doubt but that the suborder is monophyletic. In particular the phrase 'haemochorial placenta' actually covers a considerable complexity of interlinked features of the foetal membranes, which as Luckett (1976) has shown are strongly derived. The anatomy of the eye has been less investigated; but as no haplorhine has a reflective layer (tapetum) [unless *Aotus* has a tapetum fibrosum (Walls 1939)], while all investigated strepsirhines, as well as a variety of non-primate mammals do, I take it to be most plausible that (1) its presence or absence is diagnostic for the two suborders and that (2) primitive mammals had it, and strepsirhines have retained it so that its absence is a derived state of haplorhines. Again, no strepsirhine that has been examined has a true macula or a fovea (though a histologically different type of fovea is found in some lemurs—Pariente 1970), nor as far as I am aware has any other mammal; but all haplorhines do. In addition, the fact that the two nocturnal haplorhines, *Tarsius* and *Aotus*, seem to have the histological features of a macula while lacking actual cones, suggests very strongly that the ancestral activity rhythm of the suborder was diurnal, as argued by Martin (1968).

There are comparatively few dental or skeletal features in these lists; it is clearly incumbent on primatologists to try to find more than there are, so as to be clearer about the affinities of Eocene fossil groups. At present it appears that (better-known members of) the Omomyidae display derived haplorhine character states, and for this reason if no other the Adapidae, Notharctidae, and any other families that can be recognized are assigned at present to the Strepsirhini. Indeed, Schwartz and Tattersall (1979) are inclined to associate the Adapidae closely with the present-day lemurs, attributing character reversal (especially, loss of dental comb) to at least some genera of which it is known that they did not have a dental comb.

## 4.2 THE STREPSIRHINI

As I have insisted throughout, in the case of living forms the full range of evidence is, at least potentially, open to us, so that we should begin by classifying the living forms and afterwards fit in the fossils, utilizing whatever evidence is available. Put like this, the outstanding problem in the taxonomy of the Strepsirhini is not the relationship of the Adapidae to the living forms, but the position of *Daubentonia*, the aye-aye.

As mentioned above, all living strepsirhines apart from *Daubentonia* have a dental comb and a toilet claw. *Daubentonia* has no dental comb; nor, however, does it have an alternative condition which could be described as primitive: instead, the permanent incisors are reduced to a single pair in each jaw, are labiolingually expanded, have enamel only on their labial surfaces, and grow from persistent pulps. The whole skull shape is modified: as shown so effectively by Cartmill (1974), the function is not rodent-like but woodpecker-like, and the rounded, shortened skull is ideally contoured to absorb the stresses of this activity.

Nor does *Daubentonia* have a toilet claw; in this case, however, the evidence strongly suggests that its alternative state is primitive. The typical mammalian claw has two layers, a deep and a superficial stratum, whose laminar convergence gives the claw its resistant point. In insectivores the deep stratum accounts for some 80 per cent of the total thickness of the claw (Thorndike 1968). Among primates, a deep stratum is retained in the New World monkeys among the haplorhines, and here forms a mere 20 per cent of the thickness of the falcula, whether claw (as in Callitrichidae) or nail (in other taxa). The deep stratum is lost altogether in catarrhines; and in all strepsirhines except *Daubentonia*, regardless of nail or claw (i.e. toilet claw).

If, therefore, a falcula with a deep stratum—one, moreover, accounting for well over half its thickness—is correctly interpreted as the primitive mammalian condition, then the ancestor of the strepsirhines and haplorhines might well have reduced the thickness of the deep stratum, but its loss would have to have been convergent in (1) the catarrhines and (2) most strepsirhines. Some degree of homoplasy is therefore unavoidable in reconstructing the evolution of the appendage here.

*Daubentonia* has claws on all digits except the hallux, which is nailed as in all living primates. All falculae, whether claws or hallux nails, have a deep stratum which forms about 60 per cent of the thickness (Thorndike 1968). They would thus appear to be more primitive than any other strepsirhine.

Three interpretations of this state of affairs are possible:

1. That *Daubentonia* is the sister group of all other Euprimates. This would reduce homoplasy in this feature to a minimum: reduction of the deep stratum would have characterized the ancestor of the other Euprimates, although its total loss would still have occurred twice. In deference to the findings of Szalay and Decker (1974) that in its foot structure *Daubentonia* shares the derived features of other strepsirhines, this interpretation may be rejected.

2. That *Daubentonia* is a true strepsirhine, but the sister group of all other strepsirhines. This is the next most parsomonious interpretation: the deep stratum is reduced independently in members of the two suborders, but only once in each.

3. That *Daubentonia* is not only a true strepsirhine but has a special

*A theory of human and primate evolution*

relationship to one of the other strepsirhine groups, such as the Indriidae as insisted by Schwartz and Tattersall (1985). This means that, within the Strepsirhini, either the deep stratum would have had to have been reduced independently several times (in the Lorisoidea, perhaps including the Cheirogaleidae; in the Lemuridae; and in the Indriidae after the separation of *Daubentonia* from the proto-indriid stem), or else it was reduced in the ancestral strepsirhine and redeveloped in *Daubentonia*. Either of these alternatives is not impossible, but the first is prolix in the extreme (and involves convergence, not mere parallelism), and the second seems on the face of it unlikely because, along with the deep stratum, the germinal layer is also lost.

It is necesary, therefore, to examine other anatomical regions to see if there is supporting evidence for any of these possibilities. It has been seen that the evidence of the foot seems to exclude the first possibility above. We must now ask, does *Daubentonia* either (a) share any derived states with a particular other strepsirhine family as maintained by Schwartz and Tattersall (1985), or (b) lack any derived states which all the other strepsirhines share?

Schwartz and Tattersall's (1985) evidence for an indriid–daubentoniid sister-group relationship is, in fact, not cladistic when examined closely, but rests on a putative functional sequence. The Indriidae have thickened lower incisors and a rounded skull for their implantation; *Daubentonia* has this taken to an extreme (see Fig. 4.1). *Daubentonia*'s almost featureless but square surfaces are taken to be logical extensions of the rectangular indriid

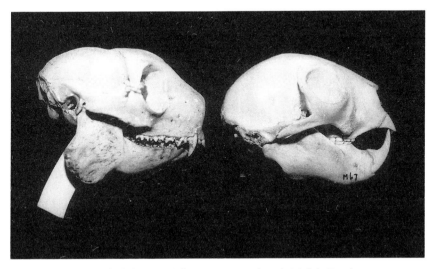

**Fig. 4.1.** Skulls of (left) *Propithecus verreauxi* and (right) *Daubentonia mada-gascariensis*. The Aye-aye skull is not satisfactorily explained as merely an Indriid skull taken to its logical extreme.

molars, though these latter have well-developed cusps, styles, and connecting cristae. This is the totality of the argument: there are no identifiable shared derived features.

A few other characters were mentioned by Groves (1974*a*): *Daubentonia* shares certain muscle attachments with the Haplorhini; for example, the occipital extension of the origin of the trapezius muscle, likely to be a primitive character, whereas all other strepsirhines lack an occipital attachment. These arguments have never been countered—even mentioned—by the proponents of an indriid–daubentoniid sister-group relationship. *Daubentonia* also lacks a complex, derived specialization of the foot that is shared by Lemuridae and Indriidae (Gebo 1985). For what it may be worth, too, the karyotype of *Daubentonia* shows no likeness to that of any other strepsirhine (Petter *et al.* 1977).

The obvious question to ask now is: Does *Daubentonia* show any sign of ever having had a dental comb? Schwartz and Tattersall (1985) consider the reduced and somewhat strengthened lower incisor–canine row of the Indriidae to be a morphological percursor to the incisor morphology seen in *Daubentonia*, but there is no close approach. Le Gros Clark (1959) described the deciduous dentition of *Daubentonia* as having two pairs of incisors in each jaw, but Luckett and Maier (1986) have identified three pairs, i.e. more primitive than other lemurs. I would say that there is no indication, from the permanent dentition or any other feature of *Daubentonia*, that the ancestors of this genus ever had a dental comb, and some suggestion that they did not.

MacPhee (1981, p. 82, Table IX) lists *Daubentonia* as having an ectotympanic that is less inclined to the transverse plane of the basicranium, and so less altered ontogenetically, than any other strepsirhine, and comparable to Lipotyphlan insectivores (these being forms with little connection between the ectotympanic and any bullar ossification that may exist). Put like this, the *Daubentonia* condition could be a primitive one, although a neotenic process could alternatively be invoked.

With regard to the molecular evidence Dene *et al.* (1976) find that, using antisera to *Lemur mongoz*, *Propithecus*, and *Microcebus*, *Daubentonia* is much the most distinct of any Malagasy lemur, though perhaps not quite as divergent as are the Lorisiformes. Sarich and Cronin (1976) rank it as just as distinct as the loris group.

Finally there is the evidence of Oxnard (1981, 1984*b*). This evidence, based as it is on multivariate analysis which many biologists feel themselves ill-equipped to examine, has proved hard to assess. The response of Tattersall (1982*b*) to Oxnard's initial demonstration of 'the uniqueness of *Daubentonia*' was that the strong divergence of the genus, in the multivariate pattern of its post-cranial measurements, was an indication of its highly autapomorphic nature and did not reflect its relationships. Oxnard (1984b) has recently shown this response to be inappropriate. While distantly related taxa may show close functional–morphological similarity to each other in particular regions of the skeleton, the more regions are taken into account the less any individual functional resemblance affects the picture, and the more do the underlying phyletic relationships emerge. This is simply another way of looking at the superficial nature of convergence. In its whole-body comparisons, *Daubentonia* is by far the most distinct of the Strepsirhini.

In sum, the evidence from many different sources indicates that *Daubentonia* is the sister group of the other strepsirhines. The recent discovery (W. von Koenigswald 1979) of a beautifully preserved hind limb skeleton of a lemur, complete with toilet claw, from Messel, shows that the divergence of these two stems dates from at least the middle Eocene. I here reiterate my conclusion, which remains unrebutted (Groves 1974*a*), that the prior division within the Strepsirhini is between two infra-orders, Chiromyiformes (with a single known family Daubentoniidae) and Lemuriformes (for all other strepsirhines).

### 4.2.1 Chiromyiformes

The single known family, Daubentoniidae, has in turn only one known genus, *Daubentonia*. The two species are the apparently (Hill 1953) recently extinct *D. robusta* and the still extant *D. madagascariensis*. The struggle to save the latter from extinction is one of the most desperate in the modern field of wildlife conservation. The conclusion that it is the sole representative of a unique infra-order merely reinforces its scientifically priceless status and the urgency of the fight to save it.

### 4.2.2 Lemuriformes

The position of *Daubentonia* does not affect the questions of the inter-relationships among the other Strepsirhini, as no polarities are thereby altered, at least none that are of significance in the present context. There are two competing views:

(1) that the sister-group relationship is between the non-Malagasy group ('Lorisiformes', family Lorisidae) and the rest;

(2) that the Malagasy family Cheirogaleidae are the sister group of the Lorisidae, and together these form the sister group of the other (all Malagasy) lemurs.

The anatomical basis for the first view has been surveyed by Groves (1974*a*), MacPhee (1981), and Schwartz and Tattersall (1979), and found wanting. All the characters defining the Lorisidae, when critically reviewed, are reduced to the structure of the bony ear: the phaneric rather than aphaneric ectotympanic, and the concentration of pneumatization in the central and posterior parts of the tympanic roof rather than in the tympanic floor (MacPhee 1981). Groves (1974*a*) and, independently, Szalay and Katz

(1973) drew attention to similarities of an apparently derived nature between the Lorisidae and the Cheirogaleidae: the presence, in both, of an anterior carotid artery (more correctly, an anastomosis between the ascending pharyngeal and promontory arteries) and differences in the venous drainage as well as in the associated foramina. The importance of this finding, in a geographical sense, is that the Malagasy lemurs would not be monophyletic (*Daubentonia* aside of course!): either the Mozambique channel had been crossed more than once, or the proto-Lorisidae had got into Africa from Madagascar, or the whole group had begun to diversify in Africa before the separation of Madagascar.

Although the lorisid/cheirogaleid affinity looked good at the time, and is still accepted by Schwartz and Tattersall (1985), its basis has come under closer inspection with not entirely expected results. Cartmill (1975) has shown that close analysis of the bony ear region reveals that the cheirogaleids show differing degrees of resemblance to the Lorisidae: if the polarities of character states as argued by Szalay and Katz (1973) are accepted, then *Allocebus* shares derived features with the Lorisidae which the other cheirogaleids do not have. This would challenge the monophyly of the Cheirogaleidae: a not impossible concept, but one which would need further study before acceptance. The relatively simple nature of the carotid difference, and the fact that it is but one of a number of variations in this region among lemurs (MacPhee 1981), suggests that it may indeed have arisen in parallel. Schwartz and Tattersall (1985) add some characters of the dentition which they see as having been present in a common cheirogaleid/lorisid ancestor. The features are, however, in no sense diagnostic: hypocones on the upper molars, for example, are usual in the strepsirhines; the upper incisors are not separated by the incisive foramina in all lorisids, and conversely are so separated in *Hapalemur* and other genera; and so on. The blood proteins of the Lorisidae and Cheirogaleidae are not alike (Dene *et al.* 1976), nor are their chromosomes (Petter *et al.* 1977).

Dene *et al.* (1976) propose to rank the Cheirogaleidae as a distinct superfamily, Cheirogaloidea, and Sarich and Cronin (1976) find that they are approximately equidistant from both lorisids and the lemurid/indriid group. For the moment I regard the inter familial affinities of the Lemuriformes as unclear, and rank them all as equal families.

### 4.2.2(a) Cheirogaleidae

The Cheirogaleidae share the following characters:

— anterior carotid artery (ascending pharyngeal anastomosis with distal promontory artery);
— lumen of promontory and stapedial arteries partially non-patent, or these arteries absent altogether;
— somewhat elongated calcaneus and navicular;

— unreduced upper incisors (a primitive feature);
— dental features listed in Tattersall and Schwartz (1974);
— primitive foot structure (Gebo 1985).

I propose the following classification of the Cheirogaleidae:

Genus *Phaner*
          Species: *P. furcifer*                    Forked-crowned lemur
Genus *Microcebus*
          Species: *M. murinus*                  Grey Mouse-lemur
                    *M. rufus*                        Red Mouse-lemur
Genus *Mirza*
          Species: *M. coquereli*              Giant Mouse-lemur
Genus *Cheirogaleus*
          Species: *C. medius*                    Lesser Dwarf-lemur
                    *C. major*                        Greater Dwarf-lemur
                    *C. crossleyi*                    Crossley's Dwarf-lemur
Genus *Allocebus*
          Species: *A. trichotis*                Hairy-eared Dwarf-lemur

1. Genus *Phaner*. Petter *et al.* (1977) stress the distinctiveness of this form and separate it as subfamily Phanerinae, which indeed Von Hagen (1978) aligns with the Indriidae on the basis of its possession of a throat gland. In fact all the distinctive features of *Phaner* appear to be uniquely derived; Schwartz and Tattersall (1985) make a good case that it shows many derived stages with *Allocebus*, while Sarich and Cronin (1976) find it to be closer to *Cheirogaleus* than to *Microcebus*.

The autapomorphic features of *Phaner* include the unique long, caniniform anterior upper premolar ($P^2$), giving it a double-canined look; the long, procumbent $I^1$, separated from $I^2$ by a gap; the sudden constriction of the snout in the posterior premolar region; and the sinuous facial profile; but the obliteration of the stapedial artery, often cited as diagnostic for this genus, may occur in *Microcebus* too (MacPhee 1981). The pelage pattern, with dark eye-rings connected by prongs (hence the 'fork-crowned' name) to a thick crown-to-rump stripe, is in effect an intensification of a common strepsirhine pattern.

Though previously reckoned monotypic, *Phaner* in fact shows well-marked geographic variation; a number of subspecies are recognizable (Groves and Tattersall, in preparation):

2. Genus *Microcebus*. The smallest living primates are included in this genus. It can be distinguished from its closest relatives, *Mirza* and *Cheirogaleus*, by its pointed molar cusps, the development of hypocones on $M^1$ and $M^2$, the curved tooth-rows, the preservation of the primitive median gap between the upper incisors, and the enlarged tympanic bullae which

posteriorly reach behind the level of basion. It shares with them, however, the presumably derived features of enlarged molars relative to its general size, convergent tooth-rows, and the molariform condition of $P^4$.

Petter *et al.* (1977) award this genus three species of which one is now separated as *Mirza* (see below). The remaining two species are *M. murinus*, larger, grey above, white below, with large outstanding ears; and *M. rufus*, slightly smaller, dark red above, yellowish below, with much shorter ears (on their nomenclature, see Tattersall 1982). Broadly, the grey species lives in the dry zone of the west and south; the red species in the humid eastern forests. However, both species appear to turn up again outside these general ranges. A grey form lives in the Angavo River district of the central eastern region: this, unlike the western grey species, has no reddish overwash, but is red around the eyes and under the head. Again, a red form occurs (sympatric with *M. murinus*) in the Morondava and Ankarafantsika districts of the west; it is still smaller than the eastern *M. rufus*, the tail is longer, the ears are bigger, and the red tone is lighter. Petter *et al.* (1977), who give these details, refer the outliers to the two well-known species, but are clearly uneasy about doing so. It is evident that further study of them is required.

Specimens in the British Museum collection show that both species fall approximately within the same head and body length range (115–140 mm), with the tail about the same; but there are two rather divergent populations of *M. rufus* represented. A series from Analamera, in the far northeast of Madagascar, are very large (152–157 mm) with short tails; and a series from Anaborano, in the northwest, are similar in size to most mouse-lemur but with extremely long tails.

3. Genus *Mirza*. Tattersall and Schwartz (1974) mention, but do not defend, their theory that the species traditionally called *Microcebus coquereli* is in fact a *Cheirogaleus*, not a *Microcebus*. Subsequently (Schwartz and Tattersall 1985) they resurrected the generic name *Mirza* for it. This step is followed here, although the evidence on balance still favours a sister-group relationship between it and *Microcebus*: they share a shortening of the anterior premolar ($P^2$), making it shorter than $P^3$, development of hypocones on $M^1$ and $M^2$, and a concave facial profile. *Mirza* lacks the greater bullar development, the obliteration of the interincisal gap, and the tooth-row curvature of *Microcebus*; its enormous size—the weight is some five-times that of *M. rufus*—is almost its only autapomorphic feature.

4. Genus *Cheirogaleus*. The dwarf-lemurs are distinguished from the mouse-lemurs by their bulbous, rather than pointed, molar cusps and the convexity of the facial profile, these two features being presumably derived; they lack the hypocone development and the reduced anterior premolar.

Petter *et al.* (1977), while adopting the traditional division of this genus into two species, again express some misgivings. *Cheirogaleus medius*, the smaller species, lives in the dry western and southern forests; it is silvery grey in colour, washed with beige, white below; its tail stores food enabling it to survive in torpid state in the dry season. *C. major* is characterized as a larger species from the eastern rain forest; the tail is longer than the head and body; the colour is reddish, with white underside as before. Petter *et al.* then go on to recognize a subspecies, *C. major crossleyi*, whose range extends from the Amber Mountain in the north down the high crests as far as Lake Alaotra, and at least in one place coexists, with no signs of interbreeding, with the coastally distributed *C. m. major*. They say that *crossleyi* is smaller and redder than *major*, with pointed muzzle, whereas *major* is more grey-brown and had a rounded muzzle.

The evidence here seems quite clear. For whatever reason, Petter *et al.* (1977) hesitate to take the step of recognizing *C. crossleyi* as a distinct species, but their own evidence—sympatry with *C. major*—requires it. Study of museum specimens is needed to define and delimit all three species; particularly as Petter *et al.* mention yet another form (from Bongolave), which they think may be intermediate between *C. major* and *C. medius*!

5. Genus *Allocebus*. The poorly known species that used to be called *Cheirogaleus trichotis* was shown to be generically distinct by Petter-Rousseaux and Petter (1967). Subsequent studies (Schwartz and Tattersall 1985) tend to indicate that it is most related to *Phaner*, and that these two together are the sister group of the other three genera. They lack the following derived states which the other three share: enlarged molars; convergent tooth rows; molariform $P^4$. It has several uniquely derived states of its own: both $P^2$ and $P^3$ are caniniform, the latter not as elongated as the former; $I^1$ is greatly enlarged, and somewhat curved forward; and the tympanic bullar region is deflated, hardly defined from the mastoid inflation. It has small hypocones, converging thus on *Microcebus*. Its caniniform $P^2$ and enlarged $I^1$, in particular, relate it to *Phaner*, and the keeled nails form a further similarity.

*A. trichotis* is reddish in colour and has engagingly tufted ears. it is known by a few specimens only; even its precise habitat is unknown.

### 4.2.2(b) Lemuridae

Traditionally (Schwarz 1931*b*; Hill 1953) this family has included three genera: *Lemur, Hapalemur, Lepilemur*; to which Petter (1965) added *Varecia* for a species previously placed in *Lemur*. Rumpler (1974) proposed to separate *Leiplemur* at subfamily level, while Petter *et al.* (1977) made it the type of a separate family. On the basis of resemblances in certain cranial and dental features, Tattersall and Schwartz (1974) associated *Hapalemur* with *Lepilemur*; but there are no other resemblances (and even the cranio-dental ones are tenuous), and the arrangement of Petter *et al.* is adopted here; more recently, Schwartz and Tattersall (1985) followed the same course.

Even thus restricted, the family Lemuridae is hard to define. All members lack the anterior carotid artery of the Cheirogaleidae; and they have the full dental complement, unlike the Indriidae. *Lemur* and *Varecia* are long-snouted, *Hapalemur* short-snouted: the latter condition would be considered dervied in haplorhine primates, but in strepsirhines the variation between taxa in facial length is so diverse that little can be made of such an argument: indeed it will be suggested (below) that facial elongation might well have occured independently in different taxa of this family. Schwartz and Tattersall (1985) mention 'very enlarged infra-orbital foramen', which is not unique to the family, and a few dental features which are similarly non-diagnostic.

Von Hagen (1979) has proposed to divide up the family (or rather, as he places all Malagasy lemurs except the Aye-aye in one family, it is his system of subfamilies that is at issue). As far as the present Lemuridae are concerned, he places the members in three groups: (1) *Varecia*, (2) *Prosimia* (i.e. all species of *Lemur* except *L. catta*), and (3) *Lemur* (for *L. catta* alone) plus *Hapalemur*. It is true that each of these three groups can be defined by the possession of derived character states. He moreover aligns *Varecia* with the Indriidae, which is more questionable but has a certain justification (see below).

Eaglen (1983) and Eaglen and Groves (in press) agree that the family ought to be split up, and that Von Hagen's groupings are in the right direction. For the moment, I will be conservative and retain the Lemuridae as recognized by Petter *et al.* (1977).

The following classification of the Lemuridae is proposed:

Genus *Varecia*
    Species: *Varecia variegata*       Ruffed Lemur

Genus *Lemur*
    Species: *Lemur catta*       Ring-tailed Lemur

Genus *Hapalemur*
    Species: *Hapalemur griseus*       Grey Gentle-lemur
             *Hapalemur simus*       Broad-nosed Gentle-lemur
             *Hapalemur aureus*

Genus unnamed
    Species: '*Lemur*' *macaco*       Black Lemur
             '*Lemur*' *fulvus*       Brown Lemur
             '*Lemur*' *mongoz*       Mongoose Lemur
             '*Lemur*' *coronatus*       Black-crowned Lemur
             '*Lemur*' *rubriventer*       Red-bellied Lemur

1. Genus *Varecia*. Though formerly included in *Lemur*, the Ruffed Lemur in fact has little in common with Ring-tailed and Brown Lemurs beyond its elongated snout. Its autapomorphic features are its large size and heavy build; the low-cusped, wide basined molars; and the panda-like colour pattern. But it is primitive, compared to both the other 'Lemuridae' and the Indriidae, in a number of features, especially those connected with the

reproductive system: there are three pairs of mammae (as in the Cheirogaleidae); the gestation period is short, about 100 days; the female builds a nest for the young; scent-marking is simple, only a throat gland being present in addition to the urine/faeces marking seen in Cheirogaleidae (and even *Phaner*, in the latter family, has such a gland). In the skull, there are no presphenoid or palatine sinuses unlike Brown Lemurs or Indriidae, and the frontal sinus is small. The malar foramen is very large, as in *Lemur catta* and *Hapalemur*. The reproductive and scent-marking characters suggest primitive retentions; the large malar foramen, unique to *Varecia* and the other taxa mentioned, might argue for a different placement. Von Hagen (1979) referred it to its own subfamily, Vareciinae, which he proposed is related to the indrids because of (1) the presence of the throat gland, as in *Propithecus* and *Avahi*, (2) the extensive fusion and flattening of the plantar pads, (3) the sharing of a similar, mainly black-and-white colour pattern, with black ventral surface, and (4) the pattern of the retina. Such data illustrate the difficulties of applying strict parsimony.

Hill (1953) recognized four subspecies in the one and only species, *V. variegata*. Petter *et al.* (1977) confirmed all of these, and found that there are at least four other colour forms, some of which seemed however to be merely the intermediates between the four subspecies; but locating the available observations and museum specimens suggests that there are anomalies in their distributions, and that they are probably only colour types within one subspecies, although there is an admitted cline of increasing white to the south:

a. *Varecia variegata rubra*, the well-known red form, restricted to the Masoala peninsula. The back and limbs are red; there is a lighter area at the tail-root, and a white spot on the nape; the face, tail, and underparts (including the inner surface of the limbs) are black, as in all forms of the species.

b. *V. v. variegata*, 'typical' form. From the Mananara–Mahambo region, bordering the range of *V. v. rubra*, according to Petter *et al.* (1977); but there are two specimens in the Smithsonian collection from the Faraony River, far to the south. The body is white instead of red, except the upper part of the back which is black, but bisected by a white stripe often continued onto the (black) crown; and there is a white collar between the black areas of the crown and the shoulders.

c. *subcincta* form. From west of Maroansetra, and the Island of Nosy Mangabe. The back is black except at the tail-base (in most specimens); there is a white 'belt' around the middle, in front of which in some examples the black may be replaced by grey. Petter *et al.* (1977) 'variety 6', from Mananara, is intermediate between this subspecies and the previous one, having a white wedge into the anterior black zone; behind the white belt the back tends to be yellow–brown instead of black.

d. *editorum* form. From the region south of Lake Alaotra and from Manakara, well to the south. This is like nominotypical *variegata* but lacks the longitudinal white stripe. Petter *et al.*'s (1977) 'variety 5' is intermediate, having a wedge of varying size into the anterior black zone; it is known from Tamatave, south of Mahambo, and from Maroansetra along with the 'typical' form.

e. 'Variety 7' of Petter *et al.* (1977). A curious unlocalized form (known only from a skin from 'south of Madagascar'), in which the wedge into the anterior black zone, and the dark zone behind the belt, are both dark brown, with only two lateral black spots.

f. 'Variety 8' of Petter *et al.* (1977). The back is almost completely white. This form has been observed in the extreme south of the species' distribution.

It is difficult to see any regularity in this pattern of subspeciation. Von Hagen (1979) points out that on the theory of metachromism, *V. v. rubra* is the primitive form; the others are more and more white, so metachromatically advanced, in the sequence *subcincta–editorum–variegata*.

2. Genus *Lemur*. It has been hinted many times (Andramiandra 1972; Mahe 1976; Petter *et al.* 1977) that *Lemur catta* is probably more closely related to *Hapalemur* than to other species commonly assigned to the genus *Lemur*. This point of view has recently been strongly urged by Eaglen and Groves (in preparation). The synapomorphic features of a group containing *L. catta* and *Hapalemur* are: possession of brachial and antebrachial glands; enlarged malar foramen; P4-3-2 sequence of premolar eruption (all other strepsirhines, except *Lepilemur* and Indriidae, have the premolar sequence 2-4-3); chromosome features; structure of facial skeleton and inter-orbital region (Fig. 4.2); clitoral glands, and a clitoral urethra; Wolffian vestiges in the ovary; retinal features; two haemoglobin components. That is not to say that they are congeneric: there remain profound differences between them, and it is not productive to combine them.

If *L. catta* is separated from the *fulvus* group, then the name *Lemur* remains with *catta*, which is its type species. *L. catta* has a wide range in the southern arid and semi-arid regions of Madagascar, but no known geographic variation. Von Hagen (1979) draws attention to the fact that, striking as it is, in many respects the facial pattern of this species is quite plesiomorphic.

3. Genus *Hapalemur*. Lacking the highly derived colour pattern of *Lemur, Hapalemur* has a primitive-appearing greenish brown fur with a pale facial mask. Carriage of the young in the mouth is presumably a derived condition; certainly the strongly molarized $P^4$ and the shortened, deep muzzle are derived.

Two species are universally recognized: *Hapalemur griseus* and *H. simus*. The latter would appear to be more derived than the former in its larger size,

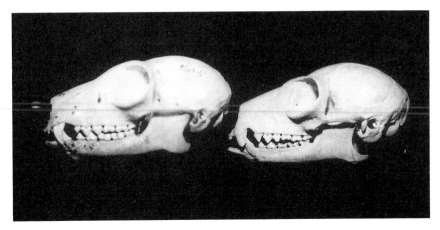

**Fig. 4.2.** Skulls of (left) '*Lemur' fulvus* and (right) *Lemur catta*. The differences, in for example the orbit/muzzle relationships, suggest that facial elongation may have occurred independently in the two groups.

extremely squared-off snout, molarized P³ as well as P⁴, the presence of a light sacral patch, and the large throat gland. The chromosome number is higher, however ($2n$ = 60, against 58 or 54); fusion is in general more frequent than fission, but as has been seen earlier this is certainly not universal. The clearly agouti-banded fur is certainly primitive, so relations between the two species are not simple. Finally the nature of the arm glands is different from in *H. griseus*: they are shifted up to near the elbow, though they seem to be as well developed as those of *H. griseus* and *L. catta*. Vuillaume-Randriamanantena *et al.* (1985) show that a subfossil species. *Prohapalemur gallieni*, is in fact the same as *Hapalemur simus*; adding the two distributions together, the species must not so long ago have been much more widespread than it is today.

*Hapalemur griseus griseus* is still widespread through the eastern rain forests, especially in areas of abundant bamboo. The very closely related *H. g. occidentalis* has been found in two widely separated areas in the west; it has a more elongated muzzle and more prominent ears, is slightly paler in colour, and has 58 chromosomes instead of 54—a supposedly more primitive condition. This is a simple case of allopatric subspecies; but more unusual is the case of the third described subspecies, *H. g. alaotrensis*. This is much larger than the other two (head and body length 40 cm, as against 25–30 cm; weight about double), and has a shorter tail, a more pointed snout, and a less-defined facial pattern; the karyotype is the same as that of *H. g. griseus*. Remarkably, it is found *within* the range of the latter, not in bamboo country but in the reed beds of Lake Alaotra. There can be no question of allopatric speciation here; rather, it is the first clear-cut example we have met with of centrifugal speciation.

There are several interesting features of this case to be noted, apart from the obvious one that the derived taxon is plumb in the middle of the range of its ancestral form. First is that its much greater size would act as a good pre-mating isolating mechanism should the ranges of the two forms meet (which is unclear from published accounts); they have been hybridized in captivity (Petter *et al.* 1977), but this does not of course indicate what would happen in the wild. Next is that there is no chromosomal difference; statispatric processes are not involved. Third is that the habitats are different: a process of divergence such as that postulated by Maynard Smith (1966) is suggested. Finally is the point that, if the *occidentalis* karyotype is correctly identified as the primitive one, then *alaotrensis* and *griseus* are sister groups: much the most highly derived of the three is the one which has arisen from within the range of one of the others (heteromorph speciation).

Whether one taxon whose range is entirely surrounded by that of another can justifiably be ranked as a subspecies is moot. I would be tempted to recognize it as full species: *Hapalemur alaotrensis*; certainly a case that flouts cladistic assumptions. Very recently, a new species, *H. aureus*, has been described; its distribution likewise fits a centrifugal model (Meier *et al.* 1988).

4. *Genus unnamed*. The lemurs of the *fulvus* group lack the numerous synapomorph features of *Lemur* and *Hapalemur*, and have a few that unite them: the presence of pericones (cingular cusps anterior to the protocones on the upper molars); large frontal and palatine sinusus; glandular perineal skin; and some degree of sexual dichromatism. Groves (1974*a*) used the name *Prosimia* (as a subgenus of *Lemur*) for the group, and Von Hagen (1979) raised the group to generic rank under this name; but Elliot (1913) had in fact fixed *Lemur catta* as the genotype, so a new name was required (Eaglen and Groves, in press).

Petter *et al.* (1977) and Tattersall (1982) recognize the following species in this genus: *Lemur macaco*, from the northwest; *L. fulvus*, widespread over the whole island except the southwest; *L. mongoz*, from restricted localities in the northwest; *L. coronatus*, from the far north; and *L. rubriventer*, from the western rain forests. I will deal briefly with each of these species, then comment on the overall picture.

*L. macaco*. Strongly sexually dichromatic: males are black, females reddish yellow. There are 44 chromosomes. Two subspecies:

*L. m. macaco*: north of the River Andranomalaza, as far as Ambilobe. Has prominent ear-tufts and cheek whiskers; eyes are reddish-orange.

*L. m. flavifrons*: south of the river Andranomalaz, as far as Befotaka; lacks ear-tufts and cheek fringes, and has blue–green eyes. Both Groves (1974*a*) and Tattersall (1982) doubted that this was more than morph of *macaco*, but its status as a genuine subspecies has recently been confirmed by Koenders *et al.* (1985).

*L. fulvus*. Sexual dichromatism varies from weak to strong; muzzle is always black. Chromosomes are $2n$ = 60 or 52 to 48. Six subspecies are recognized:

*L. m. fulvus*. Greenish brown, head blackish in male, brown in female; light to white spots above eyes. Mainly found in the east between the Bay of Antongil and the Mahanoro region; also in the northwest at Ankarafantsika and east of the Galoka Mountains, where it is sympatric with *L. macaco*, and at Mayotte in the Comoros, where it is certainly introduced.

*L. m. sanfordi*. Male dark grey–brown with light coloured head, white ear-tufts; female greyer. From the Vohemar district to the northern tip.

*L. m. albifrons*. Male browner, with white head (except for facial mask); female greyer. From the Vohemar district south to the Bay of Antongil.

*L. m. rufus*. Male greyish, with varying amount of reddish wash, red patch on crown, hands and feet red; female red, with grey crown; both sexes have white forehead. Ranges in the west from about the Mahanoro region south to Manakara; in the east from Majunga south to Tabiky.

*L. m. albocollaris*. Dark brownish olive above, the female redder-toned than the male; head black in male, redder in female; crown black in male, grey in female; cheek whiskers and throat white in male, grey in female. Diploid chromosome number is 48: all four previously described subspecies have $2n$ = 60. This is restricted to a small range around Farafangana, in the southeast.

*L. m. collaris*. Entirely resembles the last, except that the cheeks and throat of the male are red, not white; and the diploid chromosome number is 50, 51 or 52. It lives in the far southeast, the Fort Dauphin area.

*L. mongoz*. Smaller, shorter eared, with white muzzle; no pre-sphenoid sinus unlike other species. Both sexes are grey–brown; the male usually has a red wash over the back, head, and cheeks, while the female has the head grey, the cheeks and neck white, but some males resemble the female in colour. $2n$ = 60. Found in a small region of the northwest, in a wedge from Ambatondrazaka to the northwest coast between Soalala and Analalava, it is sympatric with *L. f. fulvus* and, marginally, with *L. f. rufus*; and (introduced) on Anjouan and Moheli in the Comoros.

*L. coronatus*. The smallest species; like *mongoz* [with which it was united in a single species by Schwarz (1931*b*)], but the male is dark grey–brown with crown black, forehead and cheeks red; the female is light grey, with a light red V-shaped band on the forehead. $2n$ = 46. Lives in the far north: from the northern tip of Madagascar southeast to the Bay of Antongil; it is widely sympatric with *L. f. sanfordi* and narrowly with *L. f. albifrons*.

*L. rubriventer*. Has short ears, hidden in the fur; muzzle is black; colour is chocolate or maroon brown, with tail black; male lighter reddish below and has a white spot on the inner angle of the eye; female has a whitish underside.

The palatine sinus is enormously inflated. From almost the whole of the eastern coastal rain-forest zone, where it is sympatric with *L. f. fulvus*, *rufus*, *albocollaris*, and *collaris*. $2n = 50$; it has the same number of acrocentric chromosomes as the 50-chromosomes form of *L. f. collaris*, but there appear to have been some translocations between some of the metacentrics and submetacentrics.

Groves (1974*a*) and Tattersall (1982) remark that a revision of these lemurs, especially the forms assigned to *L. fulvus*, is long overdue. There are, for example, some anomalies of distribution: Groves (1974*a*) maps several apparent overlaps between *fulvus* and *rufus*, for example, which is confirmed would mean that they must be distinct species. Some, at least, of these cases of overlap could well be due to errors in placement of localities: for example, Ambohimanga is mapped in my 1974 paper at 18°S, but I have found another place of the same name at about 21.30°S, which does seem a more likely locality for *rufus*. In another instance, I was until the publication of Petter *et al.*'s book in 1977 unaware of the very considerable variation within *albifrons*, so that a specimen from Ambatondradama (in the Maroansetra district) was attributed to a *rufus* × *albifrons*; thus the indications of *rufus* up the east coast, within the range of *fulvus*, seem to be disappearing.

There is however still one case of overlap, which seems to be difficult to explain away. The Smithsonian collection has several specimens of *rufus* and one *fulvus*, all in numbered series, from the Ambodiasy Valley, Faraony River: this is north of Manakara, and well within *rufus* country. Tattersall (1982) briefly discussed this case, suggesting an error; and clearly field investigation is needed.

Tattersall (1982) also mentions *albifrons* specimens in the American Museum of Natural History, collected in the Manombo/Vondrozo region, 23°S. In this case, I think they have simply been misidentified: the specimens are simply females of *albocollaris*, and similar to others from the same region in the British Museum.

I have remarked above that my supposed case of intergradation between *rufus* and *albifrons* was in error. I now think that, even though I had emphasized the amazingly broad variability within *rufus*, I had still under-estimated. Specimens from the Tabiky region which I thought were *rufus/albocollaris* intermediates are probably just extreme examples of *rufus*. An undoubted *albocollaris* comes from Lokosy, near Tabiky, and on the opposite side of Madagascar from its recorded range; but this could be another case of several places with the same name.

So we can make the following statements about the taxa assigned to *L. fulvus*: (1) there is some evidence of sympatry between *fulvus* and *rufus*; (2) all others appear to be strictly allopatric; (3) there are no real cases of the interbreeding in the wild between them.

The chromosomal evolution of the genus is interesting. The primitive number is surely 60; both on the usual high-is-primitive assumption, and because both the majority of the *fulvus* forms and the very different *mongoz* have this number, all with apparently similar morphology. From this state there are reductions: *collaris* with 52, 51, or 50, and *rubriventer* with 50; *albocollaris* with 48 (apparently a simple Robertsonian advance on *collaris*); *coronatus* with 46; *macaco* with 44.

A metachromatic series is also possible in the genus (Fig. 4.3). *L. f. fulvus* is on several counts the most primitive: hairs have a grey base, followed by several alternating yellow and grey bands, then long yellow–brown tips; sexual dichromatism is minimal; differentiation between different chromatic fields is also minimal. The females of *L. f. collaris* and *albocollaris* have the same grey base, then a light red–yellow band, a black band, and a reddish tip. All other taxa have a maximum of three zones per hair: a grey base, a grey to black shaft, and a straw-coloured to red tips (*sanfordi, albifrons, rufus*, males of *collaris* and *albocollaris*, females of *coronatus*). Most of the remainder have just the grey base and yellowish or reddish tips: *mongoz*, *rubriventer*, males of *coronatus*, females of *macaco*. Finally, males of *macaco* have uniformly black hairs. It is interesting in this series how males of some taxa are metachromatically more advanced than the females: this is the case with the banding patterns just as with the overall colour patterns.

Females are in fact rather alike, at least in the *fulvus* group, *mongoz* and *coronatus*. The yellow–brown of *fulvus* becomes yellow–grey in `collaris` and *albocollaris, sanfordi*, and *coronatus*; browner and redder in *mongoz*, *albifrons*, and *rufus*. The black mask is more marked in *fulvus* than in others; the cheeks and supra-orbital regions are more differentiated in *rufus*, *coronatus*, and *mongoz* than in any of the others. The males, chromatically more differentiated, are much more different from each other. The relatively undifferentiated (female-like) *fulvus* pattern could be developed through several pathways: (1) The bands are reduced to one eumelanin, one phaeomelanin; females become paler, greyer, males also greyer with the underside going to white and the cheek pattern becoming more differentiated, paler, and more tufty then whitish and spreading round the head. This route goes via *sanfordi* to *albifrons*; at an early stage along this route, a divergent stock losing the eumelanin band would develop into *mongoz*. (2) The bands reduced (but not as much as in the previous route: to two phaeomelanin with one eumelanin) and the colour paler, gives us the female *collaris/ albocollaris*; further reduction of the agouti banding, and differentiation of the cheek pattern, gives us the males, with *albocollaris* (white-cheeked) being more advanced than *collaris* (red-cheeked). (3) The bands reduced to one of each, as in route (1), but both sexes highly differentiated in a distinctive way, gives us *rufus*; it is possible to take this still further to an extremely pale form, with further differentiation of the coronal pattern, to arrive at *coronatus*; and loss of the eumelanin band in both sexes, but retention of the red tone, to give first *rubriventer*, then the female of *macaco*, and finally (by switching to

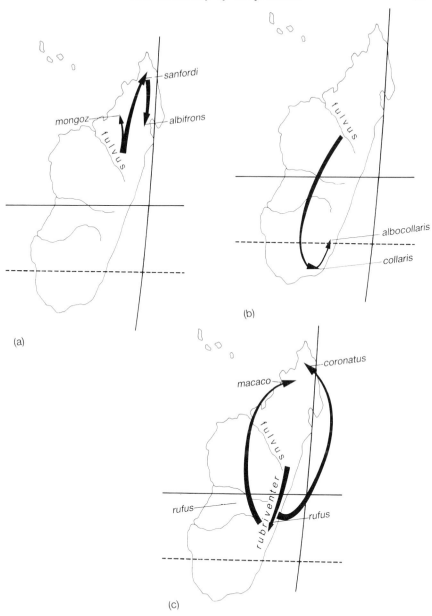

**Fig. 4.3.** Metachromatic evolution in species referred to *Lemur*. (a) extreme reduction of hair banding; paler, greyer tone of female; increasing differentiation of male's head pattern: *fulvus—sanfordi—albifrons*, with a side branch to *mongoz*. (b) reduction of banding, with development of cheek pattern in male: *fulvus—collaris—albocollaris*. (c) extreme reduction of banding as in (a), but both sexes differentiated: *fulvus—rufus—rubriventer—macaco*, with side branch from *rufus* to *coronatus*.

the eumelanin pathway and complete saturation) the male of *macaco*.

These are minimum pathways: that is to say, although the assignment of *mongoz* to the first, *coronatus* to the second, and *rubriventer* and *macaco* to the third, is open to question, there seems no possibility of reducing the number of pathways. Each one begins with *fulvus*, which really fits the bill of a surviving primitive form very well, and then passes through other supposed races of *L. fulvus*. All that the taxa assigned to the 'species' *L. fulvus* have in common is that they retain a primitive position on the metachromatic scale; or, to put it another way, they lack the particular derived states of any of the other four species. Cladistically, they are in fact older than the other four species, and it remains to be seen, by investigation in the field of border areas, whether they are also good biological species.

### 4.2.2(c) Family Megaladapidae

Schwartz and Tattersall (1985) indicated their agreement with Petter *et al.* (1977) that *Lepilemur* should be separated at family level from other extant lemurs: but they demonstrated that it has close affinities with the recently extinct *Megaladapis*, and moreover that the correct name, by priority, of a family containing them both is Megaladapidae. Citing such features as the similarly specialized mandibular condyle, these same authors proposed that the Megaladapidae and Indriidae are sister groups, and such a position (supported by the molecular data of Dene *et al.* 1976) does seem more plausible than the one proposed by Von Hagen (1979), that *Lepilemur* retains primitive features and is, if anything, a sister group to the Cheirogaleidae. Quite out of court, apparently, is the view formerly held by Tattersall and Schwartz (1974) that its closest relative is *Hapalemur*.

Derived in common between *Lepilemur*, at least, and the Indriidae are characters of the jaw joint, the shearing crests of the molars, the loss of the carpal os centrale, the loss of carpal vibrissae, and locomotor specializations. Another possible similarity is in the order of permanent premolar eruption. In *Lepilemur*, as in *Hapalemur* and in *Lemur catta*, this is 4–3–2, whereas in Cheirogaleidae, '*Lemur*' *fulvus* group, and *Varecia* it is 2–4–3 (as it is, indeed, in the Lorisidae). The Indriid 4–3 would fit into either of these patterns—assuming that the two premolars in this family really are the third and fourth of the original set. It must be admitted, however, that morphologically Schwartz (1974) is quite correct; they are strikingly like 2 and 3 (respectively) of the normal set, not 3 and 4. If it really is the posterior one that has been lost, not the anterior, then of course the Indriidae at once slot into the *Lepilemur* pattern of back-to-front eruption.

It must be admitted that the *Hapalemur/Lemur catta* group do to some extent fit into the same pattern. Not only their 4–3–2 premolar eruption, but their long gestation (over 130 days, as in *Lepilemur* and the Indriidae), their lingually opening mandibular molar talonid basins, and their incipiently vertical posture and locomotion: such characters might identify them as the sister group of the Indriid/Megaladapid line. This is a rather radical inter-pretation of lemur interrelationships, but it has been stressed throughout that current orthodoxy is totally unsatisfactory.

Genus *Lepilemur*. Petter and Petter-Rousseaux (1960) reviewed the diversity in the genus and recognized several subspecies, within a single species. With the development of karyological work in Madagascar, unexpected diversity between described forms was revealed, and a number of different species came to be recognized, culminating in the announcement of a new species, *Lepilemur septentrionalis*, based entirely on the number and morphology of its chromosomes (Rumpler and Albignac 1975). Petter *et al.* (1977) recognized seven species. Tattersall (1982) put them back into a single species, with six subspecies.

It seems inescapable that chromosome differences, if they go beyond simple Robertsonian processes or perhaps small-scale inversions, will act as post-mating isolating mechanisms. In *Lepilemur*, the karyotype of *L. microdon* is unknown, while *L. dorsalis* and *L. leucopus* have identical karyotypes; all other species are known to be unique in their chromosomes, with differences of some complexity between them. Hybrid sterility would therefore seem to be assured in those cases; captive data are lacking, the problem being to persuade these animals to breed at all in captivity; and field data on their border zones are also lacking. The gross morphological differences seem, in general, clear enough to support a multispecies model [though, as we shall see, not necessarily quite as many species as are recognized by Petter *et al.* (1977)].

*Lepilemur mustelinus*: a species from the northern half of the east coast rain forest, south from about Vohemar. Colour is medium red–brown, with the tail dark on the terminal half; $2n = 34$.

*L. microdon*: differs from the last in being less reddish, the eyes (irises) light yellow instead of brown, and in having a light yellow collar on the neck. The karyotype is unknown. It inhabits the southern half of the eastern rain forest, its limits with *L. mustelinus* being unknown. Tattersall (1982) places this in the synonymy of *mustelinus*, but the differences [as described by Petter *et al.* (1977), and observed by me on a few museum specimens] seem real enough; pending description of its karyotype, it would probably be better to place it as a subspecies of *L. mustelinus*.

*L. leucopus*: this is a throughly distinct species, from the southern xerophytic bush zone. It is a very light beige–grey in colour with white underside, hands, and feet; it is much smaller in size, and the tail is as long as the head and body, instead of shorter. $2n = 26$.

*L. ruficaudatus*: from the southern half of the western dry forest zone. It is beige–grey, darker than *leucopus*, and without the white areas; the tail, as in *leucopus*, is fairly long, and reddish, the tip often white. Size is intermediate between *mustelinus* and *leucopus*. Diploid chromosome number is 20.

*L. edwardsi*: from the northern half of the west coast dry forest zone, the exact boundary with *ruficaudatus* being uncertain. It is merely somewhat darker, often washed with red; chromosome number is 22, there being some morphological differences among the metacentric/submetacentric set as well as an acrocentric pair which *ruficaudatus* lacks. Tattersall (1982) provisionally recognizes this form, but doubts that it is really distinct from *rudicaudatus*; in this case, because the chromosome differences amount to more than a simple Robertsonian fusion, I would be inclined to maintain the species separation—but the two are almost sibling species.

*L. dorsalis*: from Nosy Be Island, and the small area of rain forest on the northwest coast. This seems another highly distinct species in its gross morphology. Small in size (only slightly larger than *leucopus*), with very small ears, almost hidden in the fur, a very short muzzle, a grey head, and red–brown body, and (usually) a dorsal stripe. There are 26 chromosomes, whose morphology appears to me [from the photos in Petter al. (1977); and without the benefit of banding studies] to be exactly the same as in *L. leucopus*. An interesting case of parallellism?–or, just possibly, could this karyotype be primitive?—in which case we might be in the presence of a case of centrifugal speciation.

*L. septentrionalis*: from the far north, the Mont d'Ambre district. This was described on the evidence of its karyotype alone: the diploid number varies from 34 to 38, with all numbers in between, and all chromosomes differ in morphology from all other species (including, in the case of the 34-chromosome morph, from *L. mustelinus*, which has the same number). Only later (Petter *et al.* 1977) was it shown that it is externally very close to *L. ruficaudatus* and *edwardsi*, but somewhat paler in colour and smaller in size than either; a virtual sibling species. At the time of the initial description of the species, four subspecies were ascribed to it: *L. s. septentrionalis* ($2n = 34$), *andrafiamensis* (38), and *ankaranensis* and *sahafarensis* (both with 36, but with different sets of Robertsonian fusions from the 38 morph). The distributions of these four, as given in the original description, are confused, and Tattersall (1982) regards them as just polymorphic variants from the same localities. A careful reading of the descriptions, however, and more especially those given by Petter *et al.* (1977), suggest that the actual position is as follows:

Sahafary forest: *septentrionalis* and *sahafarensis* and hybrids between them;
Mont d'Ambre and Ankaranana: *ankaranensis* and *andrafiamensis* and hybrids;
Andrafiamena range: *andrafiamensis* only.

So there is sympatry: between pairs, not between all four. It is evident, as Tattersall (1982) implies, that there has been misunderstanding here—perhaps about the very nature of subspecies. This is the time, perhaps, to stress yet again that subspecies are populations; that is, they can only indicate geographic entities, not mere morphs within a population. So the objective conclusion is that there can only be two subspecies of *Lepilemur septentrionalis*: one (nominotypical *septentrionalis*) restricted to Sahafary, the other (I here select *ankaranensis* to have priority) in the other localities.

### 4.2.2(d) Family Indriidae

Here, for once, is a taxon about whose monophyly there can be no doubt. They share dental reduction (though whether the mandibular dental formula is 1123 or 2023 there is dispute), a characteristic shortening of the facial skeleton and specialization of the post-cranial skeleton, enormously lengthened (and specialized) gut, and some synapomorphies with the Megaladapidae as discussed in subsection 4.2.2(c) above.

The Indriidae are not in fact as isolated among the Lemuriformes as has sometimes been claimed. In particular, it has recently been shown that they share with Lemuridae and Megaladapidae a complex structural—functional remodelling of the foot skeleton and musculature, termed the I–II adductor grasp by Gebo (1985); the Cheirogaleidae, Daubentoniidae, Lorisidae, Adapidae, and all Haplorhini show a more primitive condition, the I–V opposable grasp. This character affords quite decisive evidence of intra-strepsirhine affinities.

The three genera are quite distinct. Schwartz and Tattersall (1985) refuse to speculate on which two might be sister groups; Von Hagen (1979) suggests that *Propithecus* and *Avahi* are sister groups because of their more advanced facial shortening than *Indri*, whereas his metachromic analysis seems to indicate, on the contrary, that *Propithecus* and *Indri* could be sister groups.

1. Genus *Avahi*. The pelage consists, remarkably, entirely of underwool; it is soft, crimpy, and agouti on the dorsum with more saturated orange tones on the tail and rump. The throat gland is paired. There are 66 chromosomes. Von Hagen (1979) perspicaciously remarks that the genus is somewhat convergent on *Aotus* in the neo-tropical region: both are undoubtedly secondarily nocturnal, probably dwarfed, and woolly coated.

There is a single species, *A. laniger*, with two well-marked subspecies: *A. l. laniger* from the eastern rain forest, and the white-browed, white-thighed *A. l. occidentalis* from the small northwestern rain forest area.

2. Genus *Indri*. The largest of the living lemurs; tailless, with less lengthened hindlimbs than other indriids (intermembral index 63, cf 58 in *Propithecus*, 57 in *Avahi*: Jouffroy and Lessertisseur, 1978), no throat gland, and a basically black colour with white on the nape, flanks, and limbs. It has a small range in the northeastern rain-forest in hilly country.

3. Genus *Propithecus*. Intermediate in size, long-tailed, with median throat glands and an advanced metachromic pattern, blocks of saturated fur being white, black, maroon, red, or orange.

The two species are readily distinguished. The eastern (mainly rain forest) species, *P. diadema*, is larger, with the tail shorter than the head and body, long silky hair, thickly haired underside, and a diploid chromosome number of 42. The western, mainly dry forest species, *P. verreauxi*, is smaller, the tail longer than the head and body, short-furred, with thin fur on the underside through which the black skin can be seen; chromosome number is 48. Their distinctness has never been doubted.

Each of the two species has five ascribed subspecies. These, from north to south, may be briefly characterized as follows:

*P. diadema* (from north to south):

*P. d. perrieri*: jet black, small in size.

*P. d. candidus*: white; the head, back, and limbs sometimes grey.

*P. d. diadema*: grey; crown black, with white facial rings; a black–brown mantle on shoulders, tapering to a point at the midback; limbs yellow-tinted.

*P. d. edwardsi*: chocolate brown head, upper back, limbs, and tail, grading via brown on flanks into beige tones on underside.

*P. d. holomelas*: black–brown, a light triangular mark at tail-base; paler below.

*P. verreauxi* (again, from north to south):

*P. v. coquereli*: white tinted with yellow; lower back grey; maroon zones on upper thighs and upper arms; continuous with same colour on underside.

*P. v. coronatus*: white, tinted with pale gold or grey on back; whole head black; limbs red-tinted; chest red. Nasal root convex: 'ram-faced'.

*P. v. deckeni*: white, or with pale golden or silvery tints on back. Ram-faced like *coronatus*.

*P. v. verreauxi*: white (sometimes slightly tinted with yellow), becoming pale grey on back; chest white; crown dark brown; forehead white.

*P. v. majori*: white, with back, thighs, and upper arms chocolate brown; head black except for ears and forehead.

As both Petter *et al.* (1977) and Tattersall (1982) point out, there are problems with many of these so-called subspecies. The form called *majori* is in fact based on dark individuals within the range of *P. v. verreauxi*; these occur predominately in two 'pockets' within the latter's range, and are connected by intermediates with dark bellies, red chests, etc. to typical examples of *verreauxi*. Within the range of *P. v. deckeni* occur some individuals that are not solely white, but have reddish areas on the limbs like *coquereli*. Blackish- or reddish-toned individuals also occur within the range of *coronatus*.

Similarly, within *P. diadema* there is doubt whether *holomelas* is really a subspecies, or rather the dark southern end of *P. d. edwardsi*. Tattersall (1982) reports the discovery of an apparently new subspecies, close to *candidus*, near Vohemar.

What is particularly interesting about the geographic variation of the two species is Von Hagen's (1979) analysis of the metachromatic progression (Fig. 4.4). The sequence of saturation–dilution–bleaching in *P. diadema* goes *perrieri/holomelas–edwardsi–diadema–candidus*: there is progressive

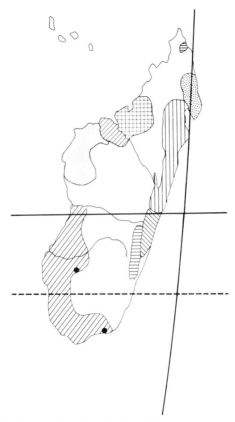

**Fig. 4.4.** Centrifugal metachromism in *Propithecus*.
(1) on east coast, *P.diadema*. Sequence runs from the peripheral, primitive forms, *perrieri* (north) and *holomelas* (south) (horizontal hatching), via south-central *edwardsi* (diagonal hatching) and central *diadema* (vertical hatching) to north-central, highly derived *candidus* (stippled).
(2) on west coast, *P. verreauxi*. Sequence runs from the peripheral '*majori*' (actually a localized morph of *verreauxi*) (south: black dots) and *coquereli* (north: cross-hatched) via south-central *verreauxi* and north-central *coronatus* (diagonal hatching) to highly derived, central *deckeni* (open dot stippling).

greying, then whitening, of the upper parts, with orange zones on the limbs and tail. The sequence in *P. verreauxi* runs *majori–coquereli–verreauxi/coronatus–deckeni*, with progressive whitening of first the upper parts then the underparts. Both sequences, which are to some extent parallel, are centrifugal: the primitive forms at either geographic end, and progressively more derived forms towards the middle. As we have seen, there are difficulties with the division of the two species into subspecies: what can still be seen is the spreading out of metachromatically more and more advanced types from the middle. That in some cases fairly clear subspecific boundaries exist, breaking the cline, whereas in others they do not—and in one case the primitive black morph (so-called *majori*) has all but been swallowed up by more derived stages—does not disturb the broad picture but only adds interest to it.

### 4.2.2(e) Lorisidae

As MacPhee (1981) finds, the real differences of the Lorisidae from the Malagasy lemurs lie in the auditory region alone: pneumatization of the tympanic floor is poor (so that ectotympanic is phaneric) but very marked in central and posterior parts of the tympanic roof (so that the mastoid region and the bone medial to the bulla are strongly inflated). The ascending pharyngeal artery anastomosis with the promontorial is present, as in the Cheirogaleidae, but this forms in addition an extrabullar rete of varying degree of ramification.

The difference between the slow-climbing Lorisinae and the long-legged (in part, vertical clinging) Galaginae is very marked; if we could be sure that the old Lorisid/Malagasy split (i.e. former Lorisiformes/Lemuriformes split: the latter being minus the Daubentoniidae) does result in monophyletic groups, two families could easily be made within the Lorisidae. Dene *et al.* (1976) find, however, that the African slow-climbers differ immunologically from the Asian ones as much as from the Galaginae, or even more, and they propose to recognize them as a new family, Perodicticidae: so the possibility of parallellism must still be taken into consideration: indeed, despite the common possession of a rete mirabile in the lower limb segments in nongalagines the development of just such locomotor specializations is foreshadowed in *Otolemur* (Bearder and Doyle 1974).

I propose the following provisional classification of the Lorisidae:

Subfamily Lorisinae
   Tribe Lorisini
      Genus *Loris*
         Species: *L. tardigradus*
      Genus *Nycticebus*
         Species: *N. coucang*
              *N. pygmaeus*
   Tribe Perodicticini
      Genus *Perodicticus*

Species: *P. potto*
Genus *Arctocebus*
Species: *A calabarensis*
   *A. aureus*

Subfamily Galaginae
Genus *Galago*
Species: *G. senegalensis*
   *G. moholi*
   *G. gallarum*
   *G. matschiei*
   *G. alleni*
Genus *Galagoides*
Species: *G. demidovii*
   *G. zanzibaricus*
Genus *Euoticus*
Species: *E. elegantulus*
   *E. pallidus*
Genus *Otolemur*
Species: *O. crassicaudatus*
   *O. garnettii*

1. Genus *Loris*. Schwartz and Tattersall (1985) propose that the two thin slow-climbers, *Loris* and *Arctocebus*, form one sister group, while the two fat ones, *Nycticebus* and *Perodicticus*, form another. This association had previously been rejected by Groves (1972a), who showed that in skull characters it is the two Asian genera that belong together and the two African genera; while as mentioned above Goodman *et al.*'s (1974) immunological data suggest that the two geographic groups might even be independently derived from an *Otolemur*-like ancestor.

Within *Loris* a single species, *L. tardigradus*, is recognized. It is clear, however, that in Sri Lanka three sharply distinct forms occur: these are the small, red form from the southwestern rain forest (*L. t. tardigradus*); the large, blackish form from the dry country and from intermediate altitudes in the central montane district (*L. t. nordicus*); and a very large form with thick, almost woolly brown fur (*L. t. nycticeboides*: and it does indeed superficially resemble a small *Nycticebus*). It would not be unexpected were these three so different forms found to be reproductively isolated; investigations at the margins of their ranges would be of great interest.

There are also small red and large black forms in (respectively) the Western Ghats and the Eastern Ghats of southern India; specimens of these are much fewer, in collections, than from Sri Lanka.

2. Genus *Nycticebus*. Groves (1972a) showed that there are two species of this genus, *N. coucang* and *N. pygmaeus*, which are sympatric in the Indochinese region east of the Mekong. Of the four subspecies within *N. coucang*, the one which most resembles *N. pygmaeus* is the small Borneo

form, *N. c. menagensis*; the resemblances are in the main symplesiomorphic, but the occurrence in both of a character—absence of $I^2$—which is certainly an apomorphous one suggests that a ring-species effect may be involved, although it is also possible to imagine that *N. c. bengalensis* has simply lost the capacity (allele?) for loss of that tooth, and that it was therefore present in the common ancestor of the genus.

3. Genus *Perodicticus*. The single species, *P. potto*, is found through the West and Central African rain forest belt, and shows some, but not very marked, subspecific variation.

4. Genus *Arctocebus*. The polarity of most of the differences between this genus and the preceding one are difficult to work out. It occupies a more specialized niche (Charles-Dominique 1971), and as a consequence the small size and narrow, pointed snout, ranked as primitive as a rule, would be as readily interpretable as correlated with this; and unquestionably the reduction to small tubercles of both second and third digits of the hand (only the second being reduced, and that to a lesser degree, in *Perodicticus*), and the loss of the tail, are autapomorphous; while *Perodicticus* has, as its only indisputable autapomorphy, the dermal shield and associated structures of the nape of the neck (Charles-Dominique 1971). Unlike *Perodicticus*, this genus is not spread throughout the African rain forest belt, but is restricted to the small central portion of it, from the Niger to the Zaire Rivers. This may in itself be significant in the light of earlier observations about the likely locus of diversification: here from a presumed pre-*Perodicticus* ancestral stock we might envisage the new taxon arising and very rapidly differentiating so that it and its sister group have achieved the status of separate genera.

A further point of interest is that *Arctocebus* has split into two well-differentiated allopatric taxa, which are in fact classed as separate species by Maier (1980); they differ in their dental characters, and also in cranial and external features as my own (unpublished) studies show. This is in fact quite comprehensible from the ecological conclusions of Charles-Dominique (1971): *Arctocebus* occupies a specialized niche, living at high densities in its optimum habitat, but at very low densities—or is absent—elsewhere. It seems to follow from this that it is a poor disperser; there is no evidence that it has ever existed outside its present range (though its habitat is there waiting for it), and where it has dispersed it has undergone local differentiation (in this case across the Sanaga River: *A calabarensis* is found north and west of the River, *A. aureus* south of it). If such a chain of argument is justified, it would not be restricted to the present case; we may expect that, as a general rule, where an African rain forest taxon has thrown up a widespread species and a restricted one:

— the widespread species will be more plesiomorphic and ecologically tolerant, the restricted one more apomorphic and ecologically specialized;

— the widespread species will have a more even population density, the restricted species a more patchy one;
— the widespread species will show comparatively minor geographic variation over its wide range; the restricted species, much more striking diversity over its much smaller range;
— the restricted species will be located in the centre of the widespread species' range: perhaps even in the Niger-to-Zaire area like *Arctocebus*, which is about as central as one can get.

5. Genus *Galago*. The Galaginae, or bush babies, have long hindlegs, which to some degree (but certainly not to the degree shown by living representatives) is a primitive condition; they retain the primitive characters of a long tail, a light body build, unreduced fingers, lack of thumb opposability, and lack of rete mirabile in lower limb segments; and possess the derived states of tarsal elongation, extra hindlimb elongation, long elongated digits with terminal disc-pads, and large mobile ears. The whole subfamily was assigned to one genus by Napier and Napier (1967); the same authors have recently (1985) admitted the existence of four genera, a course which is followed here.

In *Galago* I include, with some misgivings, the rather divergent species *Galago alleni*, as well as the cluster of species that were formerly (Schwarz 1931a) all included in *G. senegalensis*.

*G. alleni* (Allen's Bushbaby), a large, grey–brown species with rusty-toned arms and legs, stands well apart from the other species in many respects (see especially the recent survey of locomotor anatomy and behaviour by Jouffroy and Günther 1985). Like *Arctocebus* it is confined to the Niger-to-Zaire forest zone, is ecologically specialized with a patchy population distribution, and shows strong geographic diversity within its small range; very dark on Bioco Island (*G. a. alleni*); browner in Gabon and southern Cameroun (*G. a. gabonensis*); and very light, very small (head and body under 200 mm in length), with a long pale-coloured tail, in Nigeria and Cameroun north of the Sanaga River (*G. a. cameronensis*). There is, however, no corresponding and more widespread plesiomorphic sister group, at least within the genus *Galago* as recognized here. This raises the question of whether the relationships of *G. alleni* are correctly assessed. T. R. Olson (personal communication) is of the opinion that in fact the species should be referred to *Galagoides*; if this is so, then of course such a sister group is supplied, and Allen's Bushbaby fits precisely into the Angwantibo mould as outlined in subsection 4.2.2(c).

The other species of *Galago* are:

i. *G. senegalensis*: widespread through the thornbush and tree–savannah country of west, east, and southeast Africa, over which vast range it varies very little: it is iron-grey in colour, whitish below, with the hindlimbs yellowish; hindlimb elongation is extreme, the intermembral index being about 55 (Groves 1974*a*).

ii. *G. moholi*: from the semi-arid thornbush country of western Zimbabwe, Botswana, Angola, Namibia, and the northern part of South Africa. This is smaller than *G. senegalensis*, shorter legged, with paler grey colour without marked contrast on the hindlimbs, and a grey–white face with prominent dark eye-rings. Butler (1964) and others find only two pairs of nipples in this species, unlike the three pairs usual in *G. senegalensis*; the gestation period is shorter, twin births are much commoner, and a post-partum oestrus is usual, unlike in *G. senegalensis* (Doyle *et al.* 1971). This is a mixture of character states that are more primitive and more derived than those of *G. senegalensis*.

iii. *G. gallarum*: another semi-arid zone species; but apart from its light greyish colour it is more primitive than either *G. senegalensis* or *G. moholi*: lacking the long legs, white underside, and yellow hindlimb tones of the former, as well as the facial pattern of the latter. It occurs in the Somali Arid Zone, and is marginally sympatric with *G. senegalensis* in northern Kenya and southern Ethiopia.

iv. *G. matschiei*: Olson (personal communication and in preparation) has found that this is the prior name for what has previously (Hayman 1937: Napier and Napier 1967) been called *G. inustus*. Hayman (1936) showed that it has 'needle claws', like *Euoticus*; Hill (1953) took this as evidence for its inclusion within the latter genus; while Kingdon (1971) found that it has, in addition, something of the enlarged orbits of *Euoticus*. Groves (1974*a*) doubted this relationship, noting that its grey colour and dental characters were those of *Galago* proper. If this assessment is correct, it implies that 'needle claws' have evolved twice, in parallel, in the Galaginae, and possibly a similar skull morphology although I was unconvinced by this. (Clearly a cladistic analysis of the Galaginae is in order; it may yet turn out that *G. matschiei* was correctly assigned to *Euoticus* by Hill and others). *G. matschiei* is a darker grey than other members of the genus, but shares the short legs and relative patternlessness of *G. gallarum*. It is a rain forest species; first identified in southwestern Uganda, it has in fact a fairly wide distribution in eastern Zaire (Rahm 1966).

It is tempting to see in this pattern of characters and ranges a case of centrifugal speciation: a highly derived central species (*G. senegalensis*) having totally replaced its primitive ancestral form which has become divided into three non-contiguous segments. This may indeed be broadly the explanation, but the ecological diversity of the three putative remnants (one rain forest, two arid country species) renders this case not nearly as clear-cut as others previously cited. If the rain forest species has, after all, to be transferred to *Euoticus*, the picture would in fact be rather simplified.

6. Genus *Galagoides*: These are small bush babies with shorter limbs and lighter build than *Galago*; the snout is longer, concave, thinner, and lower, with marked elongation of the premaxillae in front of the incisors, and cheek teeth with high, prismatic cusps. The upper first and second molars have a small metaconule, absent in *Galago*: molar hypocones are small; the nasals noticeably broaden rostrally (in *Galago*, their posterior breadth is over 85 per cent their anterior; in *Galagoides*, less); there is a diastema between the upper canine and anterior premolar; the palate is longer than it is broad (in *Galago*, it is broader than long). *G. demidoff* (this is the prior name for the species traditionally called *demidovii*: T. R. Olson, personal communication), the smallest non-Malagasy primate, is like *Perodicticus potto* distributed throughout the West and Central African rain forest belt, but in addition extends out into some montane forest isolates in East Africa: Mt. Marsabit in Kenya, and the Uluguru/Nguru chain in Tanzania. The colour is grey with varying amounts of red or maroon wash, so that it varies individually from nearly grey to quite red overall; in the Powell-Cotton Museum collection from Cameroun there are 23 red, 13 grey; 18 are intermediate though close to the grey type. Geographic variation seems poorly marked: the Uganda/ Tanzania race, *G. d. thomasi*, is noticeably larger. Vincent (1969), however, finds considerable difference in detail between his Congo and Central African Republic samples, so the question should probably be further investigated.

In the East African coastal forests is found a much larger species, grey in colour. Formerly ranked as a subspecies of *Galago senegalensis*, it was independently realized by Kingdon (1971) and Groves (1974*a*) that it is in fact quite a distinct species, *G. zanzibaricus*, with much less-elongated hindlimbs, elongated premaxillae, and other features aligning it closely with *G. demidoff*. Kingdon (1971) suggested that Lawrence and Washburn's (1936) *G. d. orinus*, from the Ulugurus, was actually based on juvenile specimens of *G. zanzibaricus* which was collected sympatrically; but the specimens have been checked and photographed for me by D. Hull (personal communication), who confirms that the case is as described by Lawrence and Washburn: *G. demidoff* and *G. zanzibaricus* are indeed sympatric at that locality, closely related though they be.

Groves (1974*a*) also found that the olive–brown bush babies from the coastal forests of Mozambique are of this type, and ranked them as *G. zanzibaricus granti*. Between them, therefore, the two species of *Galagoides* seem to inhabit all of the closed forests of tropical Africa.

7. Genus *Euoticus*. This is the most highly autapomorphic of the Galaginae, and its affinities are consequently very hard to determine. The body build is robust for a bush baby; the mammae are reduced from the three pairs usual in other genera (two may occur in *Galago* and *Galagoides*) to only a single pair; the orbits are huge and circular; the nails have needle-like

points, like *Galago matschiei*; the dental comb is exceptionally procumbent; and the anterior upper premolar is long and dagger-like. Charles-Dominique (1971) finds that it has a diet high in gum, and uses its needle claws to dig into bark while hanging upside down and piercing with its canine and caniniform premolar, and scraping with its toothcomb. This way of life is strongly convergent with *Phaner* in Madagascar, and inferentially with *Allocebus*.

The distribution of the genus is the same as that of *Arctocebus* and *Galago alleni*: like these two, it has a specialized niche and is strongly autapomorphic, as we have seen; and also, like them, has strongly marked geographic variation. *E. elegantulus* is foxy red above and sharply contrasting silvery white below; it occurs south of the Sanaga River in Cameroun, Rio Muni, and Gabon. *E. pallidus*, which occurs on Bioco Island and north of the Sanaga in Cameroun and southeastern Nigeria, is much more red–brown, becoming grey on the shoulders, arms, and nape, and on the tail; it has a darker brown dorsal stripe; it is yellowish grey–white below; and is very much smaller (head and body length under 200 mm) but longer tailed.

It is interesting that the pelage colour of *E. elegantulus* strikingly resembles that of the 11–15 day old *Galagoides demidoff*, as described by Vincent (1969).

8. Genus *Otolemur*. These are the largest bush babies, and in my view the most plesiomorphic: Bearder and Doyle (1974) note that the species is capable of almost lorisine locomotor patterns as well as more typically galagine ones; and there are some similarities—such as the presence of a heel gland—that even suggest that the genus might be referred to the wrong subfamily. In terms of uniquely derived characters *Otolemur* has a peculiarity of the finger nails [terminally concave, as pointed out by Hayman (1937)], and cranial and dental features: a characteristic mesio-distal decrease in the size of the molar series; a large occipital bone, ascending well onto the dorsum cranii; elongated muzzle; low, bunodont molar cusps; high, raised $P^2$.

It is now fairly commonly recognized that there are two species in this genus: a large (head and body length usually more than 300 mm) grey, grey–brown, or black form with large ears (55 mm or more in length), and a small (under 300 mm) brown one with small ears (less than 50 mm long). They also differ in their reproductive parameters (Eaglen and Simons, 1980). The large species is found from Angola, Natal, and Transvaal, north as far as to northern Tanzania, southwestern Uganda, and Rwanda; the small species, in northern Tanzania, and eastern Kenya. The nomenclature is contorted; Olson (1980) has proposed a neotype for the name *crassicaudatus* (the earliest name), settling that name on the southern species and making *O. garnettii* the prior name for the northern one.

## 4.3 THE HAPLORHINI

The interrelationships of the different extant haplorhine groups are quite clear, being dichotomous at any rate down to superfamily level. The names used (and convincingly justified) by Hoffstetter (1982) are: for the first level, Tarsiiformes vs. Simiiformes; for the second level, within Simiiformes, Platyrrhini and Catarrhini; for the subfamilies of Catarrhini, Cercopithecoidea and Hominoidea. If the first-level division is given infra-ordinal rank, then a new rank has to be inserted for the second-level split—Hoffstetter uses 'section', which is a reasonable if somewhat informal solution. An alternative would be to forget one of the dichotomies, and have a three-way split somewhere; i.e. to raise Catarrhini and Platyrrhini to infra-ordinal level along with Tarsiiformes, or [the strategy adopted by Simpson (1945), who did not even have infra-orders] to forget the Platyrrhini/Catarrhini split and divide the Simiiformes ( = Anthropoidea of Simpson) into three super-families, with Ceboidea (equivalent to Platyrrhini) taking its place alongside Cercopithecoidea and Hominoidea. The evolutionary relationships are simply too clear-cut for either of these solutions to be of any real value.

A third possibility is that adopted by Szalay and Delson (1979), though in a different context: to reduce the third-level split to below superfamily rank. If this is done, then for a superfamily incorporating all extant catarrhines the name that takes precedence is, unexpectedly, Cercopithecoidea. But the category Hominoidea is not only well-used but is very useful to indicate a suite of characters in common between apes and humans, and flexibility is much decreased if the ranks are pushed downward like this.

When all is said and done, freedom of taxonomic thought is the aim of the formal structures, and if we want to insert an extra rank because it is helpful, then let us do it. McKenna (1975) did so for supra-ordinal ranks, and if Hoffstetter (1982), who is the one who has laid down the taxa and their ranks in modern times, uses Section then that is the way it will be.

### 4.3.1 The Tarsiiformes

The Tarsiiformes are plesiomorphic among the Haplorhini in their retention of a large yolk sac in the embryo, and other primitive retentions in embryology (Luckett 1976); in their incomplete post-orbital closure; in their relatively small brain; and presumably in their dentition, though it is conceivable that hypocones may have been lost and cusps once again become crystalline. The features of the soft anatomy that made Woollard (1925) call it 'a lemur of the lemurs' are again primitive retentions.

The living tarsiers are also highly derived in many features. They have elongated hindlimbs and, especially, calcaneus and navicular; the tail is modified, only sparsely haired below with rat-like scales; the eye is enlarged, and lacks cones; there is only one pair of lower incisors; the ear is enlarged and mobile (which could be either a specialization for its carnivorous way of life, or a primitive character unchanged from the large ears of many strepsirhines); there are expanded, disc-like terminal pads on the digits; and small triangular nails on all digits except the second and third toes which bear toilet claws.

A word about the toilet claws is necessary. Schwarz and Tattersall (1985) suggest that the tarsier may be a 'lemur of the lemurs' after all, if the toilet claw on the second toe is homologous with that of lemurs. If: but there is absolutely no question of homology. The revealing findings of Thorndike (1968) have already been mentioned (section 4.2 above); lemurs (Lemuriformes in the present sense) lack a deep stratum in the falculae, and lack a terminal stratum in the germinal matrix, and this is the case whether in the nails or in the toilet claw. On the other hand the tarsier has the terminal matrix, and the deep stratum which is derived from it, again in nails and toilet claws alike. If, as seems commonsensical, the deep stratum supports the strong curved structure of a claw, then the tarsier's toilet claws could well be primary, while the nails would profitably be seen as reduced claws (they are reduced anyway, of course), while the toilet claws of lemurs are evidently modified nails.

The tarsier eye also deserves a short discussion. It has a fovea, a structure not unknown in lemurs (Pariente 1970), but well-developed only in haplorhines; and, if not a macula as such (which is defined as a cones-only region: and the tarsier has no cones), it certainly has a histologically macula-like region (Wolin 1974). These features, like those of the fetal membranes and external nose, place it among the Haplorhini. The apparent total absence of cones is surprising, because even *Cheirogaleus* does not entirely lack them (Polyak 1957). I suggest that because it lacks the strepsirhine tapetum, which does so much to increase light levels, the tarsier as a none the less nocturnal animal needs every rod it can get, so has completely eliminated cones in their favour. It is interesting that the only other primate known to lack cones altogether is *Aotus*: the only other nocturnal haplorhine, i.e. the only other nocturnal primate to lack a tapetum [although it now appears it may have a sort of tapetum after all: see Walls 1939 (Prof. R.W. Rodieck, pers. comm.)].

Extant tarsiers are assigned to a single genus with four species (Niemitz 1984):

> *Tarsius bancanus*: Sumatra, Banka, Belitung, Natuna Island, Borneo;
> *T. syrichta*: Mindanao, Leyte, Negros;
> *T. spectrum*: Sulawesi, Sangihe Isand, Peleng,
> *T. pumilus*: Central Sulawasi: Montane forest.

The interrelationship between these four species are not even. The two Sulawesi species seem to be quite close; indeed, they were only recently distinguished (Niemitz 1984; Dagosto and Musser 1986). The recent detailed study of Sulawesi tarsiers by Musser and Dagosto (1987) is a model of what a systematic study should be; the relationships of all four species are also discussed in the same paper. Mackinnon and Mackinnon (1980) surmise, on the basis of vocalizations, that yet other species may be found on Sulawesi; Musser and Dagosto (1987, p. 39) give reasons for doubting this. As Niemitz accepts, the Sulawesi species are much more primitive than the Sundaland and Indonesian ones, with less limb, tarsal, and digital elongation, a more hairy tail, and far less expanded orbits. As analysed by Niemitz, there would also seem to be consistent ethological differences, and the Sulawesi species are much less specialized ecologically.

The Philippine tarsier is, as was emphasized by Niemitz (1984), intermediate between Sulawesi and Sundaland species: i.e. it too is primitive, but less so than the Sulawesi species. While the differences from *T. bancanus* are none the less noticeable, *T. syrichta* is still a highly derived species, whereas for anyone whose idea of a tarsier is based on the Sundaland species, the Sulawesi ones hardly look like tarsiers at all, especially in the matter of orbital enlargement (quite lacking the spectacular enlargement seen in *T. bancanus*, and to a lesser degree in *T. syrichta*).

The various cranio-dental (Fig. 4.5) and post-cranial differences between the Sulawesi tarsiers and the others render it profitable to recognize them at generic level. A proposal to rank them as two different genera is under preparation (Groves, in preparation).

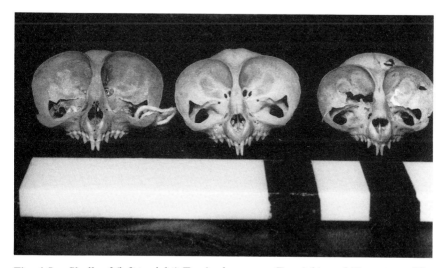

**Fig. 4.5.** Skulls of (left to right) *Tarsius bancanus, T.syrichta* and *T.spectrum*. The differences, particularly in orbit size, between *T.spectrum* and the other two species are marked.

### 4.3.2 The Simiiformes

Hoffstetter's (1982) use of this name is gaining acceptance; he explains why the perhaps more familiar term Anthropoidea cannot be used.

The two sections, Platyrrhini and Catarrhini, have sometimes been mooted to be independent derivatives from 'prosimians'. Delson and Rosenberger (1980) have finally cut away the last remnants of the foundations of this theory and laid it to rest for good, one would hope. There remains the problem of zoogeography. The assumption has been that the simiiform ancestor evolved in the Old World (Africa being the usual candidate) and dispersed across the South Atlantic to South America. If, as usually supposed (Le Pichon 1968; Cachel 1981), the break-up of Gondwanaland and the formation of the South Atlantic are mid-Cretaceous, at the latest, in age, then there is no hope of getting primates from Africa to South America dryshod. Hoffstetter (1979) has reconstructed Eocene currents (simiiform primates are emplaced in both continents by the mid-Oligocene, so an Eocene time table is favoured), and has proposed that with the aid of a raft of drifting vegetation and a *Cheirogaleus*-like food-storage system a proto-platyrrhine could survive a voyage across a South Atlantic only half as wide as today, and land safely in Brazil.

Croizat (1962) took great exception to such 'ad hockery'. For him there are such regularities in Ethiopian/Neotropical affinities that we cannot entertain any notion of chance dispersal. Either the formation of the South Atlantic was a much later event than has hitherto seemed possible, or the period of 'mobilism' (i.e. the requisite range extension) was much earlier, or orthogenesis from previously emplaced (identical?) precursors explains it; or else South America is of composite origin, and the proto-platyrrhines were carried across by a piece of Africa that separated from the main.

The earliest known catarrhines, or presumed catarrhines, are in fact Eocene in age, and found in Burma, not Africa. A Croizatian Pacific track would then have been involved: which raises still more problems than the Atlantic one.

#### 4.3.2(a) The Platyrrhini

Compared to the Catarrhini, the platyrrhines are quite plesiomorphic; indeed it has been difficult to find any characters in which they might be said to share the derived state. Delson and Rosenberger (1980) however, have, pointed out a few such characters: pattern of post-orbital bony mosaic, with malar-parietal articulation; loss of $M^3$ hypoconulid; reduction of metaconules, loss of paraconules; presence of intraplacental maternal vessels; enlarged embryonic nasal capsule.

Traditionally the platyrrhines have been divided into two families: Callitrichidae for the marmosets (with claws and no third molars), and Cebidae for the rest (with nails, and third molars). One genus, *Callimico*, does not fit into this neat division: it has claws, but third molars are present. For Hershkovitz (1978), this is no difficulty: there has been much parallelism in the New World monkeys, such that morphology can in part be predicted from the species' body size. Small ones (which are necessarily primitive) lack third molars and have claws (and the smallest of all have V-shaped mandibles), and the largest ones automatically develop prehensile tails; so it is natural that there should be one genus that, by virtue of its intermediate size, should fall between the two 'family' definitions. (As a matter of fact, *Callimico* is not the largest of the clawed group.) Hershkovitz sees no difficulty, then, in allocating *Callimico* to a separate family: Callimiconidae.

Others have seen the problem differently (e.g. Rosenberger 1980). Evolution, while it might have its predominate directions, is susceptible to apparent reversals; and callitrichid size reduction is very likely one such case, claws and all. So *Callimico* would be only the most primitive stage of the callitrichids, before third molar loss.

I suggested above that the nails of tarsiers might best be seen as modified claws. Those of platyrrhines are of a primitive structure like those of tarsiers (Thorndike 1968); so I would make the same proposal in regard to them too. I suggest that *Callimico* can be seen as the most overall plesiomorphic platyrrhine, retaining claws as well as third molars. The anatomical observations of Hill (1957) however, do, confirm its status as a callitrichid.

The monophyly of the Callitrichidae, if we except the inclusion of *Callimico*, has never been doubted; but that of the Cebidae has, and with good reason; they share no feature but that of not being marmosets and of having nails rather than claws. Rosenberger (1977) finds a major division between *Cebus* and *Saimiri*, on the one hand, and all the other genera on the other. The first pair have as derived features a very gracile zygomatic arch, reduced third molars, enlarged canines with a honing mechanism on the anterior lower premolar, a narrow inter-orbital space, and very distinct enlargement of the premolars. The remaining genera share, as their derived states, a posteriorly deepened mandible, enlarged molar hypocones and reduced paracones, and a deepened glenoid fossa. The marmosets align in most respects with *Cebus* and *Saimiri*; there are thus only two basic divisions in the platyrrhines, as before, but they are different from the traditional ones. Rosenberger therefore has two families: Cebidae (*Cebus, Saimiri,* and the marmosets), and Atelidae (the other genera).

A number of factors intervene to qualify acceptance of this promising arrangement. First, though not necessarily damning, is that the molecular data (Baba *et al.* 1979; Cronin and Sarich 1975) by no means agree with it. Second is the rather subjective nature of some of the characters. Third, and perhaps most important, is that all the lineages that are so far known as fossils go back a long way: the *Aotus, Saimiri*, pitheciine, alouattine, and marmoset lineages, at least, all go back to the Miocene (Delson and Rosenberger 1980; Setoguchi and Rosenberger 1985). Thus any special relationships that can be detected between any of these lines must have been short-lived at best. I have spoken already (Chapter 1, section 1.5) of the necessity for conservatism in classification; but where the old classification clearly will not do, but the new one is problematic as well, what will best serve the cause of communicating reality? I suggest that the first step is to agree on what the undoubted natural groups are, and then rank them all, equally, at whatever rank one thinks appropriate.

The natural groups surely are:

— marmosets, including *Callimico* (see above);
— *Cebus* with *Saimiri*: Rosenberger's (1977) analysis is convincing that these two genera are at any rate related, if not necessarily very close;
— *Aotus*;
— the Pitheciinae (*Pithecia, Chiropotes, Cacajao*: about whose interrelated-ness there is no dispute);
— *Callicebus*;
— the Atelinae (*Ateles, Lagothrix, Brachyteles*), with *Alouatta*: all analyses, including the molecular ones, agree on this association.

Considering the known antiquity of at least some of these groups, and the inferred antiquity of the others, I propose that each be recognized as a family: Callitrichidae, Cebidae, Aotidae, Pitheciidae, Callicebridae, Atelidae. [Since this was written, the very thorough and scholarly review by Ford (1986) has appeared—too late to be taken into consideration here, but its conclusions should be considered carefully by all future workers.]

*4.3.2(a)(i) Callitrichidae* Except for the amino-acid sequencing data of Baba *et al.* (1979), no modern analysis suggests that the Callitrichidae are the most divergent of the platyrrhines. They are related to *Cebus* and *Saimiri* (Rosenberger 1977), or to *Aotus* (Baba *et al*, 1979), or at most are just one line of a very ancient radiation (Cronin and Sarich 1975).

*Callimico*, compared to the other marmosets, would seem to be the plesiomorphic sister group; the others have, as their most obvious derived state, third molar loss. Sarich and Cronin (1976), however, place *Callimico* as the sister group of *Callithrix* alone, their combined lineage being part of a three-way split whose other lines are *Saquinus* and *Leontopithecus*: so the position is not quite clear-cut.

1. Genus *Callimico*. This genus, apparently monotypic (*C. goeldii*), is found in the Upper Amazon tributaries region. While rather little is known of its natural history, that it is genuinely carnivorous (in the sense of preying on vertebrates as opposed to mainly arthropods) is of considerable interest (Lorenz 1969).

2. Genus *Callithrix*. This genus has been extensively treated by Hershkovitz (1978), who adopts, however, a predominately phenetic approach to supra-

specific taxonomy. The data presented by him point, as indeed has been suggested by Rosenberger (1983), to the inclusion in the genus of the pygmy marmoset, *Cebuella*.

*Callithrix* (incorporating the Pygmy Marmoset) can now be defined by the possession of a number of uniquely derived states. These include, first and foremost, the characters of the dentition: the incisors are elongated, the canines somewhat shortened and incisiform, and the whole battery placed in a curve around the point of a rather V-shaped mandible: in which respect the genus differs from all other Simiiformes. While Hershkovitz (1968) saw a resemblance to the similarly V-shaped forms of tarsiers and many lemurs, in each case the form is associated with different specializations (dental comb in lemurs, loss of an incisor in tarsiers, incisor elongation in *Callithrix*), and is thus almost certain to be evolved in convergence. Approached from a different direction, it is clear that in all respects *Callithrix* belongs to the same (monophyletic) group as *Saguinus* and other callitrichids, which have the otherwise simiiform condition of long canines, structurally differentiated from the incisors, and placed at the corners of a rectangularly shaped mandible. Rather than postulate three (or more?) cases of convergent evolution (convergent to the point of identity!) between the other callitrichids, other platyrrhines, and catarrhines, I think it is easier to see the *Callithrix* condition as a (superficial) convergence on that of the small lemurs, as a specialization for its exudate eating habit.

*Callithrix* may also be differentiated from other callitrichids by its genitalia. As shown by Hershkovitz (1978), in *Callithrix* the glans penis is relatively undifferentiated from the penis shaft, whereas other callitrichids have the general simiiform condition of glans and shaft differentiated by the presence of a sulcus retroglandis (less developed in *Saguinus*). The unpigmented labia majora, and the position of the preputial folds mediad to the scrotal fold, distinguish the female genitalia. Rosenberger (1983) makes a case for *Callithrix* and *Leontopithecus* forming a sister group relative to *Saguinus*, although the relationship is admittedly not close; they share the derived character states of the upper central incisors enlarged and thickened, lower incisors tall, upper molars triangular, and the absence of the entepicondylar foramen of the humerus.

The differences between *Callithrix* and *Cebuella*, as given by Hershkovitz (1978), are that the latter is small in size with a multibanded agouti pelage; the penis shaft lacks spines; the scrotum is sessile; $I^2$ has a mesial as well as a distal cuspule, as do both lower incisors; and the lower premolars are at least as long as they are wide. The polarity of the dental characters is difficult to interpret; the genital characters would seem to be derived, considering the widespread possession of penis spines and pendulous scrota in primates; while the pelage would seem to be metachromatically primitive.

Within *Callithrix*, as recognized by Hershkovitz (1978), there are two species groups: the *jacchus* and *argentata* groups. In the former the pelage is agouti and the flank pattern undifferentiated from that of the lower back; and the scrotal folds of the female are sessile and papillated (as in *Cebuella*). In the *argentata* group the pelage is not agouti, the arms and mantle (forepart of the back) are pale in colour, and there are prominent white hip patches. Most striking of all, perhaps, is that the young of the *argentata* groups are like the adults, perhaps darker in colour; while the young of the *jacchus* group have a simpler agouti pattern than the adults, rather than marbled, and lack the facial pattern and the ear-tufts of the adults. In short, the young of the *jacchus* group are almost exact replicas of the Pygmy Marmoset, *Callithrix* (ex-*Cebuella*) *pygmaea*!

The data on growth and maturation of members of the genus support a conclusion that *C. pygmaea* is essentially a paedomorphic derivate of the *C. jacchus* group, derived by progenesis. The data cited by Hershkovitz (1978) record that *pygmaea* becomes sexually mature and ceases to grow at about one year of age, while all other members of the genus grow until they are well over one year old, and reach sexual maturity at 18 months or more. If we were to terminate the growth of a member of the *C. jacchus* group, and bring forward its sexual maturity, at one year, we would be left with a *C. pygmaea*: with the sole exception of the dental differences, and, externally, of the ear-tufts, which begin to develop at around six months in the *jacchus* group. The latter share, as their derived characters, the differentiated facial pattern (black mask, light or white forehead).

The *C. argentata* group are distinguished from the others of the genus by their derived states: the non-agouti pelage as described above, and their largely bare ears, which are not hidden by any tufts that may be present (the ears are hairier in the young than in the adult).

The distribution of these species groups is noteworthy: the agouti *C. pygmaea* lives in the Upper Amazon forests, its range (according to Hershkovitz (1978) ) bounded by the River Purus and the Caqueta–Japura–Solimoes river system. The agouti *C. jacchus* group lives on the east coast, from the southern bank of the Lower Amazon to the Rio de Janeiro district. The metachromatically derived *C. argentata* group lives in the middle, between the two primitive types, among the southern tributaries of the Amazon approximately to its estuary. Inferentially, centrifugal speciation has been at work.

*C. pygmaea* seems to be monotypic, and there are no taxonomic problems beyond its generic allocation.

In the *C. jacchus* group, Coimbra-Filho and Mittermeier (1973) recognize five species, which approximately replace each other down the eastern seaboard of Brazil, with narrow areas of marginal sympatry (or parapatry). They are:

a. *C. jacchus*: from Piaui, Ceara, Pernambuco, and Paraiba; its southern limit is the River Sao Francisco. The ear-tufts take the form of white, or greyish, fan-like tufts from in front of the ears themselves; the pelage is marbled, and there is a white frontal blaze.

b. *C. penicillata*: with a very wide range, from Bahia south to the River Jucurucu (17°21′S), then inland and north to the River Tocantins at 9°14′S, 48°12′W, south to the Serra do Itatiaia at 22°33′S, 44°38′W. It resembles *C. jacchus* overall, but the ear-tufts are black and more drooping, less fanlike; the crown and temples are also blackish.

c. *C. geoffroyi*: occupies the lowlands south of the coastal range of the last, from the River Mateus (18°38′S) to about 21°S. It looks strikingly different from the other species by virtue of its totally whitish crown, temples, and face, set off against the black ear-tufts and darker body tones.

d. *C. flaviceps*: lives at high altitudes inland of the range of *C. geoffroyi*, from 20′–21°S and about 40°36′ to 41°40′W. Unlike the previous three species the ear-tufts grow from the inner surface of the ear pinna itself; the fan-shaped tufts are yellowish, as are the whole head and throat (yellow-agouti).

e. *C. aurita*: inhabits the coast from 21°S to about 23°30′S, and some way inland. The tufts resemble those of *C. flaviceps* but are whitish; the crown is modified agouti, ochraceous, or black; the cheeks, temples, and throat are black.

There is a controversy over the status of these five taxa. Coimbra-Filho and Mittermeier's (1973) claim that five species are involved is challenged by Hershkovitz (1978), who disputes the evidence of marginal sympatry and ranks them as five subspecies of a single species. Specimens from the River Mateus and other localities are suggested by Hershkovitz to be hybrids between the two taxa: the evidence seems equivocal, but even were it correct it would be suggestive, not conclusive; and in fact Rosenberger (1983) has pointed out that these so-called hybrids are not intermediate, but closer to *C. geoffroyi*, nor are they from parapatric localities: they tentatively propose that instead a further species is involved, for which the available name is *C. kuhli*. At Aracruz (19°55′S, 40°36′W) both *flaviceps* and *geoffroyi* have been collected: Hershkovitz considers this to be evidence not of marginal sympatry but of altitudinal replacement about this locality. Of five *flaviceps* from Santa Teresa, one has dark ear-tufts, suggestive according to Hershkovitz of intergradation with *geoffroyi*; another (much more convincing) *geoffroyi* × *flaviceps* hybrid, with agouti forehead and crown but a pale frontal blaze may, as he agrees, be not from the wild at all but bred by a noted amateur naturalist living in the region.

Consequently, while the case is not clear-cut, the proposal of Coimbra-Filho and Mittermeier (1973), that there are five species of the *Callithrix jacchus* group, does seem more reasonable at present (the *kuhli* question is a separate matter). But the general picture is clear: five sharply distinct, closely

**Fig. 4.6.** Centrifugal metachromism in *Callithrix jacchus* group. Sequence runs from primitively agouti, peripheral *jacchus* (north) and *aurita* (south) (horizontal hatching), via intermediate forms to highly derived central *geoffroyi* (stippled).

parapatric taxa, the northern ones with tufts in front of the ears, the southern with intrinsic tufts; tufts white in the far north (*jacchus*) and south (*aurita*); agouti pelage in the north and south, more modified in the centre; and the most modified, conspicuous facial pattern also in the centre (*geoffroyi*). A remarkable centrifugal pattern (Fig. 4.6).

The *C. argentata* group has, according to Hershkovitz (1978), two species:

— *C. argentata*: ear-tufts absent, the ear being nearly bare;
— *C. humeralifer*: tassel-like tufts emerging from the whole pinna: inner and other surface and rim alike.

Their distributions are remarkable. *C. humeralifer* lives between the Rivers Aripuana and Tapajoz; the range of *C. argentata* surrounds it to the east, south, and southwest, the two being sympatric between the Upper River Aripuana and its tributary the River Roosevelt. It cannot be claimed, however, that such a distribution reflects centrifugal (or any other particular mode) of speciation, as the two appear to be divergently specialized: *C. argentata* would seem to have become facially and aurally dipilated (the young are hairier than the adults) as thoroughly as the ears of *C. humeralifer* have undergone hypertrichy.

Each of the two has independently, as Hershkovitz (1978) shows, gone through a bleaching sequence. The three subspecies of *C. humeralifer* are curiously distributed, the most primitive (*humeralifer*) and most bleached (*chrysoleuca*) being found on opposite sides of the River Canuma, the intermediate one (*intermedius*) further downstream. The three subspecies of *C. argentata* have a more centrifugal distribution: the most primitive (*melanura*) and the next (*argentata*) are peripheral in range to the most bleached one (*leucippe*).

3. Genus *Leontopithecus*. The beautiful, and excessively rare, lion marmosets lack the characteristic incisor–canine relationships of *Callithrix* but according to Rosenberger and Coimbra-Filho (1984) they do share a few derived features with the latter, the lower incisors in particular being relatively tall, a possible incipient stage to the *Callithrix* arrangement. The genitalia are less modifed than the latter (Hershkovitz 1978): that their genital features are shared with *Callimico* and with some, but not all, species of *Saguinus* is sufficient indication of the polarity in this case.

Though usually placed in a single species, Rosenberger and Coimbra-Filho (1984) argue that three full species should be recognized. *L. rosalia*, from the Rio de Janiero district, is a beautiful golden colour all over; cranio-dentally it is characterized by its reduced anterior dentition and very abbreviated pre-maxillae. *L. chrysomelas*, living further north, on the coast between the Rivers Pardo and das Contas in Bahia Province, has the golden tones restricted to the head, the body being black; it has a more robust physique but a small face and small teeth. *L. chrysopygus*, the southernmost species, found between the Rivers Paranaponem and Tiete, has the orange tones on the loins and rump, the rest being black; it is the largest species, more resembling *L. rosalia* in its detailed anatomy but less gracile. The authors remark that 'The two populations at the tail ends of the swath of their distribution are those that most resemble one another, counterintuitively' (1984, p. 167), and go on to identify the black colour type as primitive, so that the central species, *L. rosalia*, is the most derived in lacking any black tones. We have seen above that there is no longer any reason to regard such a pattern as 'counterintuitive'.

4. Genus *Saguinus*. This large, unwieldy genus has the callitrichine synapomorphies that distinguish the subfamily from *Callimico* but lacks the apparent synapomorphies that unite *Callithrix* and *Leontopithecus*. But whether the genus has any derived features of its own is unclear: the genital features cited by Hershkovitz (1978) that so well order the polarities of the other marmosets are here variable, and the cranio-dental characters that do certainly define the genus seem to be primitive callitrichine ones. Seeing the enormous diversity in external features among the species, the question must be asked whether some of the species might not be phyletically closer to the *Callithrix/Leontopithecus* group.

Maintaining, in the absence of contrary evidence, the unity of the genus, following Hershkovitz (1978) six species groups are recognized, ten species in all:

*inustus* group
  *Saguinus inustus*: Mottled-faced Tamarin

*bicolor* group
  *Saguinus bicolor*: Pied Tamarin

*leucopus* group
  *Saguinus leucopus*: White-footed Tamarin
  *Saguinus oedipus*: Cottontop Tamarin or Pinche

*nigricollis* group
  *Saguinus nigricollis*: Black-naped Tamarin
  *Saguinus fuscicollis*: Saddlebacked Tamarin

mystax group
  *Saguinus mystax*: Moustached Tamarin
  *Saguinus labiatus*: White-lipped Tamarin
  *Saguinus imperator*: Emperor Tamarin

*midas* group
  *Saguinus midas*: Black Tamarin

The first three groups are known as bare-faced tamarins, and are derived in their hair-tract patterns; the fourth, fifth, and sixth groups are known as hairy-faced tamarins, and are primitive, having long hairs on the cheeks, forehead, and crown. The bare-faced groups divide into an intermediate group (*leucopus* group), in which the cheeks and sides of the face are hairy, but the forehead and anterior crown have only very short, downy hairs, and two highly derived groups (*inustus* and *bicolor* groups), in which the cheeks and sides of the face are likewise sparsely haired, or even naked.

It is of interest that the three hairy faced groups are widespread throughout the Amazon Basin (with the exception of a section south of the middle/lower Amazon), and the intermediate *leucopus* group is isolated in Panama and northern Colombia; while the two advanced bare-faced groups are both central in distribution: *S. inustus* is found northwest of the Rio Negro, between the ranges of the *leucopus* group and the hairy-faced groups, and *S. bicolor* is found in the middle of the ranges of the hairy-faced tamarins (though sympatric only with *S. midas*).

The monotypic species *Saguinus inustus* is blackish or dark brown all over; only the face has peculiar mottled light and dark pigmentation which gives it its vernacular name.

*Saguinus bicolor*, the most nearly bald-looking of all species, has a unique parti-coloured pattern, the foreparts being whitish or buffy, the hindparts brown, the belly reddish or orange. Three well-marked subspecies are distinguished by Hershkovitz (1978).

The intermediate bare-faced group has two species. *S. leucopus* is pale buffy and has a short upstanding crest on the crown, contrasting strongly with the very short-haired anterior crown. *S. oedipus* has a brindled reddish brown upper side, the underside, face, and extremities being pure white; of the two subspecies, *S. o. geoffroyi* has a short crest like that of *S. leucopus*, while *S. o. oedipus* has a long sideways-flaring crest like an Iroquois warrior. The distributions form a centrifugal pattern: *S. leucopus*, mainland Colombia (north of the Rio Atrato); *S. o. geoffroyi*, Panama; *S. o. oedipus*, in between the other two in isthmian Colombia.

The three hairy-faced groups are differentiated as follows: in the *nigricollis* group the facial skin is entirely black or brown, but there is a band of short grey–white hairs around the mouth, and there is a sharp transition between dark mantle and a lighter (usually more agouti) middle body zone: in the *mystax* group there is a similar, but longer haired, white circumoral zone, but the underlying skin is itself unpigmented, and the mantle and midback zones are little or not differentiated; in the *midas* group the face, hair, and skin alike is black, as is virtually the whole body.

The *nigricollis* group has two species: *S. nigricollis*, in which the pattern is bipartite, between the mantle and the rest of the body (lower back, rump, thighs, and underparts); and *S. fuscicollis*, in which the pattern is tripartite, between mantle, saddle and rump. The two are very closely related, and it seems possible that a saddle/rump differentiation might have arisen independently, more than once; but Hershkovitz (1978) has analysed them as they stand, and even at this level the results are remarkable. *S. nigricollis* has only two subspecies: the very primitive *S. n. graellsi*, from the Upper Rio Napo, in which the mantle is agouti and the lower back, thighs, and underparts are brown; and *S. n. nigricollis*, from between the Lower River Napo and the River Ica, in which the mantle is saturate, black, and the lower back, rump, thighs, and underparts are reddish. But in *S. fuscicollis* there are no fewer than fourteen subspecies, and they show a general progression from primitive, mainly agouti-haired, forms, to brightly coloured black and red forms (*S. f. lagonotus, tripartitus*) in one series and to completely bleached white forms (*S. f. melanoleucus*) in another. (The theory of metachromism was, in fact, born largely from Hershkovitz's analysis of the chains of subspecies in this species). Along the first sequence we get subspecies that are more and more saturated in their patterns: banding is progressively eliminated from the hairs, until solid colours are reached. In the second series, bleaching progressively occurs: pigmentation drops out altogether. The exact sequence is disputable; as Hershkovitz's (1978) book went to press, a new subspecies was discovered, which in the Appendix he described as *S. f. primitivus* and placed at the very stem of the species, necessitating a slight rejuggling of the two main metachromic sequences. So it might be better simply to identify a cluster of primitive (i.e. agouti) races, rather than try to arrange them in sequences, since Hershkovitz himself has changed his mind about their exact ordering. These agouti races are *S. f. primitivus, weddelli, avilapiresi, fuscicollis*, and *fuscus*, and broadly speaking they occupy the southeastern, eastern, and northern rim of the species' distribution. The western edge is occupied by two highly derived forms, *S. f. tripartitus* and *lagonotus*; but, remarkably, they are sympatric with the other primitive member of the species group, *S. nigricollis graellsi*. There is thus, except for the southwestern strip, a complete ring of primitive taxa bordering the range of the species group. The saturating series, the one that ends in *tripartitus*, is a straight westward-travelling chain; but the highly derived bleaching series, *primitivus–cruzlimai–crandalli–acrensis–melanoleucus*, curls round in spiral and ends in the very centre of the range, in the angle between the Rivers Taranara and Jurua.

To the *S. mystax* group are ascribed three well-defined species. *S. labiatus* is the most primitive, in that the skin of the nasal alae is not depigmented (unlike that of the inner borders of the nostrils), and the underside is yellow or red. *S. mystax* has unpigmented skin entirely around the nostrils; there is a short white moustache; the underside is black. *S. imperator* more resembles *S. mystax*, but has much longer, sweeping moustaches, grey or brown underparts (with often a reddish tone on the belly), and the tail is reddish instead of black. The three live parapatric in an area south of the Amazon around the Rivers Purus and Jura: *S. labiatus* to the east of the Purus, *S. mystax* further west, *S. imperator* to the south around the headwaters of the Rivers. The most sharply patterned of the group is a red-crowned subspecies, *S. m. pileatus*, which lives in a central range between the Purus and Jurua. As an anomaly, a marmoset ascribed to *S. labiatus* (*S. l. thomasi*) lives north of the Amazonas/Solimoes, opposite the range of *S. m. mystax* and so about as far from its alleged conspecific as it is possible to be; if it is correct, as I have suggested above, to regard the characters of *S. labiatus* as primitive for the group, then clearly the race *thomasi* is allocated to the latter only by the common possession of symplesiomorph features and ought to be awarded species status. [The differences are the less extensive crown pattern (red line in front, grey spot behind; very well-marked in *labiatus*, only barely developed in *thomasi*), and the wholly black throat and chest instead of the chest being reddish yellow like the belly.]

The last of the three species groups of hairy-face tamarins is the *S. midas* group. This has a single species, with an all-black race, *S. m. niger*, and the golden-handed *S. m. midas*.

The diverse and widespread tamarins illustrate very well many of the evolutonary ideas introduced in previous sections of this book. Twenty years ago Hershkovitz (1968) introduced the 'principle of metachromism', and despite disbelief at the time it has come to enjoy considerable acceptance; many analysts have found it useful as a general method, and consistent in its results, so that it stands today as one of the best-tested examples of directionality in evolution. *Saguinus fuscicollis*, the type case of the principle, is today as good an example as ever it was.

Second, we have seen how centrifugal processes can be invoked in the sub-speciation and speciation of the genus. In the genus as a whole the most apomorphic species occupy a central range; in one species group after another the central taxa are found to be central in their distribution. It is only unfortunate that so little is known of the karyotypes of marmosets; although information so far available (Chiarelli 1979) suggests that the Callitrichidae as a whole are karyotypically conservative, with all those so far tested having $2n = 46$, except for the plesiomorphic $2n = 48$ of *Callimico* and the autapo-morphic $2n = 44$ of *Callithrix pygmaea*.

*4.3.2(a)(ii) Cebidae* Hill (1970) kept *Cebus* and *Saimiri* together in his subfamily Cebinae of the Cebidae. Hershkovitz (1970a, 1978, 1984) sepa-rated them, awarding *Saimiri* its own subfamily, Saimirinae. More recently Rosenberger (1980) has reunited them; they share a reduction of the third molars (which he sees as a precursor to their loss in the marmosets), enlargement of canines with development of a honing system between upper canine and anterior lower premolars, enlargement of premolars, fore-shortened face and gracile zygomatic arches, shallow glenoid fossae, narrow inter-orbital pillar, specializations of the visual cortex and glandular external genitalia. While any one of these conditions could have evolved in parallel, taken together they do seem to support an association of the two. That they relate them to the marmosets is less clear; any such common ancestry must in any case have been brief. Perkins (1975) listed some cebine synapomorphies in the characters of the skin: presence of abundant dermal melanin (shared with the pitheciines), hairs in independent perfect lines (occasionally found in the pitheciines also), and certain features of the chemistry of the eccrine glands (again, not entirely confined to the two genera). None of these character states are shared with marmosets.

The molecular data also do not support a cebine/callitrichid association. In the analysis of Baba *et al.* (1979), *Cebus* and *Saimiri* are clearly linked to each other, though separated by a considerable distance, and together form a sister group to all the remaining platyrrhines, while the marmosets form a group with the others, being closer to *Aotus* and *Callicebus* than to the pitheciines and atelines. The findings of Cronin and Sarich (1975; Sarich and Cronin 1976) conflict with this; no special link is found between *Cebus* and *Saimiri*, which both spring from a more or less equal seven-way basic playrrhine split. This may simply indicate that at this level the method gives poor resolution; the association of the two genera seems otherwise to be so well-supported that I will follow Rosenberger's (1981) scheme, except that I keep them separate from the Callitrichidae in their own family, Cebidae.

1. Genus *Saimiri*. After a long period in which the Squirrel Monkeys were

generally assigned to a single species, *Saimiri sciureus*, with some authors separating the Central American *S. oerstedti* with a query, there is now agreement that several species should be recognized. Hershkovitz's (1984) arrangement is as follows:

a. *S. boliviensis*: black brow-band forming 'Roman arches' over each eye; lateral supra-orbital vibrissae inconspicuous; black tail pencil thin. In these three characters the species differs from all others. Ears are tufted. General colour is yellow-agouti, the hindlimbs more buffy, the forelimbs more orange. The two subspecies are *S. b. boliviensis* (Bolivian rain forest), in which the crown is black in both sexes, and *S. b. peruviensis* (from between the Rivers Huallaga and Tapiche, Peru), in which it is blackish only in the female, and blackish agouti in the male.

b. *S. sciureus*: like all the subsequent species, this has 'Gothic arches' over each eye, conspicuous lateral supra-orbital vibrissae, and a thickly black-tufted tail. Ears are tufted. Colour is greenish agouti, the hindlimbs, at least, being much the same as the body in tone. Hershkovitz (1984) recognizes four subspecies: *S. s. cassiquirensis*, with orange-toned forearms and a pale nuchal collar, its range being flanked by those of the other three, which are nearly identical phenotypically, lacking the collar; *S. s. sciureus* (from the Guyanas and northeastern Brazil) and *S. s. macrodon* (from between the Rivers Apoporis and Jura) have orange forearms, and differ only in karyotype, while *S. s. albigena* (from Colombia) has grey-agouti forearms.

c. *S. ustus*: differs from *S. sciureus* in its red–brown-agouti colour, the forelimbs being grey-agouti but the hindlimbs blackish, at least in the female, and from all other species in its bare, untufted ears. It is the most southerly species.

d. *S. oerstedti*: very close to *S. sciureus*, but the crown blackish in females, blackish agouti in males; the colour more orange; the forearms bright orange. The two subspecies are both very restricted, and separate, in distribution: *S. o. oerstedti*, in which the crown is black in both sexes, the hindlimbs are orange like the forelimbs, and there is a black cheek stripe, is found in Panama and the neighbouring part of Costa Rica; *S. o. citrinellus*, in which the crown retains agouti banding, the hindlimbs are greyish agouti, and there is no marked cheek band, is found around the Rio Panita in Costa Rica.

Thorington (1985) has a slightly different arrangement. For him, *S. ustus* (which he calls *S. madeirae*) is a distinct species, but *S. boliviensis* is not: where Hershkovitz (1984) sees sympatry between *S. b. peruviensis* and *S. s. macrodon*, Thorington sees integradation between a Roman-arch form (*S. sciureus boliviensis*) and a Gothic-arch form (*S. s. sciureus*). Clearly, there is a case for re-examination of this area, preferably by a field study; in the meantime, because the list of characters, including behavioural, purporting to differentiate the two arch types (which Hershkovitz sees as the prior division within the genus) is so very impressive, Hershkovitz's system will be adopted here.

Thorington (1985) also refuses to recognize subspecies distinguished only by karyotype, so unites *macrodon* with *sciureus*, and also considers the distinctions of *albigena* too weak to merit its recognition. The result of this is the remarkable one of the range of a distinctive, apomorphic subspecies, *S. s. cassiquiarensis*, separating the range of its plesiomorphic relative, *S. s. sciureus*, into three separate parts! While any character, karyotype included, is surely available for use by taxonomists, so that (at least as far as the recognition of *S. s. macrodon* is concerned) I adopt Hershkovitz's (1984) system, the way Thorington lumps the subspecies brings out in a very striking manner their centrifugal arrangement.

In the genus as a whole, it seems likely that, as Hershkovitz (1984) opines, the Roman arch is primitive, so that *S. boliviensis* here retains primitive characters; on the other hand, the bare ears of *S. ustus* seem likely to be primitive, and so does the high chromosome number (with as many as seven acrocentric pairs) of *S. s. sciureus*. The distribution of primitive characters, but not of overall primitive taxa, is therefore peripheral in South America; on the other hand it is clear that *S. oerstedti* is in the main the most derived of the four species—anything but centrifugal. (Hershkovitz suggests that this species originates from the hybridization of feral pets of different species; but it is hard to see how this could be so given the homogeneity of the characters.)

Very recently, a new species, *S. vanzolinii*, has been described (Ayres 1985). Occupying a very small range right where the ranges of *S. sciureus*, *S. ustus*, and *S. boliviensis* meet, it appears to a considerable extent to combine the derived conditions of the other three, except that it retains the primitive Roman arch.

2. Genus *Cebus*. The capuchin monkeys are the intelligent short-faced organ grinders' monkeys; they have prehensile but fully haired tails, and dextrous hands with a pseudo-opposable thumb. Their taxonomy is complex; four species are generally recognized, divided into a tufted and an untufted group. Tufted Capuchins all belong to one species, *Cebus apella*, characterized by its frontal tufts, dark pre-auricular stripe, narrow narial opening, deep mandible, and shorter broader molar teeth: all these are almost certainly derived conditions. The three untufted species, without true frontal tufts but with a dark wedge from the crown into the light forehead, are:

a. *C. capucinus*: black above; the forehead, throat, chest, shoulders, and forelimbs are whitish. The other two differ in being brown or grey, not black; and the pale areas are never so extensive.

b. *C. albifrons*: the crown cap is rounded, with a broad buffy or whitish forehead band in front of it; the limbs are often yellowish or reddish; the hands and feet are the same colour as the limbs.

c. *C. olivaceus*: the cap is wedge-shaped as in *C. capucinus*, and restricted to the vertex; the forehead band is narrow; there is often a well-defined dorsal stripe; the tail is black distally; the limbs are dark brown, with lighter tipped hairs, and the hands and feet are contrastingly darker. There is often a dark line connecting the point of the crown wedge to the nose. The build is more robust than the other two untufted species, but less so than the Tufted Capuchin, which is often very thickset indeed. For the correct nomenclature of this species, see Husson (1978): Hershkovitz (1949) called it *C. nigrivittatus*, Hill (1960) *C. griseus*.

The polarity of characters of these three species is not entirely clear-cut, but it seems probable that *C. olivaceus* is closest to the common ancestor, with its agouti pelage and less differentiated crown pattern; but by virtue of its robust build it could be the sister group of *C. apella*. The other two untufted species would be derived separately from a pre-*olivaceus* form; certainly, *C. capucinus* is extremely highly derived in its sharply disruptive colour pattern and extensive white areas. *C. capucinus* is found in Central America and northwestern Colombia; *C. albifrons* in the rest of Colombia and in eastern Ecuador: *C. olivaceus* in Venezuela, the Guyanas, and Brazil north of the Lower Amazon. The range of *C. apella* overlaps that of *C. capucinus* and *C. olivaceus* and extends south to Bolivia, southeastern Brazil, and Paraguay. There is no centrifugal pattern to be seen here; it is, perhaps, of some interest that, as in *Saimiri*, a highly apomorphic species is found in Central America.

A vast number of subspecies have been described, especially for *C. albifrons* (surprisingly, given its restricted range) and *C. apella*. I would hate to give an opinion on the validity of the named forms. In *C. capucinus* the central forms seem to have truly white light areas, whereas in the peripheral forms these areas are more yellow or buffy. In *C. albifrons* the central forms tend to have orange or reddish limbs, and often a reddish mid-dorsal zone, while the peripheral forms have these regions coloured more like the rest of the body. In *C. olivaceus* the centrally distributed (east Venezuelan) *C. o. olivaceus* has the most contrasting colours (dark grey body, grey–white forehead, blackish dorsal stripe).

Taken together, these three vicariant forms have an interesting distribution of frontal fringes. (These tufts above the brows, found in females only, are to be distinguished from the true frontal tufts of *C. apella*.) In the central species, tufted females are normal: only a few appear to lack these fringes. In each of the other two, it is the generally derived central subspecies (*C. olivaceus olivaceus*, *C. capucinus imitator*) which possess the superciliary tufts. If I have correctly identified the possession of the tufts as derived, then we have here an interesting example of parallelism (orthogenesis, in the true sense), in which the same feature re-emerges in three central regions.

*C. apella's* geographic variation is very difficult to handle. A group of subspecies from north of the Amazon are all dark in colour, with black hands and feet, and usually only a poorly expressed supra-orbital band. Those from the Upper Amazon region, from western Colombia south to Bolivia, are very contrasty in colour, the red–brown body contrasting with the blackish spinal stripe, limbs, hands and feet, and tail; the white supra-orbital band is well-developed. In central–eastern Brazil and Paraguay are some forms with enormously developed, horn-like frontal tufts, either contrasty in colour or else wholly dark. The southeastern races are overall the most primitive, with poorly developed tufts, little colour contrast, and poorly developed crown caps, Regional subspecies groups seems to be the main pattern here, although the central races (*C. a. nigritus, C. a. cay*) could be seen as in some respects more derived than the rest.

*4.3.2(a)(iii) Aotidae*    The genus *Aotus* is of admittedly uncertain affinities. That it is the only nocturnal monkey means that it has a number of highly derived states which could confuse its relationships: its loss of a well-developed retinal fovea, for example, could be misinterpreted as primitive, and the large post-orbital fissure could be seen as either primitive or as derived (a result of orbital enlargement), though Cartmill (1981), after detailed consideration, still opts for the former—a potentially significant conclusion, if subtantiated. Rosenberger (1980) places it in his family Atelidae, and together with the *Pithecia* group and *Callicebus* in the subfamily Pitheciinae; but the second of these decisions, at least, is taken with some misgivings. The structure of the skin is very similar to that of *Callicebus* (Perkins 1979), and this in turn is a primitive type not unlike that of strepsirhines; for example, in its grouping of follicles in independent circular clusters. Cronin and Sarich (1975) suggest that, from the albumin evidence, *Aotus* may have been the first lineage to separate from the platyrrhine stem. Baba *et al.* (1979) tentatively align the genus with marmosets and, still more tentatively, with *Callicebus*, but remark that these interrelations seem highly questionable. Taken overall, the separation of *Aotus* seems to have been very early, and it could indeed be, as Sarich and Cronin (1976) propose and as Cartmill (1981) implies, the sister group of all other platyrrhines.

The intrageneric taxonomy of *Aotus* has undergone a revolution in the past decade. Mounting evidence of extreme karyotypic diversity led De Boer (1983) to propose a nomenclature for the recognized seven karyomorphs and four or more phenotypes; but his suggestion was quickly superseded by the 'preliminary revision' published by Hershkovitz (1978), in which nine full species (two of them new) were recognized. The nine species were divided into a red-necked group and a grey-necked group. The former are clearly derived, in both phenotype and karyotype. There is no central/peripheral pattern to this: the derived species occur south of the Amazon, the primitive ones north of it. It is possible within the northern group to see the central species, *A. nigriceps*, as most derived, with its entirely orange throat; in the southern species the most derived is certainly *A. trivirgatus*, with its orange mid-dorsal band and extensively orange inner limbs, but this is not central in range, except within the genus as a whole (and, interestingly, its range borders on that of the northern group, including *A. nigriceps*).

Other characters show a mosaic distribution. *A. infulatus* and *A. azarae* (red-necked) and *A. vociferans* (grey-necked) have an interscapular whorl, and *A. brumbacki* (grey-necked) has an interscapular crest. The orange tone of the underside extends to the wrist and ankle in *A. trivirgatus* (grey-necked) and in *A. nigriceps*, *A. nancymai* (correctly, *nancymaae*), *A. infulatus*, and *A. azarae* (red-necked), but only to the knee and elbow in other species. The whole picture is complex in the extreme; and, as Hershkovitz (1983) notes, there are still plenty of gaps in the data (the karyotypes of some of the species, for example). Galbreath (1983) has essayed a phylogeny of those species whose karyotype is known; the differences between even related karyotypes are often considerable, implying several successive mutations between the primitive and the more derived forms.

*4.3.2(a)(iv) Callicebidae*   The genus *Callicebus* (titi monkeys) is another which is generally supposed to be primitive in many respects; Hershkovitz (1963) objects to such an assessment, but it is supported by the tegumentary analysis of Perkins (1975). Rosenberger (1980) points out that it shares with the pitheciines the characters of tail, narrow lower incisors, stout canines, enlarged molar hypocones, and cheek teeth which in general have rather little crown relief. These characters seem convincing, and were there no contrary evidence his taxonomic association would be followed; but no independent evidence supports it. Cronin and Sarich (1975) take the *Callicebus* line back to the initial platyrrhine radiation. Baba *et al.* (1979) more or less agree, except that the *Cebus/Saimiri* line splits off before the three-way *Callicebus*/marmoset/pitheciine-plus-ateline split. Sarich and Cronin suggest that there may be some slight evidence to associate *Callicebus* not with the pitheciines but with the atelines.

Under these circumstances judgement must be suspended—at least until such time as two or more independent sources of evidence can be found to agree! Placement of the genus in a family of its own is, therefore, by way of being an interim solution.

Like *Aotus*, *Callicebus* has an unexpected degree of karyotypic diversity; but at least this seems so far to follow species lines as already laid down by Hershkovitz (1963). Hershkovitz recognized three species, but they are clearly ready to be reassessed, because one (*C. torquatus*) is based on a restricted set of derived characters (sharply contrasting colour of hands and feet; white or orange throat patch or collar), a second (*C. personatus*) has a different set of derived characters (pale colour, black hands and feet, black ear-tufts), while the third (*C. moloch*) is based on a residue of more primitive subspecies and so is virtually undefinable. Within *C. moloch*, as recognized by Hershkovitz, there are three subspecies-groups: a north eastern (*C. m. ornatus* and *discolor*), in which the forehead has a grey—white band; a central (*C. m. cupreus* and *brunneus*), in which the forehead is black; and an eastern and southeastern (*C. m. moloch, hoffmannsi,* and *donacophilus*), in which forehead, crown, and nape alike are grey to red–brown like the back. Other characters cut across this division somewhat: the pale grey digits generally found in the northeastern group may foreshadow the light hands and feet of sympatric *C. torquatus*; *C. t. donacophilus* has ear-tufts like its neighbour *C. personatus*, but they are white instead of black; and so on.

Hair banding patterns are rather simplified in the central and northwestern *C. moloch* and in *C. torquatus*, with a pair of bands (black/brown or some variant) after a grey base. *C. personatus*, however, has a series of alternating rings, and far northwestern *C. torquatus medemi* has a terminal black zone. This is perhaps the only sense in which central taxa may be said to have more derived states.

It is clear that the last word has not been said about the taxonomy of *Callicebus*.

*4.3.2(a)(v) Pitheciidae*   After reviewing a number of taxonomically isolated genera, it is a pleasure to turn to the survey of one of the few clusters about whose interrelatedness there is no disagreement. Hill (1960) and other specialists class *Pithecia, Chiropotes,* and *Cacajao* together as the subfamily Pitheciinae. They agree in the possession of characteristically tall, narrow, procumbent incisors in both jaws, large stout canines, flattish low-cusped cheek teeth, expanded ribs, and exceptionally broad nasal septum ('ultra-platyrrhine'). Rosenberger (1977, 1980) associates them with *Callicebus*; reasons for some reluctance to accept this affiliation are mentioned under that genus. Perkins 1975 finds the structure and chemistry of the pitheciine skin primitive, but less so than that of *Aotus* or *Callicebus*, and difficult to associate with any other form unless one is prepared (as nobody really would be!) to split them up and place *Pithecia* on the callitrichid stem, *Cacajao* on the cebid/atelid. Cronin and Sarich (1975) derive the pithecine stem from the very base of the platyrrhines; Baba *et al.* (1979) relate them, very distantly, to the ateline stock; and Dutrillaux *et al.* (1981) find that, according to chromosome banding, *Cacajao* and *Lagothrix* are closer to one another than either is to *Cebus*. Two independent sources of evidence do, therefore, align the pitheciines and atelines; but before definitely accepting this verdict one would wish to identify morphological features which the two share to the exclusion of *Callicebus* and *Aotus*.

This is the reason why, again as an interim measure, I recognize a distinct family, Pitheciidae.

There are three genera in this family, whose distinctiveness has never, to my knowledge, been seriously questioned; though their exact interrelationships are still unclear.

1. Genus *Pithecia*. The sakis, distinguished by the lank cowl of hair which surrounds the face and the white para-nasal streaks, were considered by Hill (1960) as falling into two species, *P. pithecia* with extreme sexual dichromatism and *P. monachus* without; but Hershkovitz (1979) showed that there are in fact (at least?) four, with differing degrees of sexual dichromatism. As we have seen above, weak sexual dichromatism occurs in *Saimiri*; but in some taxa of *Pithecia* it is extraordinary, comparable to some members of the *Lemur fulvus* group.

a. *P. hirsuta* is dark, blackish, or brown (with a subterminal buffy bands) in both sexes; black-bearded; the shoulder whorl and crown ruff are well-differentiated; the underparts blackish; the hands and feet pale, buffy agouti. It is widespread, south of the Caqueta–Japura–Amazonas system.

b. *P. albicans* is buffy in colour, except for the blackish back and tail; the whorl and ruff are extreme in their development. It is found within the range of *P. hirsuta*, to the east of the Rio Jurua, but not east of the Rio Madeira.

c. *P. monachus* has a certain degree of sexual dichromatism. The female is like *P. hirsuta* but has a pale forehead, buffy beard, orange underparts, and pale hands and feet. In the male the whole crown is buffy, and the hairs there are short and stiff, tending to baldness. This species too lives within the range of *P. hirsuta*, west of the Rio Jurua; its range does not however extend so far either north or south, and does not overlap that of *P. albicans*.

d. *P. pithecia* is extraordinarily sexually dichromatic. The female is black or brown agouti, with orange underparts, and the hands and feet are agouti; the whorl and ruff are weakly differentiated. The male is blackish, with no agouti banding at all; the head and face are whitish to reddish; the crown hairs are all short and stiff. It is allopatric to the other three species: north of the Amazon, east of the Rio Negro and Orinoco.

The relations between these four species are actually fairly simple. The most widespread species, *P. hirsuta*, is the most primitive: it lacks sexual dichromatism, has the least saturated fur, the least bleached hands and feet, the least differentiated ruff and crown hair. Within its range it has given rise to (1) the paler *P. albicans*, with its more developed ruff, in the eastern part, and (2) the sexually dichromatic *P. monachus*, in the western part. The only species which is not sympatric with any other, *P. pithecia*, must have originated from a *P. monachus* population which crossed the Caqueta–Japura and then the Rio Negro: the females are almost indistinguishable, while the male *P. pithecia* is simply a still further differentiated stage of that of *P. monacha*.

What we have here is two cases of sympatric speciation and one allopatric. It is interesting that the two derivatives of *P. hirsuta* have arisen in different places, and are derived in different directions, and that the one which has speciated again has simply accentuated the features that already differentiated it. The chromosomes of this group are unfortunately entirely unknown, so the mechanics of speciation, whether stasipatric or not, cannot be elucidated: but the two sympatric derivatives look quite different from their ancestral form, yet the basis of the differences seems very simple, and there is no reason why a single step, perhaps a recombination, maintained by assortative mating, could not have been responsible in each case.

2. Genus *Chiropotes*. The skull of these bearded sakis differs from that of *Pithecia*, in its prognathism and other characters, but externally they are rather similar. The two species, indeed, could have diverged in different directions from a *Pithecia*-like stock which lacked the white para-nasal streaks and had the whorl on the vertex instead of on the occiput or nape. The two species are *C. satanas*, in which the beard is greatly enlarged and the crown tufts are hypertrophied into a 'bouffant' hair-style; and *C. albinasa*, in which the nose is white, sharply set off from the black fur. The two are allopatric; subspecies have been identified within *C. satanas*, but their status is uncertain.

3. Genus *Cacajao*. In this genus, the uakaris, the skull more resembles that of *Chiropotes*; the fur is long and lank; the face is remarkably lacking in subcutaneous fat; and, unique among New World monkeys, the tail is shorter than the head and body. The two species are allopatric. *C. calvus* has a red facial skin, the anterior crown is bald, the pelage is fine and silky, and the underparts are light toned, sparsely haired; *C. melanocephalus* has a black face, the crown is wholly haired, the pelage is coarser, and the underside is well-haired and black. In overall colour, *C. melanocephalus* is black on the foreparts and tail, chestnut on the trunk and hindlegs. *C. calvus* has four well-marked subspecies; *C. melanocephalus* two (Hershkovitz 1985).

*4.3.2(a)(vi) Atelidae*   The one association of two subfamilies in Hill's (1960, 1962) system that is now well-supported is of Atelinae (spider and woolly monkeys) and Alouattinae (howler monkeys). Both groups have a prehensile tail; so does *Cebus*, but in the present group the underside of the terminal fifth or so is naked, ridged, and covered with dermatoglyphics: such a remarkable and unique specialization that this alone, it would seem, would suggest very strongly that they are related. Perkins (1975) finds similarities in the skin: strong epidermal monoamine oxidase activity, thick epidermis with much epidermal but no dermal melanin, and eccrine glands throughout the skin, not merely confined to the friction surfaces.

Whether this group might be closer to some other platyrrhines than to others is arguable. While Dutrillaux *et al.* (1981) reconstructed the chromosome morphology of *Lagothrix* as closer to that of *Cacajao* than to that of *Cebus*, and Baba *et al.* (1979) tended to agree on the basis of their immunological and sequencing data, Cronin and Sarich (1975) took the ateline lineage right back to the initial diversification of the platyrrhines, adding however that there is some evidence that they might be somewhat closer to *Callicebus*. Rosenberger (1977, 1980) also associated the atelines with the pitheciines, but as *Callicebus* and, with misgivings, *Aotus* were placed along the pitheciine lineage this cannot be used as independent evidence of the interrelationships of specifically atelines and pitheciines.

Accordingly, I provisionally restrict the Atelidae to the present group alone. The four genera are *Lagothrix*, *Brachyteles*, *Ateles*, and *Alouatta*; each is highly distinct. *Brachyteles* and *Ateles* share the loss of a thumb, a highly characteristic shared derived feature; but as Zingeser (1973) found the teeth of *Brachyteles* to retain primitive characters like *Alouatta*, it may be that *Ateles* and *Lagothrix* are in fact sister groups, and thumb loss is an example of orthogenesis.

1. Genus *Lagothrix*. Woolly monkeys are restricted in distribution, rather uncommon; the two species are very distinct, and it is questionable whether they do in fact share derived characters, since the unique wiry, crinkled pelage is in fact found only in *L. lagotricha*.

*Lagothrix lagotricha* is divided by Fooden (1963) into four subspecies, differing in colour of body, whether or not the crown contrasts in colour with the body, and the presence or absence of a pale mid-sagittal crown streak. None of them is really 'central' or 'peripheral' in distribution; in its overall brown colour, with no contrasts on the crown, the northeastern *L. l. lagotricha* ranks as most primitive. Interesting is the clinal variation of colour within *L.l. lugens*, distributed along the eastern margin of the Cordillera in Colombia: from nearly black at high altitudes, it grades in the lowlands from pale grey to darker, browner, then paler again from south to north.

*Lagothrix flavicauda*, the high altitude Peruvian species, is a deep mahogany colour with a buffy nose and upper lip, and yellow underneath the terminal half of the tail. It was rediscovered about ten years ago (Mittermeier *et al.* 1975) after being unrecorded for half a century, and is still a grave conservation concern.

2. Genus *Brachyteles*. This, the most endangered genus among all of the Haplorhini, has a single species confined to the now largely eliminated Atlantic forests of southeastern Brazil. *B. arachnoides* is light grey-brown in colour with a light facial ring. Despite its popular name, Woolly Spider Monkey, its fur is not wiry like that of *L. lagotricha* but rough and thick. The difficulties of assessing its relationships have been mentioned above.

3. Genus *Ateles*. Spider monkeys proper have a spindly body build, setting off their ungainly pot bellies; arms, alone among platyrrhines, are longer than the legs; the thumb, as in *Brachyteles*, is absent; the cheek teeth are small, low-crowned, bunodont.

How many species there are remains uncertain. The four-species arrangement of Kellogg and Goldman (1944), accepted by Hill (1962) and Konstant (1985), is rather peculiar by reason of the crossing distributions of two of them, and is in pressing need of re-examination. Provisionally, I will split each of the two 'crossing' species into its isolated components, thereby raising the total of species to six, as follows:

a. *A. paniscus*. This form has long (75–100 mm) silky pelage, black all over; the face skin is pinkish red. The tail is curiously thick at the base and tapers towards the tip. It is found north of the lower Amazon, in Brazil and the Guyanas.

b. *A. belzebuth*. The pelage is shorter, less than 70 mm long, black or dark to medium brown; of the facial skin, only the muzzle is somewhat light coloured, but the chin and cheek hair is often light, and there is a whitish triangular forehead band; the underside is contrastingly golden brown to white. The tail is uniformly slender; the feet are less elongated than in *A. paniscus*. This species is found north of the Upper Amazon, along the Cordillera into Colombia.

c. *A. chamek*. Placed by Kellogg and Goldman (1944) as a subspecies of *A. paniscus*, this form does bear a general resemblance to the latter in its all black fur and long feet; also, there are often flesh-coloured rings round the eyes. The tail, however, is slender; the hair is shorter and coarser; and there are buffy or golden hairs on the inner side of the thigh and the pubic region. It occurs south of the main Upper Amazon system.

d. *A. marginatus*. This form was placed as a subspecies of *A. belzebuth* by Kellogg and Goldman (1944) and indeed resembles the latter in its white forehead band and cheeks, though the chin is dark; eye-rings and muzzle are light; the underside is black like the upper. The tail is also considerably shorter [less than 770 mm long, compared to (usually) above this length in the previous three species], but not robust like *A. paniscus*. The pelage is long, but not silky. It lives in Brazil south of the Lower Amazon.

e. *A. fusciceps*. From Ecuador, western Colombia, and Panama: another black species, sometimes with pale eye-rings; the underside is dark, the chin is often white but the cheeks are dark, and there is no forehead band, although the front of the crown is lighter, browner than the rest; the lips and chin are sprinkled with white hairs. This is another relatively short-tailed species.

f. *A. geoffroyi*. In this species the black colour may extend over the whole upper side, or be restricted: in the extreme case, to the crown, hands and feet, and upper side of the tail, the rest of the upper parts being brown, reddish, or buffy. The underside is reddish, buffy, or whitish, generally contrasting somewhat with the upper side. There are often flesh-coloured eye-rings; the chin may be white; there are white hairs around the lips; the forehead may have a light band. This species is smaller than the others. It is found from Panama through the isthmus into southern Mexico, the furthest north of any platyrrhine.

There is, thus, rather than a restricted number of clear-cut species, a larger number of taxa with a confusing mosaic of characteristics. As the only known sympatry is between *A. geoffroyi* and *A. fusciceps* (see Konstant 1985), this test for species status cannot be used.

Distinctive subspecies exist in *A. belzebuth* and *A. fusciceps*; and an extra-ordinary diversity (nine named subspecies) in *A. geoffroyi*. Within the latter the most centrally distributed form, *A. g. geoffroyi* from southeastern Nicaragua, is the least black of all, and the two peripheral ones, *A. g. vellerosus* from south–central Mexico and *A. g. grisescens* from north-western Colombia, are the blackest. In the central forms the forehead hair stands up in a triangle, and may be light.

4. Genus *Alouatta*. Whatever the inter relationships of the other three genera, there is no doubt that *Alouatta* is the sister group to them, with its primitive dentition (Zingeser 1973), skull, and brain (Falk 1979). It has, of course, the noteworthy specialization of its enlarged, hollowed-out hyoid bone and associated cartilages, giving it perhaps the loudest voice in the animal kingdom.

The six species of howler monkeys arrange themselves in a very convincing centrifugal pattern. The largest hyoids occur in the three centrally distributed species (*A. seniculus, belzebul*, and *fusca*); they are smaller in the outlying *A. palliata* and *caraya*, and probably (though not definitely known) in the marginal northern *A. villosa*. *A. seniculus* is strikingly red, not black or brown like the others; it and *A. belzebul* alone lack light banding or tipping on the hair; and *A. belzebul* has contrastingly red or yellow hands, feet, and tail-tip, whereas the others are uniformly coloured. They may be briefly described as follows:

a. *A. villosa*. A completely black, soft-coated species; ears are hidden by fur; cheek fur is buff or brown. The hyoid is not precisely described, but from the description of the voice is probably relatively small. Restricted to the Yucatan peninsula, Guatemala, and Belize.

b. *A. palliata*. This species has often been deemed conspecific with the previous species, but is marginally sympatric with it in Guatemala. It has silky black hair, with brownish to buffy tones especially on the flanks where it forms a short buffy mantle. The hyoid is small, with a rudimentary cornua and a large posterior opening into a poorly inflated tentorium. It lives in Central America, and into Ecuador and western Colombia.

c. *A. seniculus*. A red to golden, soft-coated species; ears are not hidden in the fur; the hyoid is enormous and strongly sexually dimorphic, that of the male being more than twice as large linearly than that of the male *A. palliata*, while that of the female is only a little larger than the latter. There are several subspecies, with varying degrees of colour contrast between the darker, more maroon, head, limbs, and tail, and the lighter, more golden body. The more central subspecies, *A. s. sara* from Bolivia and, to a lesser extent, *A. s.*

*stramineus* from southern Venezuela and northwestern Brazil, have more contrasting patterns than the more peripheral (Peru–Colombia–North Venezuela–Guyana–East Brazil) forms.

d. *A. belzebul*. A black or brownish species in which the hyoid is again very enlarged, though the structure is not quite as specialized as in *A. seniculus*. The hands, feet, and tail-tip are reddish or yellow, contrasting with the body. There are a grading series of subspecies from eastern and southern Brazil, the more westerly ones being more saturated in colour than the more easterly.

e. *A. fusca*. The hyoid in this species is large but less inflated than in the two previous species. The colour differs between the two sexes: males are dark brown, females a duller, less-yellowed tone. The hair is long and dense, and the ears are hidden in the fur. The species lives in southeastern Brazil. A taxon, *A. f. beniensis*, from northern Bolivia, has been assigned to this species; but not only is the range completely separate, but the usual sexual dichromatism is lacking, and it is difficult to see what there is in common beyond the possession of characters a stage more primitive than *A. belzebul*.

f. *A. caraya*. A more strongly sexually dichromatic species with a still less-specialized hyoid, from the dry forests of southern Bolivia, Paraguay, and southern/southeastern Brazil. Males are black, with red–brown tones, and females are yellowish to olive buff. The hair is long and wiry.

This picture is, both overall and (to some degree) within some of the species, a typical centrifugal one. Certain taxa are in need of revision, but the overall picture is none the less clear.

### 4.3.2(b) The Catarrhini

Whereas the search for derived character states common to all platyrrhines has only comparatively recently turned up convincingly successful results, the derived features of the catarrhines have long been obvious. Most noticeable are the elongation of the tympanic ring into a tube, associated with a deflation of the auditory bulla; and the reduction of the premolars to two in (each half of) each jaw, associated with a well-developed honing system for the canine, which I have dubbed the canine/sectorial system (Groves 1974*b*). Many platyrrhines also have a canine hone system, but the honing premolar is of course $P_2$, and the morphology of the tooth is not precisely similar to that of the sectorial $P_3$ of catarrhines. Only in *Homo* and *Hylobates* among extant catarrhines is the canine/sectorial complex not sexually dimorphic, while only in *Homo* is the sectoriality lost altogether.

Delson and Rosenberger (1980) have added other features to the list of derived catarrhine conditions: the presence of facet 'X' on the lower molars (but see Chapter 5): the presence of hypoconulids on $M^1$ and $M^2$ in addition to $M^3$; the loss of the lateral orbital fissure (i.e. the most complete post-orbital closure of any primates); the reduction of the pre-sphenopalatine lamina of the palatine; the villous placental disk; the strong development of the cytotrophoblastic shell; the narrow nasal septum and reduction of wing cartilages and olfactory scrolls; and the loss of the vomero-nasal (Jacobson's) organ.

Within the catarrhines a division has long been recognized between the Cercopithecoidea (Old World monkeys) and Hominoidea (apes and humans). Crudely, the former are more derived in their dental characters, primitive in post-cranial; the latter, more primitive in dentition, derived in post-cranium.

*4.3.2(b)(i) Cercopithecoidea*    The superfamily Cercopithecoidea are above all characterized by their bilophodont molar teeth: on first and second molars, in both jaws, the cusps are reduced to four in number, and joined transversely in pairs by lophs. Between mesial and distal cusp-pairs, each molar is constricted ('waisted'). On $M^3$ the form is the same, except that the hypocone may be reduced; on $M_3$ it is again the same, except for the retention in many taxa of the hypoconulid. (Hypoconulid loss is of course an evolutionary reversal where it occurs.) Bilophodonty is variably developed in different cercopithecoid taxa, but is always present to some degree.

A further derived character is the extreme form of the canine/sectorial complex: the upper canine, in the male, is not only slender and dagger-like in most taxa, but the mesial groove is continued uninterrupted onto the root; and $P_3$ is strongly compressed and unicuspid, the enamel line being bowed down below the general alveolar margin of the tooth-row.

To the dental characters Martin and Gould (1980) add characters of sperm morphology. Old World monkeys have sperm characterized by an unusually small head and small nucleus—a clearly derived pair of traits when compared to other primates.

Although morphological variation within the Cercopithecoidea is less than that within other primate superfamilies (Schultz 1970), the clarity of the divisions within the superfamily render it profitable to recognize two full families: Cercopithecidae and Colobidae:

*Cercopithecidae.* The Cercopithecidae are distinguished from the other cercopithecoid family, Colobidae, by:

— the (primitive) lack of foregut fermentation and complex stomach;
— the presence of cheek pouches;
— the lacrimal fossa is contained entirely within the lacrimal bone;
— the upper lateral incisors are relatively small, pointed, much reduced compared to the centrals;
— on the mandibular cheek teeth, the trigonids are relatively elongated, and the median notches, especially on the lingual side, end high above the crown base.

Within the Cercopithecidae, there are clear differences between *Cercopi-*

*thecus* and *Erythrocebus*, on the one hand, and a group of genera centred around *Papio* on the other. (The *Papio* group includes *Theropithecus, Mandrillus, Cercocebus,* and *Lophocebus*.) The differences are:

1. The *Papio* group (Papioninae) all have a diploid chromosome number of 42; the number in the *Cercopithecus* group (subfamily Cercopithecinae) is variable, but always greater than 54.
2. In the Papioninae the lingual surfaces of the lower incisors lack enamel.
3. In all the Papioninae the females' sexual skin, incorporating vagina, perineum, and anus, undergoes cyclical enlargement, being maximally swollen around the time of ovulation; there is no sexual swelling in the Cercopithecinae.
4. All the Papioninae have some degree of facial elongation.
5. Papionine molars have variable degrees of 'flare' (Delson 1973), a convexity of the buccal and lingual surfaces. This flare is greater in larger toothed forms.
6. $M_3$ always has a hypoconulid in the Papioninae; never in the Cercopithecinae.
7. Lower canines of females are slender and resemble smaller versions of the males' in Cercopithecinae; but are incisiform in Papioninae.
8. All Papioninae have a rich repertoire of facial gestures, which are quite lacking in the Cercopithecinae (Van Hooff 1967).
9. In the Papioninae the ischial callosities of adult males are fused across the midline; they are smaller and quite discrete in Cercopithecinae (as in female papionines).
10. In cercopithecine skulls there is an orifice halfway along the occipito-mastoid suture.
11. The palatine foramina are reduced in the Cercopithecinae.
12. The digastric insertion scars on the mandible are large in the Papioninae, but reduced in the Cercopithecinae.
13. The axes of the auditory tubes are transverse in the Cercopithecinae, but slant postero-laterally in Papioninae.

The polarities of the character states involved in this list are not always easy to deduce. Chromosome number in other catarrhines is nearly always above 42, but is rarely as high as 54; probably both states of character 1 are derived. In character 2, the papionines are clearly derived; and the same is almost certainly the case for characters 5, 7, and 9. Sexual skin is found in two other catarrhines: in the chimpanzee it takes the same general form, whereas in *Procolobus* it is totally different; with some misgivings, therefore, I suggest that character 3 is also derived for papionines. Character 4 is probably also derived for the Papioninae; in this subfamily it is positively allometric, but in no colobid or hominoid, however large, is there facial elongation to a degree remotely comparable to that in a large baboon. Character 5 is again derived for the Papioninae, and for reasons exactly analogous to character 4 (degree of flare depends on molar size, not on body size as such). In character 6, on the other hand, the Cercopithecinae are almost certainly derived: as noted earlier, hypoconulids on all molars are part of the morphotype of the Catarrhini as a whole. Character 8 is arguable: well-developed facial gestures are seen in most hominoids, too, but as gibbons and colobids lack them their development is most plausibly seen as parallel derived in papionines and hominids. Characters 10 to 13 (which were first pointed out by Verheyen 1962) are without question derived for the Cercopithecinae: in every case the papionine condition is shared by all the other catarrhines.

These thirteen characters, in any event, differentiate readily and absolutely between the two groups of genera concerned, and if the Cercopithecidae and Colobidae are to be awarded full family rank then these two groupings within the former should evidently be ranked as subfamilies. But there are three other genera that do not fall so clearly into the scheme. These are *Macaca, Allenopithecus*, and *Miopithecus*.

The genus *Macaca* falls clearly into the Papioninae on charaters 1,2,7,8, and 10 to 13. Some species-groups have sexual swellings; some do not; either the genus is hopelessly para- or polyphyletic, or some species have lost them, or others have gained them. Large species have considerable degrees of facial elongation, which small species lack; large-toothed species have some degree of molar flare, while small-toothed ones do not; these two conditions confirm and extend the conclusion that positive allometry is involved, in that small-sized and small-toothed macaques are noticeably smaller and smaller-toothed than any of the central group of papionines. Small macaques also often lack $M_3$ hypoconulids, and even within one species (e.g. *M. fascicularis*) there is variability, suggesting that here there may be simply a size threshhold operating (but the discrimination between the subfamilies remains: even a very large cercopithecine, such as *Erythrocebus patas*, lacks all trace of a hypoconulid). Finally, as for character 9, one species, *Macaca sylvanus*, has ischial callosities fused in the male; but this is not a size-threshhold effect, as this species, though large as macaques go, is none the less smaller than others (such as *M. thibetana*) in which there is no fusion.

I would see no difficulty, therefore, in accepting *Macaca* as a member of the Papioninae also. Only in character 9 might there be some indication of a special status, perhaps paraphyly for the genus; this will be discussed below.

*Allenopithecus* possesses papionine characters 3,5,7, and 9; there might also be said to be some development of facial elongation, as much as is concomitant with its small size. It lacks cercopithecine derived characters 10 to 13, but it is cercopithecine in its lack of an $M^3$ hypoconulid—yet its cheek teeth are extremely large (and its molar flare correspondingly well-developed). It is therefore either a papionine with one character developed in parallel to the Cercopithecinae, or a cercopithecine with four parallelisms with the Papioninae; the former hypothesis is more parsimonious. If it is truly a papionine, then it is the sister-group to the other genera in that its mandibular incisors are well-enamelled on their lingual surfaces; likewise its diploid chromosome number is 48, which would be primitive whichever subfamily it belonged to.

The case of *Miopithecus* is more difficult. The exceptionally small size of this monkey render size-dependant characters (4,5,6,9, and perhaps 11 and 12) unusable. It has sexual swellings, well-developed facial gestures, and tympanic axes that are not transverse, like the Papioninae, but characters 7 and 10 are decidedly cercopithecine. The diploid chromosome number is 54, and Dutrillaux *et al.* (1981) see its general chromosome morphology as decidedly cercopithecine in type, whereas that of *Allenopithecus* is somewhat equivocal. Overall, it would be difficult to place *Miopithecus* with confidence in either subfamily (a point made as long ago as 1966 by Jolly); on balance I would place it in the Papioninae, but the placement is provisional.

*Cercopithecinae.* As Verheyen (1962) has shown, there are quite a number of cranial characters available for discrimination of the taxa within the Cercopithecinae. The problem is to decide polarities, but given the above decisions about the affinities of *Macaca*, and especially of *Allenopithecus*, a plausible scheme can be reconstructed.

Most members of the subfamily have crania in which the inferior suborbital region is regularly curved towards the dental arcade, the pyriform aperture is round–oval (not angled in the middle), the temporal lines anteriorly follow the posterior borders of the orbits, and the nasal bones usually run straight across inferiorly instead of being pointed in the midline; dentally, $I^2$ is small and pointed, and does not bite in a continuation of the occlusal line of $I^1$; externally, all these species have extensive black areas on the limbs. The exceptions to this are the genus *Erythrocebus*, and the species (or species group) currently called *Cercopithecus aethiops*. While lacking the above characters—all of which, using the Papioninae as the out-group, are evidently derived—these two share a few derived states of their own: in side view, the orbits do not slope forward inferiorly, but their lower borders are situated behind the upper margins; the auditory tube has a V-shaped lower margin (a feature often seen however in the juveniles of the other species); and the orbits themselves are angular instead of oval. There is thus a possibility that these two together form the sister group to all other species; at any rate, it is clear that *aethiops* can no longer continue to be referred to the genus *Cercopithecus*. Fortunately the type species of *Cercopithecus* is *C. diana*; for *aethiops* the prior available name is *Chlorocebus* Gray, 1870.

1. *Erythrocebus.* The Patas Monkey, *E. patas*, has always attracted special attention because of its large size, its unusual red colour, and its limb specializations (Napier and Napier 1967). Its placement in its own, mono-typic, genus is perfectly justified as long as the various species referred to *Cercopithecus* can safely be shown to have their own suite of synapomorph features. If it is desired to unite *patas* and *aethiops* in a single genus the name

*Chlorocebus* would have priority by 27 years (*Erythrocebus* dates from Trouessart, 1897); which would make a good enough reason for not so uniting them, although the relative fewness of their shared derived states would make this in any case inadvisable.

2. *Chlorocebus*. Although Dandelot (1959) divided the *aethiops* group into three species, and later on (1971) into four, Napier (1981) has shown that these groups do in fact grade into one another, and are not reproductively isolated. Consequently, distinct though they are phenotypically, they would simply be subspecies groups within a single species. The four groups show an interesting mosaic of characters; overall the centrally distributed one, the *tantalus* group (which ranges from Ghana to northwestern Uganda) is the most highly derived, with the most discrete brow-band, a white tail-tip, and sky-blue scrotum like the northeasterly *aethiops* group, orange tufts at the base of the tail, and contrasting stiff yellow cheek whiskers like the westerly *sabaeus* group.

3. *Cercopithecus*. The skull characters as described by Verheyen (1962) sort the various species of this genus into five groups: (a) the *mona-cephus* group, (b) the *hamlyni-lhoesti* group, (c) the *mitis-nictitans* group, (d) *neglectus*, and (e) *diana*. Not included in Verheyen's craniological descriptions, for obvious reasons, were the subsequently described *C. salongo*, the poorly known *C. erythrogaster*, and the enigmatic *C. dryas*, still known only from a single juvenile specimen. His five phenetic groups can be justified on other grounds as well, with one exception; that Dutrillaux *et al.* (1981) find that the karyotype of *C. lhoesti* places this species quite apart from the others, and along the *Erythrocebus–Chlorocebus* lineage. It is difficult to know what to make of this finding; except that, as these authors' analysis of the karyotype of *Miopithecus* places this, too, on the same lineage—a placement which would seem impossible to justify on any other grounds—it seems possible that the polarities of karyotypic change in this genus, susceptible to parallelism as it is, have been misinterpreted.

The *C. mona* and *C. cephus* species groups share a reduction of sexual dimorphism; pale facial skin; bicoloured tail, light underneath; temporal lines slightly bowed out behind the coronal suture before converging again; short, triangular nasal bones; low, broad choanae; lower margin of mandible slightly curved; and a diploid chromosome number of 66. That the two groups are so obviously closely related is interesting, as they are widely sympatric, sharing in fact almost the whole of their distributions.

The *mona* group is conventionally divided into five species: four of them vicarious, the fifth (*C. mona*) overlapping with two of the others. The West African *C. campbelli* is a generally grey colour, the foreparts browner; the arms are black, the legs usually dark grey; the underside and inner surfaces of the limbs white or grey; there is a pale brow-band, yellow cheek whiskers, and a black band on either side from eye to ear. The two subspecies are quite distinct: *C. c. campbelli*, found from Senegal to the Cavally River (on the Liberia/Ivory Coast border), has little contrast between the tones of the fore- and hindparts, the thighs are grey, and the brow-band is white; while *C. c. lowei*, found from the Cavally to the Volta Rivers, has more fore/aft contrast, black thighs, and a yellow brow-band—except for the last feature, a metachromatically more advanced pattern. Just to the west of the Volta, it occurs sympatrically with *C. mona*, which is browner, more contrastingly coloured, and has a long oval patch on either side of the rump: a strongly derived pattern.

The range of *C. mona* extends eastward into Cameroun, mainly north of the Sanaga River, but reaching south of this river in the Ongue district near the mouth; it occurs also on Principe Island. In its Cameroun range it is sympatric with *C. pogonias*, which is in some respects less metachromatically advanced, in others more so: it is grey with (often) a black saddle; the legs are grey; the underside is orange; the tail-tip is black; there are orange ear-tufts; and there is a black sagittal crest of hair between the usually black eye–ear streaks. The four recognized subspecies have a peculiar distribution: *C. p. pogonias*, from north of the Sanaga River (also Bioco Island) resembles *C. p. nigripes*, from the Gabon coast, in that the saddle is sharply defined, black, while *C. p. grayi*, which is found between the ranges of these two, and *C. p. schwarzianus*, Mayombe, has a red saddle that blends with the colour of the flanks.

*C. denti*, which represents the group in northern and eastern Zaire, lacks alike the striking colours of *pogonias* and the rump-patches of *mona*, but has white ear-tufts. South of the Zaire River it is replaced by *C. wolfi*, which is somewhat more contrastingly coloured, with red or yellow underside, black tail-tip, and again white or red ear-tufts, and often a reddish saddle.

Kingdon (1980) sees *C. pogonias* as the most highly derived of these five species; not only in colour pattern, but in behaviour too, making the most emphatic, highly ritualised signals. This, of course, is what the probabilities of centrifugal speciation would predict as well. *C. mona* is also highly derived, but in a different direction, and is again a centrally distributed species.

The *C. cephus* group have also been interestingly analysed by Kingdon (1980), who divides the group into five species, vicariant in distribution. From west to east these are *C. petaurista*, *C. sclateri*, *C. erythrotis*, *C. cephus*, and *C. ascanius*. As the evidence suggests that *sclateri*, *erythrotis*, and *cephus* intergrade where their ranges meet (see, for example, Struhsaker 1970), I would reduce these five species to three; but as far as the overall pattern is concerned it makes no difference. The central taxa are much the most highly derived, with elaborations of the white cheek whiskers and restricted white nose-patches of *C. petaurista* and *C. ascanius*; especially the bright blue face and white V-shaped moustache-stripe of *C. c. cephus*, and the yellowish cheek whiskers, separated by a strong black stripe from the white throat, of all members of *C. cephus* (including *sclateri* and *erythrotis*). The patterns are complex, and for a full account of subspecific variation see Kingdon (1980).

It seems likely that the poorly known *C. erythrogaster*, recently reviewed (with field observations) by Oates (1985), is a member of the *cephus* group. It has a very restricted range in the vicinity of Benin City, southern Nigeria west of the Niger Delta. The white throat, characteristic of all the *cephus* group, is developed in this species into a bushy ruff. It is possible, however, that it may be an outlying member of the *C. lhoesti* group.

Although the karyotype analysis of Dutrillaux *et al.* (1981) separates them completely, craniologically *C. lhoesti* and *C. hamlyni* are undeniably close. In both the pterygoid fossa is very deep; the internasal suture persists into adulthood; the ascending ramus is low; the symphysis is relatively prominent; the molar occlusal surfaces are broad; incisor occlusion is edge-to-edge rather than overbite; and, apparently reversals from the basic cercopithecine condition, the palatine foramina are big and the digastric insertions are extensive.

*C. hamlyni* is a peculiar grey species with a white brow-band, a white stripe down the centre of the nose, and a cape of back-swept hair from the forehead back over the nape. It is large and strongly sexual dimorphic; the range is restricted to eastern Zaire.

*C. lhoesti* is also large and sexually dimorphic, and found in much the same region as *C. hamlyni*. It is dark grey, more orange on the back, with a black brow-band, nose, and chin, and white throat and cheek whiskers. Generally associated with it taxonomically is *C. preussi* from Bioco Island and the Cameroun/Nigeria border region; the colour pattern is superficially alike, but the tail is white below, the cheeks are not white, and the facial structure seems different. Verheyen (1962) did not study its skull, assuming—as was usual at that time—that it is merely an isolated western population of *C. lhoesti*. But not all authors have accepted this assumption: both Pocock (1907) and Eisentraut (1973) considered it related to *C. mitis* (see below), and Kingdon (1980) also raised this possibility. The chromosome number is 66, whereas that of *C. lhoesti* is 60.

A third species seems about to be added to this group (if it indeed is a homogeneous group). Kingdon (1980) presents an artist's impression of it, and calls it *C. solatus*; but no description is given, and the name remains a *nomen nudum* (deliberately, as the formal description of the new species is presently being undertaken by M. Harrison (personal communication), and is not yet published).

The *C. mitis* group is, overall, the most primitive species group of the genus, with multi-banded hairs and lack of sharp colour contrasts on most of the body. Cranially, members have a presumably derived condition of strongly developed masticatory apparatus, with generally a sagittal crest, and the temporal lines closely following the posterior margins of the orbits before turning backward. The two species are easily distinguished: *C. nictitans*, spread from Sierra Leone east to the Central African Republic, is dark grey with black arms, hands, and feet, and a white nose-spot; and *C. mitis*, widespread from Zaire through East Africa north to Somalia, south to Angola and Zambia, is variable but never has a white nose and has bushy, speckled cheek whiskers and a stiff brow-diadem. *C. mitis* divides again into two subspecies groups: the *mitis* group west of the Rift Valley (southwestern Ethiopia to northwestern Zambia, Angola), is darker, with the diadem pale; and the *albogularis* group east of the Rift Valley (Somalia to northeastern Zambia) is paler with less dark on the limbs, and generally with whitish throat and ear-tufts. Patterns of subspecific variation within each of these two groups are centrifugal: in the eastern group the centrally distributed subspecies (Kenya, northern Tanzania) have the white of the throat developed into a collar which nearly encircles the neck; and in the western group the central race, *kandti*, differs strikingly from all the rest in its contrastingly orange back.

The remaining two well-known species, *C. neglectus* and *C. diana*, resemble each other somewhat in the development of a white beard and of haunch stripes, and in the lower margin of the mandible being strongly curved. It seems doubtful whether these similarities are more than convergent, however, as in other respects they are totally different. *C. neglectus* is strongly agouti greyish with a reddish brow-band, rump, and tail-base, and black tail and limbs; the scrotum is blue. It is a large terrestrial (forest-floor-living) species. *C. diana*, a canopy dweller, is dark grey with a dark red saddle; the chest and throat are white and the face jet black. It is confined to West Africa; the two subspecies differ conspicuously in colour pattern.

Quite unexpected was the description, at the end of 1978, of a new species, *Cercopithecus salongo*, on the evidence of an incomplete skin from Zaire. Kuroda *et al.* (1985) have now provided the first more complete description of it, including some field notes. Externally there are some resemblances to *C. diana*, as its original describer (Thys van de Audenaerde 1977) had noted; but the resemblance seems little more than superficial, and such features as the bicoloured tail (white below with a black transverse stripe near the base of the underside) set it far apart, as well as its habitat (terrestrial, forest floor). The figure of the skull makes it clear that it is indeed a *Cercopithecus*, not a *Chlorocebus*, with its oval orbits, nasals apparently not pointed, and so on; the nasals seem short and somewhat triangular, which would ally it to *C. cephus*, but its exceptionally small $I^2$ would place it near *C. neglectus*. On the other hand the inter-orbital pillar is not constricted unlike in *C. neglectus*; the premaxilla seems to be narrow, unlike in the *C. mitis*, *lhoesti*, and *hamlyni* groups; and the internasal suture is closed in the figured skull. The species is very small in size, perhaps the smallest of the genus, and with very little sexual dimorphism.

Still mysterious is *Cercopithecus dryas*, of which the type, and only known, specimen is a juvenile skin and skull in the Tervuren Museum. The inferior mandibular border is not curved in the *C. diana* fashion, although its describer (Schwarz 1932) likened it to the latter species. There are a few features recalling *Chlorocebus*, such as the less-oval shape of the orbits than *Cercopithecus*, but at its immature age it is difficult to tell whether this is a true resemblance or only apparent. It does not have the V-shaped auditory tube, unlike *Chlorocebus*.

The diversity and complexity of this genus deserve far more space than I have here devoted to it, but I have tried to bring out some of the fascination of these beautiful monkeys, and to suggest some of the dominant patterns of their speciation and diversification.

*Papioninae*. It should be clear from the discussion above that there is a central group of genera—*Papio, Theropithecus, Lophocebus, Cercocebus, Mandrillus*—and three other genera which lack some of the synapomorphies of the central group. These are *Miopithecus* (perhaps not a member of this subfamily at all), *Allenopithecus*, and the perhaps paraphyletic *Macaca*.

1. *Miopithecus*. The talapoins are the smallest Old World monkeys, with a suite of characters remarkably convergent on the New World *Saimiri*: small size, green colour, large bisexual social groupings, high frequency of insectivory, reproductive seasonality (Gautier-Hion 1971). The genus ranges from the Rio Muni/southern Cameroun region south to Angola, but it appears that representatives from north and south of the Zaire River are specifically distinct (Machado 1969). The differences, as described by Machado, are striking and concern ethological as well as morphological features. All the available names in the genus, as Machado has shown, refer to the southern form and are therefore synonyms of *Miopithecus talapoin*; a formal description of the (in fact, much better known!) northern species was said by Machado to be already in progress as long ago as 1969, but unfortunately has not yet appeared.

2. *Allenopithecus*. The sole known species, *A. nigroviridis*, known as Allen's Swamp Monkey, has recently become better known as a result of field observations by Gautier (1985). It appears to live in swamp forest on both sides of the Zaire River, in both Zaire and Congo, but higher up the River than the talapoin with which it is not, as far as is known, sympatric. No diversity has been described within the species.

3. *Macaca*. This genus, of 19 species according to Fooden (1976), is still a taxonomic problem: less because of internal problems, concerned with the interrelationships of its component species and subspecies (largely as a result of the work of Fooden) than because of the question of its relationship to other genera of the Papioninae. The most detailed consideration (see, for example, Szalay and Delson 1979) has failed to uncover a single derived feature shared by all species of the genus. I will return to this problem below.

The species of the genus are divided into four species groups by Fooden (1976): these are the *sylvanus-silenus, fascicularis, sinica*, and *arctoides* groups, and they are distinguished above all by their male genital morphology. Delson (1980) notes that these are in effect phenetic groups, and that *M. sylvanus* is not closely related to *M. silenus* and its relatives, while on the other hand *M. arctoides* seems to be simply a specialized derivative of the *sinica* group.

*Macaca sylvanus*, the Barbary Ape, is the only species of the genus which is not found in Asia: its range is the Atlas region of Morocco, Algeria, and Tunisia, although in the Pliocene and Pleistocene it lived also in Europe (and indeed has been re-introduced there, on Gibraltar, in historic times). It lacks some of the specializations of other macaques: the crown hair is simple and erect, the sexual swelling of the female is restricted to a circular area including some of the rump fur, the facial skeleton is relatively short, and the pelage is agouti-banded. It is the only species in which the ischial callosities in the adult male fuse across the midline like those of most other papionines (*Miopithecus* alone excepted). Its extremely short, nubbin-like tail is a uniquely derived character, convergent (or parallel) with tail reductions in other macaques of different species groups.

The *silenus* group has a distribution split into two discrete areas by virtue of the species' restriction to broad-leaf evergreen forests. *M. silenus*, the Lion-tailed Macaque, lives in southwestern India, while *M. nemestrina*, the Pig-tailed Macaque, is continously distributed from Burma south to Sumatra and Borneo. A third species, *M. pagensis* from the Mentawai Islands, is recognized by Wilson and Wilson (1976) but had been included by Fooden (1975) in *M. nemestrina* as a subspecies. All three have a distinctive crown pattern of short central hair and elongated lateral hair continuous with cheek whiskers, with generally a central crown whorl or parting. The two peripheral species are both black, with sharply defined areas of light grey or white about the face, while the central *M. nemestrina* is a more primitive brown.

Related to the *silenus* group are the very short-tailed Sulawesi macaques, placed in seven species by Fooden (1969) but in four by Groves (1980) as a result of field observations. Like the peripheral species further west, the Sulawesi species tend towards black pelage, with additional adornments: *M. maura* from southwestern Sulawesi is the least modified, *M. ochreata* from southeastern Sulawesi has grey legs, *M. tonkeana* from the center has grey–white rump and cheek whiskers (southern subspecies) or brown limbs (northern form), and *M. nigra* from the far north has extremes of facial elongation and tail reduction, and has a crest on the crown.

In the *silenus* group, therefore, the reverse of centrifugal processes have evidently been at work. There would appear to have been extensive parallelism, or orthogenesis, from a primitive *nemestrina*-like morphotype.

The *fascicularis* group contains four species distributed down eastern Asia from Japan (*M. fuscata*) to Taiwan (*M. cyclopis*), and from China and northern India (*M. mulatta*, the Rhesus Monkey) to southeast Asia (*M. fascicularis*, the Crab-eater or Long-tailed Macaque). The Crab-eater is distributed along the Lesser Sunda Islands to Timor—probably by human agency—and into the Philippines. The scalp hair is directed backwards in the species of the group; the glans penis is small and narrow, the baculum short. Tail length increases from north to south.

The *sinica* group have four species, found in Sri Lanka (*M. sinica*, the Toque Macaque) and southern India (*M. radiata*, the Bonnet Monkey), and again in far northern India (*M. assamensis*) and south–central China (*M. thibetana*). They all have scalp hair radiating from a point (except some specimens of *M. assamensis*), and the glans penis is elongated and pointed. As in the previous group, tail length increases from north to south.

The distinctive species *M. arctoides*, known as the Stumptail or Bear Macaque, is confined to the Burma/south China region. It has an extremely long, pointed glans penis, and thin lank body hair; the forehead and anterior scalp are thinly haired and become progressively bald during adult life. Delson (1980) is certainly correct in seeing these peculiarities as extreme expressions of the characters of the *sinica* group.

Fooden (1982) has most interestingly analysed the distribution patterns of the Asian macaques. Ecologically they divide into those inhabiting broad-leaf evergreen forests (*silenus* group and northern representatives of the *sinica* group) and those in other forest types (*fascicularis* group and southerly representatives of the *sinica* group). Within each of these two types the different species are vicariant, but between the two types there is wide sympatry. The only exception to this is *M. arctoides*; a broad-leaf evergreen forest species, it is none the less widely sympatric with both *M. nemestrina* and *M. assamensis*.

From the point of view of present concerns, the macaques throw up a number of interesting problems. The three widespread species groups—the *silenus, fascicularis*, and *sinica groups*—seem to show basically uncontroversial evolutionary patterns, in which chains of subspecies have simply achieved species rank. [It should be added that species rank is, in most cases, inferential only, as there is but a single case of parapatry. In this case, hybridization at a point along the species border between *M. mulatta* and *M. fascicularis* at first inclined Fooden (1964) to unite the two in one species; later (1971*a*), although hybridization proved to be occuring all along the border whether in Burma, Thailand, or Vietnam, he decided to keep them separate, although it was only field observations by Eudey (1981) that showed that the hybridization is indeed casual and sporadic, and does not represent a full breakdown of RIMs.] These chains of taxa show north–south geoclines, corresponding to ecological 'rules' ('Bergmann's rule', size increase with reduced mean temperature; 'Allen's rule', reduction of extremities—in this case, tail length—with reduced temperature), and not centrifugal patterns.

There is however one striking case of apparent centrufugal speciation: that of *M. arctoides*. Here the most highly autapomorphic species in the genus has arisen in the centre of the range of the genus in Asia. It does, however, span the ranges of members of two different species groups, so that it would be possible to see it as having speciated allopatrically from one of them—almost certainly the *sinica* group—and then gone into partial sympatry with it. Whatever the verdict on this species, however, the general 'orthodoxy' of evolutionary patterns in macaques seems evident. One cannot help noting that macaques are part of a subfamily in which all members bar two (*Allenopithecus* and *Miopithecus*) have 42 chromosomes, and stand out in sharp contrast to most of the primate groups we have so far been considering, in which chromosome diversity is the rule. The very existence of this glaring exception underlines White's (1978) insight, that chromosomal evolution is the rule in non-allopatric modes of speciation.

In conclusion the question of the monophyly of the genus *Macaca* must be returned to. There is little in common between the species groups, at least as reorganized by Delson (1980). There is no derived skull pattern either; indeed, there seems nothing to differentiate a skull of a macaque from that of the African Pliocene *Parapapio*, and even little to distinguish it from the most primitive living definitive papionine, *Allenopithecus*, beyond the latter's large teeth and strong molar flare. Are, then, the different species groups of macaques independant lineages from the papionine stem, orginating at a point above the separation of *Allenopithecus* but below the beginning of the diversity of the 'central genera'? A very tentative suggestion would be that the Asian macaques share at least a short common stem, with a few derived conditions such as reversal of the coalescence of the male's ischial callosities and a specialization of the scalp hair pattern, while *M. sylvanus* is a little-altered version of the ancestral type of both the 'central genera' and the Asian macaque stem. The African, though not subsaharan, distribution of the Barbary Ape could be relevant in such a scenario.

4. *Mandrillus*. The 'central group' of genera of the Papioninae are extremely closely interrelated, but it seems worthwhile to continue to distinguish them if only for purposes of communication. As pointed out by Groves (1978), viewed in the context of the other *Papio*, the mangabeys are so clearly polphyletic that, unless they are to be subsumed under *Papio* along with *Theropithecus* and *Mandrillus*, two genera have to be recognized. The compromise of dividing the magabeys into two subgenera while retaining but a single genus (Szalay and Delson 1979) does not address this problem.

If mangabeys are generically distinct from *Papio*, then so are mandrills. Some lines of evidence (Barnicot and Hewett-Emmett 1972; Dutrillaux *et al.* 1981) suggest that the mandrills and the *Cercocebus* mangabeys together form a sister group to the other genera (*Papio*, *Theropithecus*, and the *Lophocebus* mangabeys). This hypothesis, however, needs to be tested on morphological evidence. At any rate, the mandrills are strongly differentiated from the other baboons, and their facial elongation could be an independant development.

The two species of the genus are well diffferentiated; though their ranges are largely separate, they do overlap south of the lower Sanaga River (Grubb 1973). *Mandrillus leucophaeus*, the Drill, is found only in a small area around the Cameroun–Nigeria border, and on Bioco Island where a highly distinctive subspecies occurs; *M. sphinx*, the Mandrill, more derived in its pattern differentiation but with more agouti-banded hair, is widespread from southern Cameroun to the Zaire River.

5. *Cercocebus*. Groves (1978) restricted this genus to the following species: *C. torquatus* (Sierra Leone to Gabon), *C. agilis* (Central Africa), and *C. galeritus* (Tana River in Kenya). Napier (1981) regards *C. atys* as separable from *C. torquatus*, and this seems acceptable if somewhat less of a lumping attitude is adopted than in my 1978 paper. But it seems inescapable that *C. galeritus* is at least as distinct from *C. agilis* as is *C. torquatus*, for reasons given in that paper. Metachromatically *C. agilis* is the most primitive species, with strongly agouti-banded fur, a variably but never strongly developed frontal parting, and eyelids that are slightly paler than the facial skin, if that. *C. galeritus* is more bleached in its pelage, has a striking contrast between a jet-black face and white eyelids, and has a fully developed crown-hair parting from which long hairs diverge. *C. torquatus* has subdued or suppressed the scalp parting but developed a contrasting red cap; the pelage is more saturated, iron-grey, and the facial pattern is dark with white eyelids as in *C. galeritus; C. atys* resembles it but lacks the reddish cap, and the face is pale.

The pattern here is rather like that in *Macaca*: the most primitive species is central in position, but there has been parallelism between its peripheral offshoots. A centrifugal element has however been superimposed on this, in that *C. torquatus* is more highly derived than either of the two species on either side of it. A further observation is that, as with *Galago, Euoticus, Arctocebus*, and *Cercopithecus*, the locale of this most derived taxon is the Gabon/Cameroun region of Central Africa.

Recently, mangabeys were discovered in the Mwinihana Rain Forest of Tanzania (Homewood and Rodgers 1981). Although Wasser (1985) refers to the local taxon as *Cercocebus galeritus sanjei*, this name would appear to have no status, and no new taxon has yet been described. From photographs of the sole available specimen, presently in captivity at Arusha, Tanzania, kindly made available by K. Homewood and R. Wirth, the Mwinihana mangabey would appear to be referable to *C. agilis*, making no approach in any respect to *C. galeritus*. There would seem to be an interesting parallel here to the interrelationships among East African *Procolobus* (see below).

6. *Lophocebus*. Groves (1978) shows that this genus differs from *Cercocebus* in the form of the nasal bones and inter-orbital region, cranial flexion, the form of the suborbital fossa, smaller cheek teeth and reduced molar flare, orbit shape, malar foramen size, and characters of the mandible.

There are also pelage differences, and *Lophocebus* lacks paling of eyelids. Rollinson (1975) notes a difference in the sexual swellings. Electrophoresis of blood proteins separates them clearly (Lucotte 1979*a*). Whatever the position of *Mandrillus, Lophocebus* is phyletically closer to *Papio* than is *Cercocebus*.

In my 1978 paper, only a single species was recognized in this genus. I now incline to think this was overlumped, and despite the evidence of very slight gene-flow at points across the Zaire River, it is appropriate to recognize *L. aterrimus* from southern Zaire as a species separate from *L. albigena* (from Cameroun, across northern Zaire, to Uganda). It is difficult to see any noteworthy pattern in the geographic arrangement of the subspecies within either species, although it may be significant that the only all-black form, *L. a. aterrimus*, could be said to occupy a more or less central range.

7. *Theropithecus*. This genus, extensively reviewed by Jolly (1970), has a single living species, *T. gelada*, known as the Gelada (or Gelada Baboon), but a number of fossil species. It is probably the sister group of *Papio* (see the review by Cronin and Meikle 1979). Present-day geladas are confined to the high plateaux of Ethiopia.

8. *Papio*. Baboons are widespread over the savannah and thornbush country of Africa, and into the margins of the forest belt. There is much controversy over their taxonomy; wherever relevant field observations have been made, hybridization has been found between neighbouring taxa. What this implies in terms of reproductive isolation, or its breakdown, has been much discussed; the latest commentator (Sugawara 1982) considers that the morphologically distinctive taxa *hamadryas* and *anubis* are on these grounds conspecific.

If all savannah baboons are united into a single species, the name *Papio hamadryas* has priority. There are at least five strongly differentiated subspecies: *P. h. hamadryas* (Hamadryas or Sacred Baboon: Horn of Africa, and western Arabia), *papio* (Guinea Baboon: Horn of Africa, and western Arabia), *papio* (Guinea Baboon: Senegal), *anubis* (Olive Baboon: Mali to Ethiopia, south to northwestern Tanzania), *cynocephalus* (Yellow Baboon: southeastern Kenya, south to the Zambezi and southwest to Angola), and *ursinus* (Chacma Baboon: south of Zambezi). Whether some of the subtypes within these, especially the last two, are also valid subspecies or only clinal variants, is still unclear.

The three northern subspecies are externally highly sexually dimorphic, with heavy manes in the males; but in *anubis* the mane grades into the fur on the body whereas in the other two it is sharply defined from the short body hair. The two southern forms are maneless. Both *hamadryas* and *papio* are red-faced; the former, in addition, is red-rumped; other baboons are black-faced. Body size increases in the series *papio–hamadryas–cynocephalus–anubis–ursinus*. Colours go from olive-toned in *anubis* to yellow in *cynocephalus*, yellow–brown in northern *ursinus*, and blackish in southern and western *ursinus; hamadryas* is grey–brown in females, silvery grey in males, and *papio* is reddish; the hairs in *hamadryas* and *papio* are agouti, with many alternating bands; the others have at most two pairs of light and dark alternating bands. The tail slopes back and up, then down in a smooth curve in most of the taxa, but in *anubis* and *ursinus* it is 'broken'; i.e. there is a sharp angulation between the basal and terminal halves. Skull characters differ between the taxa: *hamadryas* and *papio* have multiple zygomatic foramina, whereas all the others have a single one on either side. The three southerly subspecies live in multi-male troops with panmictic breeding systems, but in *hamadryas* and *papio* there are harem groups within the larger multi-male troops—in the former the harems tend to separate out spatially for daily foraging, while in the latter they apparently do not (Boese 1973).

It will be seen from this rapid summary that, even as far as the restricted range of characters cited goes, relationships between the five subspecies are not at all simple. The primitive banding of the hairs and perhaps the harem-style breeding system (if comparisons with geladas and mandrills are appropriate) might lead to the conclusion that *P. h. hamadryas* and *papio* are primitive peripheral forms between which a more derived taxon, *anubis*, has evolved; but it is hard to see the unusual type of mane as anything but highly derived, nor the red skin colour; and zygomatic foramina are single in all related genera. Two alternative hypotheses suggest themselves: that formerly *P. h. papio* was more widely distributed, between the ranges of *hamadryas* and *anubis*, since when there has been a change in distribution; and that *hamadryas* and *papio* have evolved independantly but orthogenetically from an *anubis*-like ancestor. An analogous pair of hypotheses could be proposed to explain the insertion of *P. h. cynocephalus* between the more similar *anubis* and *ursinus*.

Lucotte and Lefebvre (1980) find that in several proteins examined by electrophoresis *papio* and *anubis* are indistinguishable, and are about intermediate between *hamdryas* and the other two; *papio/anubis, cynocephalus*, and *ursinus* bear a more or less equilateral triangular relationship to each other. This implies that, as far as the northern forms go, the distributions have not changed very much (and suggests that *papio*, for all its distinctiveness and its morphological convergences towards *hamadryas*, has only originated from *anubis* rather recently). In southern Africa, however, there may have been a change in distribution.

To sum up in the Papioninae: study of the taxonomy and interrelationships of extant forms in conjunction with their geographic distributions leaves the strong impression that modes of diversification have been on the whole different from those in other groups of primates. Instead of the centrifugal patterns so pervasive elsewhere, we find a simple vicariant pattern with either stepwise change of morphological features or else parallelism. Parallelism, of course, is by no means absent elsewhere, but the otherwise starkly 'traditional' taxonomic picture in this chromosomally uniform group throws the cases of orthogenesis into relief.

*Colobidae.* The leaf monkeys differ from the Cercopithecidae primarily in features of their digestive tract. Most conspicuously, their stomachs are complex, tripartite or quadripartite, with only the distal compartment (the pyloric part) containing hydrochloric acid and digestive enzymes: the proximal parts contain bacteria and other symbiotic organisms which ferment cellulose and hemicellulose, thereby greatly increasing the available energetic component of the diet. The morphological and physiological details of the digestive system in various colobid genera are described in such sources as Hill (1952), Kuhn (1964), Hollihn (1971), and Peng *et al.* (1983). Associated with the compartmentization of the stomach is the development of characteristic longitudinal and transverse muscle bands, and a reticular groove which enables fluid ingesta to bypass the fermentation chambers. The salivary glands are hypertrophied; the liver is large, flattened, and confined to the right hypochondrium. The cheek pouches, such a conspicuous feature of the Cercopithecidae, are lacking.

Cranially, colobids tend to have a high facial angle and rounded frontal region; generally the inter-orbital pillar is broad; the ascending ramus of the mandible is high and anteroposteriorly broad. Characteristically, the lacrimo-maxillary suture runs through the lacrimal fossa; in the Cercopithecidae, it is on or in front of the fossa, which is therefore situated entirely in the lacrimal bone. In the dentition, the molar cusps of colobids are high and pointed.

Colobids are nearly always of slender build with long limbs and a shortened thumb. The tail, with one exception, is long, with elongated caudal vertebrae supported proximally by chevron bones.

Delson (unpublished thesis, 1973) separates the Colobidae, which he ranks as a subfamily (Colobinae) of the Cercopithecidae, into two subtribes, Colobina and Presbytina, for the African and Asian genera respectively. Apart from the total absence of the thumb in the former, there are two differences in the dentition: in the Colobina, $M_3$ is broader posteriorly than anteriorly (instead of narrowing distally) and the protoconid on $P_3$ is reduced in size. All three of these features of the African colobids are clearly derived, which leaves the Presbytina as a group defined only by primitive retentions.

Welker (1981) notes that *Procolobus* differs strikingly from other colobids in its mother–infant relationships, and proposes a subfamily Procolobinae to separate it from the other genera, which he places in the Colobinae. What information there is, however, suggests that *Presbytis* too may lack the colobine 'handing-round' of infants to other troop members, showing instead a pattern where the infant takes the initiative in transferring. The pattern in *Nasalis* appears not to have been described.

Among the Asian genera, *Nasalis* is clearly, as has long been recognized, outstanding in many respects. It is, for example, the only genus in which the interorbital pillar is narrow, cercopithecid-like; the nasals are long and narrow; and the muzzle is long, the ascending ramus back-tilted. Szalay and

Delson (1979) consider this morphology an evolutionary reversal, the short, broad face of the other genera being, in fact, the cercopithecoid morphotype (because of its similarity to that of gibbons). While the similarity to gibbons is interesting, it is not identical and may be only superficial; certainly the functional connection of the normal colobid facial morphology to the colobid dental specializations, which are undoubtedly not primitive, cannot be ignored (Vogel 1966). *Nasalis*, though at least one of its species (*N. larvatus*) is certainly specialized in the degree of its facial elongation, would appear none the less to be primitive in lacking the full extent of the colobid masticatory specialization: it has thinner enamel than other colobids (Kay 1981); it lacks their frequently heavy supraorbital ridges; there is commonly a sagittal crest; the mandibular angle is not enlarged, and the ascending ramus slopes back. In other features, too, *Nasalis* appears primitive: it has 48 chromosomes, whereas all other colobids—all genera except *Presbytis* being known—have 44 [although the banding patterns, in as far as they have been analysed (Dutrillaux *et al.* 1980), are far from being identical]; the intermembral index is over 90 (shared however with *Pygathrix* and some fossil genera), and the infant colouration does not contrast with that of the adult.

It is true that apart from *Nasalis* the Asian colobids do seem to form a group separate from the Africans. Cranially, they are shorter faced, and all possess a sub-orbital fossa; in the muscular system, Asian colobids as far as known differ from African in the loss of the clavicular head of M. pectoralis major (Ashton and Oxnard 1963). It should be noted, however, that Cronin and Sarich (1975) make *Colobus, Pygathrix*, and what they call '*Presbytis*' three equal lineages.

The evidence is therefore fairly good that *Nasalis* is the sister group to other colobids, and should be placed in a subfamily, Nasalinae, apart from the other genera, Colobinae.

1. *Nasalis*. Groves (1970*a*) placed the Pig-tailed Langur of the Mentawai Islands, *Simias concolor*, in this genus as *N. concolor*, noting that it retained certain more primitive features (such as less-specialized skull form) than the Proboscis Monkey, *N. larvatus*, while being highly specialized in others (such as its extremely short tail). The short but upturned nose of *N. concolor* resembles that of the juvenile *N. larvatus*; in the latter, the nose elongates rapidly, finally in the adult male turning downwards and overhanging the lips—an example of hypermorphosis.

Napier (1985) resurrects *Simias*, agreeing that the two genera are close, but considering the differences too great to permit them to be combined. Until a full taxonomic revision of the family is undertaken, I would prefer to retain the two species in the same genus.

2. *Pygathrix*. This genus contains five species; *P. nemaeus, avunculus, brelichi, roxellana*, and *bieti*. The last of these was relegated to the status of a subspecies of *P. roxellana* by Groves (1970*a*), but is here revived in

concordance with the findings of Chinese workers (see, for example, Li and Lin 1983). The Indochinese species *P. nemaeus* lacks the full development of the nasal specializations of the other four; consequently the recognition of two subgenera, nominotypical *Pygathrix* for *nemaeus*, and *Rhinopithecus* for the others, seems entirely justified: it remains possible, however, that the peculiar leaf-like upturned nose of *Rhinopithecus* has been lost in *P. nemaeus*, which does have the same nostril shape, which seems difficult to interpret as such.

*P. nemaeus*, the Douc Langur, is restricted to the southern two-thirds of the Indochinese peninsula, east of the Mekong River. The more brightly coloured, red-legged *P. n. nemaeus* occurs in central Vietnam and southern Laos; the black-legged race, *P. n. nigripes*, lives in southernmost Vietnam and the neighbouring part of Kampuchea.

*P. avunculus*, a largely black-coloured snub-nosed monkey, is known only from the highlands between the Red and Black Rivers of northern Vietnam. The three Chinese species share derived characters, such as the shortened digits, so probably together form a sister group to *avunculus*. *P. brelichi*, from the Fan Jin Shan range of south–central China, is grey and long-tailed, and lacks the facial subcutaneous fat padding of the other two: the black *P. bieti*, the largest (perhaps baboon-sized) species, from the mountains between the Mekong and Yangtse gorges in Yunnan, and the beautiful golden *P. roxellana* (NB not '*roxellanae*': Krumbiegel 1978) from the highland slopes in Sichuan, Ganssu, and Hubei.

3. *Presbytis*. On the evidence of dental, cranial, and other features (Weitzel 1983) this genus is properly restricted to a group of species from Malaysia and western Indonesia. They have exceptionally thick dental enamel, short faces, no supra-orbital ridges, and a larynx that is very reduced in size. Neonates are white in colour, with at least a trace of a black dorsal stripe, and often also a cross-stripe over the shoulders: a pattern seen also, if in muted form, in *Pygathrix*. Brandon-Jones (1977) points out that cranially the most primitive species is *P. potenziani* from the Mentawai Islands, which is black in colour like primitive forms of other colobid genera (he cites *Trachypithecus*, but *Colobus* and probably *Pygathrix* can also be instanced). Next most primitive are three grey species: *P. comata* (West Java), *P. thomasi* (northern Sumatra), and *P. hosei* (northern Borneo). Central in distribution are a cluster of red and brown taxa, which Brandon-Jones sees as being more derived. How many species are represented in this central cluster is unclear; two (*P. melalophos* and *P. femoralis*, the latter occurring in Malaya also) are recognized in Sumatra by Wilson and Wilson (1976), and three appear to be partly sympatric with each other and with *P. hosei* in Borneo (the widespread *P. rubicunda*, and the more locally distributed *P. frontata* and, again, *P. femoralis*). Some of the Malayan and east-Sumatran races ascribed by Wilson and Wilson to *P. femoralis* are separated by Brandon-Jones (1984) as *P. siamensis*.

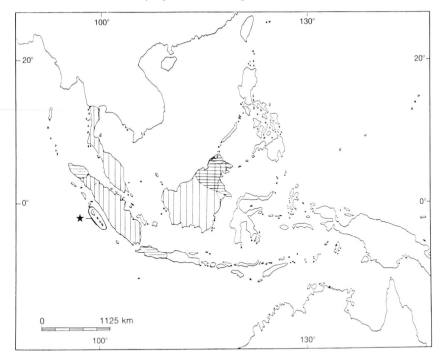

**Fig. 4.7.** Centrifugal pattern in genus *Presbytis*. Most primitive species is *P.potenziani* (marked with star); then the peripherally distributed, grey *P.comata, thomasi* and *hosei* (horizontal hatching); with the most derived, red/brown, species (*melalophos, femoralis, frontata, rubicunda*) being most central (vertical hatching).

This very clear centrifugal pattern (Fig. 4.7), as elucidated by Brandon-Jones (1977), predicts that chromosome banding patterns will be found to differ between the various species, at least between the central and peripheral ones; as is already known to be the case in *Colobus* (see below). It is also worth remarking on the multiplicity of species (each in its turn divided into well-marked subspecies) in this limited area.

4. *Trachypithecus*. The cranial and dental characters of this genus set it quite apart from *Presbytis*, with which it shares only the primitive features of the Asian group. The neonate is golden coloured, and the 'handing-round' behaviour is well known as it is in *Semnopithecus* and *Colobus*.

The distribution of this genus is much wider than that of *Presbytis*: it extends east to Bali and Lombok, north into southern China, and has outliers in southwestern India and Sri Lanka. An extremely interesting centrifugal pattern has been described by Brandon-Jones (1977). At the three corners of the distribution are found black species: in Sri Lanka the Purple-faced Langur, *T. vetulus* (incorrectly called *senex*: see Napier 1985), and in southwestern India the closely related Nilgiri Langur, *T. johni*; in the Vietnam–Laos–China border area, *T. francoisi* (which Brandon-Jones prefers to regard as a cluster of admittedly closely related species): in Java, Bali, and Lombok, *T. auratus*, which has however a red morph in some locales. In the first two of these regions, moreover, there is a tendency to develop white (or silvery, or creamy yellow) areas on the head, rump and tail, presumably in parallel. So close are these resemblances that in his 1984 publication Brandon-Jones actually includes the form commonly known as *T. francoisi poliocephalus* (from Cat Ba Island, Vietnam) in *T. johni*, and a still undescribed all-black race (from far northwestern Vietnam) in *T. auratus*.

The centrally distributed species can again be divided into a peripheral and central group. The peripheral group, basically grey in colour, with grey or black facial skin, are *T. cristatus* from Sumatra, Borneo and smaller islands, and the western Malay coast, recurring in eastern Thailand and Indochina; and *T. pileatus*, a much larger species from Assam, northern Burma, and southern China. This latter species has both monomorphic and black-capped subspecies, and a well-differentiated offshoot, usually classed as a full species, *T. geei*, in Bhutan.

In the centre of the range occur one or two highly derived species with pale limbs and a striking facial pattern of white eye-rings and circumoral zone. Usually one species, *T. obscurus*, known as the Dusky or Spectacled Langur, is mapped as occurring in Malaya and peninsular Thailand and Burma, and a second, *T. phayrei*, further north in Thailand and Burma, and into Bangladesh and Tripura to the northwest and Yunnan to the northeast, the ranges of the two being separated by a wedge of *T. cristatus* (Fooden 1976); but the two are certainly very alike, and Brandon-Jones (1984) recommends redrawing the boundary between them. The important point, however, is the double centrifugal pattern of the genus as a whole (Fig. 4.8), taken in conjunction with the extension of this pattern into other Asian genera.

5. *Semnopithecus*. To this genus belongs only a single species, the Sacred, Hanuman, or Indian Gray Langur, *S. entellus*. Brandon-Jones (1984) divides the species into two, and unites the genus with *Trachypithecus*; while it is true that the two genera share many characters, their skulls are clearly separable and they differ in other respects too: for example, the neonate colour in *Semnopithecus* is black, surely a primitive condition. As well as India, this species is found in Sri Lanka, Pakistan, Nepal, Bhutan, and Bangladesh, also in the Chumbi Valley (the wedge of Tibet between Sikkim and Bhutan). Napier (1985) has recently incisively reviewed this species, whose geographic variation is simple and predictable from geography (and, in the north, altitude).

6. *Colobus*. The Black-and-White Colobus monkeys of Africa are uniquely derived in their enormously enlarged larynx, the form of the basicranium, shape of the incisors, and other features.

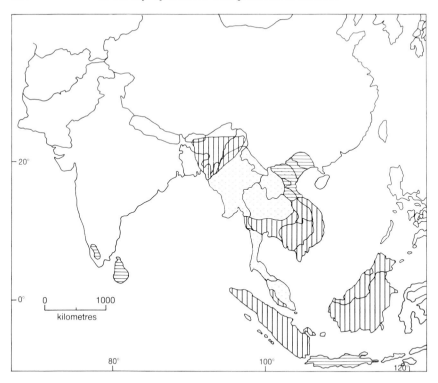

**Fig. 4.8.** Centrifugal pattern in genus *Trachypithecus*. Sequence runs from peripheral, black species *vetulus* (Sri Lanka), *johni* (S.W.India), *francoisi* (Indochina/China border) and *aurata* (Java) (horizontal hatching), via intermediate grey *pileatus* (Assam/Burma border) and *cristatus* (S. Indo-china, and Sumatra/Malay/Borneo region) (vertical hatching), to highly derived (depigmented facial pattern) *obscurus* and *phayrei* (stippled).

Hull (1979) recognizes four species in the genus: *C. polykomos* (West Africa), *C. angolensis* (from Angola to East Africa), *C. satanas* (Cameroun/Gabon region), and *C. guereza* northern Cameroun to Ethiopia and Kenya). The first three are allopatric, but *C. guereza* is sympatric in northern Cameroun with *C. satanas* and in northeastern Zaire and on the Uganda border with *C. angolensis*; but in Kenya and northern Tanzania it and *angolensis* are parapatric. The vicariance of the first three has led some authorities (Napier and Napier 1967) to unite them, retaining just two species in the genus. However, Hull's (1979) analysis has shown that the three allopatric species are readily distinguishable.

To these four, Oates and Trocco (1983) add a fifth, *C. vellerosus*. This taxon hybridizes with *polykomos* around the headwaters of the Bandama River in Ivory Coast, whence previous authors had united the two into a single species. Oates and Trocco, however, not only doubt that the hybridization is more than casual, but demonstrate that the loud-call (the characteristic black-and-white colobus 'roar') of *C. vellerosus* is identical to that of *C. guereza*, while that of *C. polykomos* is more primitive; the *C. angolensis* roar is still less derived, and that of *C. satanas* is the most primitive of all. (Investigations of the larynx of these forms is clearly in order.) They thus derive an evolutionary scenario; in terms of the present thesis, we would have a primitive black species giving rise, presumably by ordinary allopatric speciation, to a more derived black-and-white daughter species; this would then differentiate clinally, and finally centrifugally.

7. *Procolobus*. Although, on the basis of Verheyen's (1962) craniological analysis, I proposed to retain red colobus in the genus *Colobus* but separate olive colobus as *Procolobus* (Groves 1970*a*), the evidence now seems to argue the other way round: red and olive colobus belong together in *Procolobus*, the similarities between red and black-and-white colobus being probably symplesiomorph. Red and olive colobus alike lack colour contrast between infant and adult, and have no infant 'handing-round' behaviour; they have specializations of the masticatory apparatus, including a sagittal crest in most adults, enlarged supra-orbital ridges, and, in the stomach, development of a praesaccus; the larynx is modified in the opposite direction from *Colobus*, its size being reduced, the subhyoid sac lost, and the pterygoid fossa deepened. Most cogent of all is the development of a unique form of female sexual swelling, and its mimicry (the perineal organ) in adolescent males (Kuhn 1972).

The cranial differences between red and olive colobus are well-marked (Verheyen 1962) and dictate their placement in different subgenera, respectively *Piliocolobus* and nominotypical *Procolobus*.

*P. verus*, the Olive or Van Beneden's Colobus, is an inconspicuous greenish coloured species, the smallest living colobid. Known mainly from the Sierra Leone to Ghana region, it may be more widespread as a specimen is recorded from northeastern Nigeria (Menzies 1970).

There are four species of red colobus (Groves and Dandelot, in preparation). The central species, *C. pennantii*, found from Bioco Island and Congo east through Zaire, to Zanzibar Island, is in many respects the most primitive; i.e. differentiation has not been centrifugal. There are a number of distinctive subspecies, of which the two central ones, *C. p. oustaleti*, found west and north of the Zaire River, and *C. p. elliotti* from northeastern Zaire, are highly variable in colour and pattern. The Zanzibar form, *kirkii*, has usually been classed as a distinct species, but is connected to the Uganda/western Tanzania *tephrosceles* by a race, *gordonorum*, in the Uzungwa region of southeastern Tanzania, perfectly intermediate morphologically as well as geographically.

The species restricted to the gallery forests of the Tana River, Kenya, *P. rufomitratus*, looks externally very like *P. pennantii tephrosceles*, but cranially is very distinct. On the other hand it bears no relation to *kirkii* and *gordonorum*. Evidently early forest connections enabled it to disperse from Central Africa to the Tana River; these were then broken, but gene flow continued between central Africa and the Tanzanian coast, with a new suite of derived character states washing back along the forest corridor, its strength declining with distance, until connections were finally broken for good. The case is remarkably similar to that of *Cercocebus agilis* and *galeritus*, even to the same Tanzanian (Uzungwa/Mwanihana region) and Kenyan (Tana River) localities being involved; the difference is only that mangabeys have not (yet?) been recorded on Zanzibar.

*Procolobus badius* is restricted to West Africa; it differs distinctively from *P. pennantii* in its vocalizations (Struhsaker 1981) as well as in its cranial and pelage features. The very rare *P. preussi*, restricted to the Yabassi district of Cameroun, is rather more like *badius* than *pennantii*, and very distinct in its own right.

*4.3.2(b)(ii) Hominoidea*   The derived character states of the Hominoidea are listed by Groves (1972*b*), and discussed in terms of function by Andrews and Groves (1976). They mainly concern the postural/locomotor apparatus, but there are other features too: the presence of a vermiform appendix (not the functionless rudiment it has all too often been claimed: Scott 1980), interstitial placental implantation (Luckett 1976), and sperm mitochondria with few gyres (Martin and Gould 1980) are examples. Consequently there is no doubt of the monophyly of the superfamily, and such proposed schemata as the separation of the gibbons—distinct as they undoubtedly are—as a separate superfamily, Hylobatoidea (Thenius 1981), serve only to obscure this.

Traditionally the Hominoidea have been divided into either two families (Pongidae for the apes plus Hominidae for humans) or three (restricting the Pongidae to the great apes and making a separate family, Hylobatidae, for the gibbons). Such a division is by inference gradistic, but proposals to reorganize the taxonomy on a cladistic basis have become increasingly urgent and have been gaining ground. I have recently reviewed these proposals, which date from Goodman (1963), and have opted for the following (Groves 1986):

Family Hylobatidae
Family Hominidae
   Subfamily Ponginae
    *Pongo*
   Subfamily Homininae
    *Gorilla*
    *Pan*
    *Homo*

When fossil taxa are brought into consideration, the three extant genera of the Homininae will need to be referred to separate tribes: Gorillini, Panini, Hominini. Thus far, only the last of these is known to contain more than a single genus.

Schwartz (1984) has proposed that it is *Pongo*, not *Pan* and *Gorilla*, that is closest phyletically to *Homo*. Koop *et al.* (1986) have emphasized that this flies in the face of all the evidence. The derived anatomical states shared by the Homininae (in the present sense) are listed by Groves (1986); the most obvious are the modifications of wrist and ankle, reduction of premolar size, relatively delayed dental eruption, structure of the heart and aorta, specialized axillary organ, preponderance of eccrine glands over body surface, densely haired scalp, lengthened small intestine but shortened colon, presence of ear-lobe, presence of frontal sinuses, and other features of facial architecture (suture patterns, foraminal positions).

1. *Hylobates.* The gibbons, or lesser apes, have among their derived features the loss of canine/sectorial sexual dimorphism (females as well as males having long slender canines and extremely sectorial $P_3$), orbits encircled by peculiar thickened rims, very short faces and reduced jaws and cheek teeth, and both arms and legs elongated well beyond allometric trends. At the same time they lack various synapomorph conditions of the taxa here referred to the Hominidae: shortened but robust canines, $P^3$ metaconid, reduced body hair, reorganization of carpus, loss of ischial callosities, and so on.

Groves (1972*b*) referred all gibbons to a single genus, and this system has been widely accepted. This, it now appears (Prouty *et al.* 1983) should be split into four subgenera: *Nomascus, Symphalangus, Bunopithecus,* and *Hylobates.* The preferred cladogram of Haimoff *et al.* (1984) splits the lineages off in this order, but it is noted that *Symphalangus* could have separated off before *Nomascus*, or these two could have split off together: in other words, relationships of the main divisions are very even, and any dichotomy is hard to elucidate. The only extensive sympatry is between *Symphalangus* and *Hylobates*, and this is presumably made possible by the strong size difference between them; whatever the mode of the initial speciation event(s), it has been overriden by time and ecological imperatives.

*Nomascus* has been shown (Dao 1983; Ma and Wang 1986) to consist of two species rather than one. *H. leucogenys*, the White-cheeked Gibbon, in which the male (and juvenile of either sex) has white or reddish cheek whiskers, and ranges through most of Indochina east of the Mekong and into Yunnan in the Xishuanbanna rain forest. *H. concolor*, the Black Gibbon, in which males and juveniles are entirely black, seems restricted to deciduous forest; it ranges from northeastern Vietnam, and Hainan Island, well into Yunnan, encircling the range of *H. leucogenys* to the north, with a small isolated population (described as *Hylobates concolor lu*) around 20°20′N, 100°10′E on the Upper Mekong in Laos.

*Symphalangus* has a single species, *H. syndactylus*, the Siamang of

Sumatra and Malaya. *Bunopithecus* has also a single species, *H. hoolock*, the Hoolock Gibbon, living in Yunnan west of the Salween, northern Burma, and India and Bangladesh east of the Brahmaputra; the Chindwin River divides it into two distinctive subspecies.

Except for recognition of the primitive species *H. klossii*, the Kloss Gibbon or Beeloh from the Mentawai Islands, there is no agreement about the number of species in subgenus *Hylobates* (see recent review by Marshall and Sugardjito 1986). The following taxa are candidates for full species status: *moloch*, a monomorphic grey taxon from Java; *pileatus*, a sexually dichromatic form with a bubbling great-call (female intergroup vocalization) from Kampuchea and southeastern Thailand; *lar*, the White-handed Gibbon, with a simpler soaring great-call, from Thailand, southeastern Burma, Malaya, and northern Sumatra, a nonsexually dichromatic taxon; *agilis*, the Agile Gibbon, again nonsexually dichromatic, and with a simple great-call, from central and southern Sumatra; *muelleri*, highly variable in colour and with a bubbling call, from most of Borneo; and the enigmatic *albibarbis*, resembling *muelleri* in appearance but *agilis* in vocalizations, from southwestern Borneo. Where distributions come into contact—*lar* and *agilis* in northern Sumatra, *lar* and *pileatus* in eastern Thailand—low levels of interbreeding occur, so species status for most, perhaps all, of the taxa would seem appropriate.

2. *Pongo*. Although there is no doubt of the early separation of the orang utan lineage from the common human—gorilla—chimpanzee stem, there are numerous parallelisms between the orang utan and one or other of the other hominids (Groves 1986). Groves (loc. cit.) and Courtenay *et al.* (in press) discuss the taxonomy of the genus *Pongo*, concluding that a single species with two subspecies is probably appropriate in the present state of incomplete knowledge.

3. *Gorilla*. The gorilla and chimpanzee share surprisingly few derived features which are not shared with the human. The inclusion of the gorilla in the genus *Pan*, as previously recommended by me (Groves 1970*b*), therefore seems decidedly inadvisable. The three subspecies of *Gorilla gorilla* may have been distinct for hundreds of thousands of years (Groves and Stott 1979).

4. *Pan*. Although it is probable that *Pan troglodytes* and *Pan paniscus*, referred to (somewhat inappropriately, as at least one subspecies of the former is as small in size as is the latter) as Common and Pygmy Chimpanzees respectively, really are distinct species, little can be said definitively about the true relationship between them until the number and interrelationships of subspecies of Common Chimpanzee have been worked out (Shea and Groves, in preparation).

5. *Homo*. Groves (1986) lists seven shared derived character states between gorilla and chimpanzee, 12 between gorilla and human, 25 between chimpanzee and human. The favoured cladogram would therefore be

chimpanzee and human as sister groups, with gorilla as their sister group, but the large number of shared derived states between each of the pairs is a strong reason for caution. Certainly immunological comparisons, from Sarich (1967) on, have been unable to resolve the three-way split. There could actually be no real reason to doubt the reality of the three-way split if the principle of heteromorph speciation (Chapter 1, subsection 1.4.3) is borne in mind. If the proto-hominine species had, as indeed most species that are at all widely distributed do, a system of non-concordant clines running in different directions, the cutting up of its range (assuming allopatric speciation) would result in three—or more!—now isolated daughter populations, each of which had high frequencies of different character states, or shared high frequencies of particular derived states with its neighbour. Kortlandt (1972) envisages an ancestral species which became divided into three by the formation of the Nile and the Niger, cutting off the proto-chimpanzee to the west, the proto-gorilla in the centre, and the proto-human to the east. In such a model, we might imagine the frequency of fluorescent sperm and the sparsity of chest hair increasing to the east, and the frequency of absence of the deep head of M. flexor pollicis brevis and the tendency to develop a white pygal tuft in the infant increasing to the west. All these four conditions are obviously evolutionarily derived. After the division into three of the parent species by the advent of the Niger and Nile, the central and western daughter populations (proto-gorilla and proto-chimpanzee) would find themselves sharing the derived conditions of thumb musculature and pygal tuft, while the central and eastern daughters (proto-gorilla and proto-human) would share the derived sperm and chest hair states.

Now let us make the tripartite division of the ancestral species a little more complex, so that proto-chimpanzee and proto-human are also neighbours; say, to the south of proto-gorilla (the Nile and Niger will, of course, not do as barriers in such a case). If southerly populations of the ancestral species had such conditions as a tendency to $I^{1/2}$ monomorphy, rather larger penis, and reduced size of throat sac, compared to more northerly ones, then after the division of the species proto-human and proto-chimpanzee would share these derived character states.

There is really no cause for worry about polychotomies. The result is some apparent inconsistency in the assortment of derived character states, but in the real world—as opposed to the strict rules of cladism!—this is only to be expected. And parallelism has not even been considered yet.

## 4.4 SPECIATION IN THE PRIMATES

To survey the evidence for speciation processes offered by the extant primates, we should first distinguish species pairs which differ in karyotype from those which do not. Those which do seem to respond best to the centrifugal pattern, implying that the speciation mode was stasipatric. Cases where this seems indicated are:

— members of the [cf. *Lemur*] *fulvus* group;
— *Arctocebus* from *Perodicticus*;
— *Euoticus* from ?*Galagoides*;
— perhaps [cf. *Galago*] *alleni*, if this derives from *Galagoides*;
— *Callithrix argentata* group from *C. jacchus/pygmaea* group;
— *Saimiri vanzolinii* from *S.* proto-*boliviensis/sciureus*;
— *Alouatta seniculus/belzebul/fusca* (several levels of derivation) from more primitive species;
— *Cercopithecus mona* from *C. campbelli/denti*.

These cases of necessity are restricted to 'geminate' species pairs and, while others may be suspected, the speciation event was far enough in the past to make the geographic pattern uncertain.

Cases where there is apparently no karyotypic difference involved—or, better, none is known—but which still correspond best to a centrifugal pattern, are:

— [cf. *Lemur*] *mongoz* from the *fulvus* group;
— *Hapalemur alaotrensis* from *H. griseus*;
— *Galago senegalensis* from *G. gallarum/moholi*;
— ?*Tarsius pumilus* from *T. spectrum*;
— *Leontopithecus rosalia* from *L. chrysomelas/chrysopygus*;
— *Saguinus inustus* from *S. leucopus* group/hairy-faced groups;
— *Saguinus mystax* from *S. labiatus*;
— *Pithecia albicans* from *P. hirsuta*;
— *Pithecia monachus* from *P. hirsuta*;
— *Cercopithecus pogonias* from *C. campbellil/denti*;
— *Cercopithecus cephus* from *C. petaurista/ascanius*;
— *Cercopithecus erythrogaster* from ?*C. petaurista*;
— ?*Macaca arctoides* from *M. sinica* group;
— *Cercocebus torquatus* from *C. atys/agilis*;
— *Presbytis* red and brown species from grey species;
— *Trachypithecus* grey species from *T. auratus/francoisi/vetulus*;
— *Trachypithecus obscurus/phayrei* from *T. cristatus/pileatus*;
— *Colobus vellerosus/guereza* from *C. polykomos/angolensis*.

This list excludes cases where the centrifugal diversification has not resulted in speciation, but only subspeciation, or not even that. Such cases include:

— bleached morphs in *Propithecus* (both species);
— bleaching series in *Saguinus fuscicollis*;
— *Saguinus oedipus oedipus* from *S. o. geoffroyi/S. leucopus*;
— *Saimiri sciureus cassiquiarensis* from *S. s. sciureus/albigena/macrodon*;

— *Cebus albifrons, C. capucinus, C. olivaceus*: central races from peripheral ones;
— *Ateles geoffroyi geoffroyi* from *A.g. vellerosus/grisescens*;
— *Cercopithecus mitis kandti* from *C. m. doggetti/stuhlmanni*;
— *Cercopithecus mitis albogularis* and allied forms from *C. m. zammaranoi/labiatus*.

There could well be other cases which emerge on more rigorous, detailed study than has been possible here. The point is certainly not to argue that centrifugal processes are the only means whereby diversification takes place, nor even that they are necessarily commoner than allopatric modes (although I incline to think that they may in fact be so—given that much of the evidence of diversification process is simply lost in the mists of time), but to list the evidence that they really do occur. It is important, too, to realize that this diversification may or may not go as far as speciation, and may not result even in subspeciation: the same diversity-generating processes are involved. It is evident, too, that we find the same effect, namely diversification generated in the centre of a species' range, whatever the cytogenetic basis: simple mutation, recombination, chromosomal mutation, molecular drive, or orthogenesis; and that it may or may not be accompanied by heterochrony.

It is worth asking, too, whether there seem to be any correlations between the type of diversification that a taxon undergoes and some aspect of its environment or social structure. White (1978) argued that for stasipatric events to be successful there simply has to be a relatively high level, at least sporadically, of inbreeding. In the first list (above) the inbreeding requirement may be fulfilled in the cases of *Alouatta* and *Cercopithecus*, but is dubious in the other examples and is positively not the case for *Saimiri*. This goes, at any rate, for breeding structures; but it may be another matter when we consider that inbreeding may be a product of patchy distribution of a species as much as of social structure itself. This is more difficult to test rigorously; indeed, there is good reason to believe that the distribution of most species is patchy to some degree, so that all species would have some chance to experience requisite inbreeding levels. Though Vrba (1980) has indeed demonstrated some correlation between speciation tendency and distributional patchiness in antelopes, as far as primates are concerned the case must remain thus far unproven.

So, what sort of species do not diversify centrifugally? The following are the cases where allopatric processes seem to account for all diversification:

— all genera of Cheirogaleidae;
— ?*Varecia*;
— *Lepilemur*;
— species within *Propithecus*;
— *Loris*;

— *Otolemur*;
— *Aotus*;
— ?*Callicebus*;
— ?*Chlorocebus*;
— *Mandrillus*;
— ?*Lophocebus*;
— *Papio*;
— *Pygathrix*;
— *Semnopithecus*;
— *Procolobus*;
— subgenus *Hylobates*;
— *Pan*.

and, almost entirely,

— *Macaca* (origin of *M. arctoides* probably excepted);
— *Colobus* (origin of *C. vellerosus/guereza* group excepted).

Looking at this list and comparing it with the previous (centrifugal) lists, it is at once obvious what the difference is. The second list contains mainly genera (or species groups) which range over two or more major biome types: rain forest and deciduous forest; lowland and mountain forest; savannah and arid zone. A few genera, especially *Mandrillus*, are exceptions; but in the centrifugal lists there are no such cases. This, then—not social structure; not open-country habitat; not karyotypic uniformity or diversity, at least as such—is the difference. If there are habitat differences, there will be adaptation (or exaptation?); if not, then evolution will proceed just the same, by centrifugal processes.

# 5
# Fossil non-human primates

It is probably too much to hope that, except in the later stages where stratigraphic resolution is greatest, modes of evolution can be worked out in fossil primates. Gingerich and others tried it for *Pelycodus* and *Plesiadapis*, and it is still controversial. But no book of human and primate evolution would be complete without them, for obvious reasons; and we may yet be able to deduce something about what made the system run.

At the beginning of the last chapter I discussed the use of the plesion concept, using *Purgatorius* as an example. Plesia will crop up often in the present chapter, and their frequency emphasizes how much we do, after all, know about primate evolution; for a plesion is a morphotype, a branch of the cladogram which lacks autapomorphous states; and what are such branches but potential ancestors?

## 5.1 PLESIADAPIFORMES

Szalay and Delson (1979) divide this suborder (Plesiadapiformes, Paromomyiformes) into superfamilies Plesiadapoidea and Paromomyoidea, and—except that *Purgatorius* is included in the latter, indeed in a monotypic tribe of one of its families!—this scheme is defensible cladistically.

*Purgatorius*, the earliest known primate (from the lowermost Palaeocene and, somewhat dubiously, from the uppermost Cretaceous as well), is certainly the most primitive. It has remarkably primitive molar patterns, with high cusps showing only the beginnings of the bulbous form which characterizes the primates in general. It is the only primate with three incisors; it is also primitive, but not unique, in having a relatively large canine and the full complement of four premolars. The mandibular corpus is slender and elongated.

All the Plesiadapiformes retain primitive features long since lost in extant groups of primates: small endocranial volume, deep post-orbital constriction, lack of ossified post-orbital bar, no petromastoid inflation, carotid circulation in bony tubes within the bulla, retention of claws on all digits. At the same time they have developed their own derived character states, which put them right out of consideration as ancestors for later primates. Incisors, at least the central ones, are enlarged, and canines reduced; none has more than three premolars, whereas early strepsirhines and haplorhines have four.

The two superfamilies differ basically in the structure of the molars; during their evolution, there is much parallellism between them—loss of canines, reduction of other teeth, development of enormous diastemata between front and cheek teeth—but the end products are very different. The Plesiadapoidea develop thickened, subvertical incisors acting, supposedly, in a gnawing fashion; the Paromomyoidea evolve cylindrical, procumbent lower incisors.

The Plesiadapoidea are divided into three families: Plesiadapidae, Saxonellidae, Carpolestidae. The first of these survived until the late Eocene; the others were confined to the Palaeocene. The Paromomyoidea are divided by Szalay and Delson (1979) into two families: Paromomyidae and Picrodontidae. The first of these, again, survived into the late Eocene; the second, is known only from the Palaeocene. The study by Schwartz and Krishtalka (1977) suggests that the Picrodontidae are in fact closely allied to one, rather specialized, paromomyid genus in particular, *Phenacolemur*; if this is the case, then the family will have to be sunk.

There has been much discussion about the demise of the suborder. Certainly the Plesiadapidae converged in their dental specializations on rodents, and they seemed to decline in diversity as the order Rodentia radiated taxonomically; but this explanation will not do for the other members of the suborder.

## 5.2 STREPSIRHINI

It is true that the earliest fossil strepsirhines and haplorhines are often difficult to distinguish, such that some genera have been shuffled back and forth between them. (There are even two genera, *Navajovius* and *Berruvius*, which cannot be with certainty allocated to either of them or to the Plesiadapiformes.) In part this is because of the scrappy nature of some poorly represented fossil genera; in part, of course, it is the nature of the evolutionary process itself.

### 5.2.1 Adapiformes

The earliest representatives of the Strepsirhini, mostly Eocene in age, are placed in an infraorder Adapiformes. Certain features are said to characterize this infraorder: $I^1$ low-crowned, slender (buccolingually), very asymmetrical, and with a strong mesial prong (Rosenberger *et al.* 1985); but it is not clear that these are more than the primitive state from which modern lemur conditions took origin. As there is much controversy over the interrelationships of the different constituent taxa, and of their relationships to extant taxa, it is probably better to retain the infraorder Adapiformes as an interim measure, to include all those strepsirhines that are neither Lemuriformes nor Chiromyiformes. [Whether they are actually strepsirhines at all is another matter: Rasmussen (1986) has raised this question in an attempt to reconcile the neontological evidence with the reputed simiiform-like features of the adapiforms but Gebo (1986) finds shared derived features in the foot skeleton.]

In most writings, such as Szalay and Delson's (1979) book, all Adapiformes are referred to a single family, Adapidae. Recently this custom has been challenged (Schwartz and Tattersall 1985), and it seems best in the present state of knowledge to recognize a number of families. Provisionally, I propose that at least four be recognized: Petrolemuridae, Notharctidae, Adapidae, and Sivaladapidae. In at least two of these, Notharctidae and Adapidae, early Eocene forms retain the full series of four premolars; reduction to three in later forms is just one example of a number of instances of parallel evolution between them.

The Petrolemuridae have recently been proposed as a family by Szalay *et al.* (1986), to include *Petrolemur* from the middle Palaeocene (the oldest known strepsirhine), and probably also the Chinese Eocene genera *Lushius* and *Hoanghonius*. Their relationship to other primates is obscure; by inference, however, they would seem to be strepsirhines.

The family Notharctidae have commonly been thought of (in the guise of a subfamily of Adapidae) as the North American branch of the adapiform group. As more and more genera are reassigned to the group, however, it appears more and more widely distributed, with representatives (such as *Laurasia*, described by Schwartz and Tattersall 1983) surviving in Europe until the middle Eocene. In the Notharctidae $M_3$ is not reduced like-many other adapiforms, and the paraconid (the primitive mesial cusp on lower molars) remains well-developed; conules are well-developed on the molars. Recently Rosenberger *et al.* (1985) have shown that, where known, notharctids have an unusual and potentially significant occlusal pattern of the incisors; $I^1$ occludes with both lower incisors at once, and $I^2$ is excluded from occlusion, and $I^1$ has a prominent mesial prong. This arrangement makes a convincing antecedent to the occlusal pattern of the lemuriform dental comb; despite various suggestions of other comb precursors from time to time, I would suggest [though the strictures of Rasmussen (1986) may confute this] that if the same relationship can be demonstrated in the non-North American notharctids the source group of the Lemuriformes will at last have been identified.

If there is now a potential taxonomically identified precursor to the Lemuriformes as far as the dental comb is concerned, what of the other main lemuriform feature, the toilet claw? I am happy to say that there is now evidence for such an organ in the Eocene; but less happy to have to admit that it is so far taxonomically unidentified, indeed very likely unidentifiable. From the middle Eocene site of Messel in West Germany, remarkable for the state of preservation of its fossils, comes a pair of legs and feet with clear evidence of toilet claws on the second toes (von Koenigswald 1979). (And associated with the pair of legs was not only a pelvis, which might have been expected, but also something quite unlooked for: an enormous baculum.)

The Adapidae contain the residue of the Eocene strepsirhines. They may thus not be monophyletic, but as removal of the two definitely monophyletic groups (Anchomomyinae and Adapinae) would still leave a residue, it seems preferable at the moment to retain one family than to have a large number of *incertae sedis* genera.

The subfamily Anchomomyinae consists of four genera known from the middle Eocene of western Europe, sharing the derived states of loss of paraconids, presence of large hypocones on $M^1$, and compressed talonid basins on the molars. In the process of adding the fourth of these genera, which they named *Fendantia* in honour of a kind of melted cheese dip, Schwartz and Tattersall (1983) confirmed the group's essential unity; and later (1985) the same authors noted certain characters in which the group resembles the Cheirogaleidae and Lorisidae. The phylogenetic reconstruction of the strepsirhines by Schwartz and Tattersall has already been criticized (Chapter 4, subsection 4.2); and the similarities between the Anchomomyinae and the two extant families, in as far as they are genuine resemblances, seem to me to be better put down to parallelism than to a distinct cladistic lineage.

The Adapinae, a European Eocene group, have been discussed in detail both by Schwartz and Tattersall (1985) and by Rosenberger *et al.* (1985). It is quite clear that the five genera share numerous derived character states: spatulate upper incisors, with the incisor–canine rows closely ranked; $I^2$ occluding with $I_2$; no gap between upper central incisors; submolariform $P^4$; well-developed hypocones on all upper molars; buccal pillars on upper molars; all-round cingulum; premolariform lower canine; and so on. Schwartz and Tattersall would wish to restrict the family Adapidae to this group, and indeed they are such a well-defined assemblage that this is probably a sensible suggestion; the only problem, as mentioned above, would be that a number of genera would thereby be left unallocated to family. The same authors note resemblances to the Indriidae in the dentition; as with their proposed anchomomyine/cheirogaleid–lorisid clade, I feel that to take the Indriidae from a separate stock as far back as the Eocene would entail far too much convergence (three separate evolutions of the dental comb would have to be envisaged, for example).

The subfamily Protoadapinae contains those genera which, while certainly of basic adapid affinity, do not belong to either of the two well-defined subfamilies, and are only dubiously related to each other. All are European, from the middle and late Eocene. A further genus, *Azibius*, the only North African Eocene primate, is difficult to allocate even this far.

Living strepsirhines live in tropical Africa and Asia, and in particular on Madagascar. So far, there are almost no fossils—of any kind, let alone primates—from these regions of Palaeogene age. The problem is still to get the proto-lemurs into the tropics before the Miocene (when there are definite lemuriforms from East Africa: see subsection 5.2.2 below), and preferably early enough to get them into Madagascar before the Mozambique Channel became impossibly wide. The question of whether a water crossing at all should be envisaged must also be faced. There are fairly numerous, readily detectable affinities between Malagasy and mainland African mammals (Groves 1974*a*), and chance dispersal across a seaway makes a very unsatisfactory explanation for this. Preferably, we should ask plate tectonic geologists to tell us when the last dry-shod crossing was possible; was there a time, for example, when Madagascar was slipping northward but along a dry transform fault, perhaps like the San Andreas Fault of present-day California, rather than out in the Indian Ocean? Or, is it still possible that the India connection lasted later than the African one? Clearly, too, the range of potential dispersers was somewhat limited: there could, for example, have been no ungulates in the source area at that time.

It was with considerable surprise that, a few years ago, adapiform primates were identified far from their previously known range in both space and time: in the late Miocene of India (Thomas and Verma 1979). The genera concerned—*Sivaladapis*, and the long-known, but always enigmatic, *Indraloris*—were assigned to a subfamily, Sivaladapinae, of uncertain affinities. As the incisors in this group were vertical and quite un-comb-like, there is no real question of lemuriform affinities for them. Gingerich and Sahni (1984) consider them a long-lived adapid derivative living probably in an arboreal folivore niche, from which they were later outcompeted by evolving colobids. In view of their uncertain origins, and the tremendous time gap involved, I suggest that they (with their Chinese contemporary, *Sinoadapis*) be awarded family rank as Sivaladapidae.

## 5.2.2 Lemuriformes

The earliest known Lemuriformes, as identified by possession of a toothcomb, are a group of genera from the early Miocane of East Africa, assigned to the Lorisidae as they clearly possess the auditory anatomy of this family. [A lorisid has recently, however, been claimed in the Fayûm, of Oligocene age (Simons *et al.* 1986)]. *Mioeuoticus* has a very robust cranium with raised temporal lines and upwardly directed orbits, as in the living slow climbers; Walker (1974) assigns it to the Lorisinae. I argued above (Chapter 4, subsection 4.2.2(e) ) that the African and Asian slow-climbers should be assigned to separate subfamilies of Lorisidae; the characters of *Mioeuoticus* are peculiarly those of the Perodicticinae.

The other two genera, *Progalago* and *Komba* (which may not be distinct), have much more lightly built crania, and more inflated bullae. Walker (1974) assigns them to the Galaginae; indeed, in its known parts, *Progalago* is hardly distinguishable from *Galagoides*.

From Madagascar there are, alas, no Tertiary fossils. What there are, instead, are a remarkable array of early Holocene forms, some under 1000 years old, evidently—tragically—eliminated deliberately, or as a by-product of habitat change, by the first human inhabitants of Madagascar. There are large species of *Daubentonia* and *Varecia*; range extensions for *Hapalemur simus*; and several genera unrepresented by living species. Most notable is the gigantic *Megaladapis*, which, as recently shown by Schwartz and Tattersall (1985), forms, together with the extant *Lepilemur*, the family Megaladapidae. The Indriidae now have to be divided into three subfamilies to accommodate fossil genera: to the Indriinae belong, as well as the three living genera, the extinct *Mesopropithecus*; to the Palaeopropithecinae, the lemuriform brachiators *Palaeopropithecus* and *Archaeoindris*; and to the Archaeolemurinae the terrestrial forms *Archaeolemur* and *Hadropithecus*. The scientific, ethical, and aesthetic disaster of their extermination reminds us how precious and how vulnerable are their still, often precariously, surviving kin.

## 5.3 HAPLORHINI

In the absence of fossilized placentae we have to turn to other features to try to recognize early haplorhines. Fortunately there are some. In particular, the position of the carotid foramen postero-medial to the auditory bulla is diagnostic of the Haplorhini; as are the passage of the olfactory process above the interorbital septum, not below it as in strepsirhines, and the reduction of the stapedial artery.

Schmid (1983) pointed out that the well-developed vomero-nasal organ (Jacobson's Organ) of modern lemurs requires a gap between the upper central incisors for possage of olfactory membrane continuous with the rhinarium, and that this can be recognized in the fossil record. Unfortunately it is unclear whether lemurs have a primitive mammal-sized vomero-nasal organ, or a hypertrophied one; so it is not possible to say for certain whether the large central gap of the Notharctidae is primitive, or derived (and so perhaps a lemuriform synapomorphy), or whether the absence of such a gap in the Adapinae is primitive or derived, although the total absence of a gap is, indeed, unlikely to be the primitive state. The importance of this in the present context is that an Eocene haplorhine group, the Microchoeridae, have a central gap, and so fairly clearly identify themselves as primitive, the probable sister group of the other Haplorhini. Also probably primitive are their very small promontory artery, and the ventral exposure of the fenestra rotunda.

The bulk of the Eocene haplorhines are currently referred (Szalay and Delson 1979) to an infraorder Omomyiformes. Clearly if the Microchoeridae are the sister group of the other haplorhines they should not continue to be referred to the Omomyiforms, let alone to the Omomyidae as Szalay and Delson do; but there is insufficient information from other omomyiformes to be sure that all of them lack the gap. As an interim measure, therefore, the Microchoeridae will continue to be placed in the Omommyiformes.

Recently Szalay and Li (1986) have identified a haplorhine plesion, *Decoredon* from the middle Palaeocene of China. Its characters are said to be very primitive but detectably primate, and recalling confirmed haplorhines.

### 5.3.1 Omomyiformes

The Microchoeridae are confined to the Eocene of Europe. Five genera are now known: *Microchoerus, Nannopithex, Necrolemur, Chasselasia*, and *Pseudoloris*. The last of these had the lower parts of the tibia and fibula fused, as do tarsiers, whence the family was formerly referred to as Eocene tarsiids; but the more is learned about them (Schmid 1982), the less tarsier-like they seem to be, so that tibia–fibula fusion seems likely to be parallel in the two. [Very recently, however, Fleagle and Kay (1977) have proposed that the close opposition, near-fusion, of tibia and fibula is a primitive 'euprimate' feature.]

Although Szalay and Delson (1979) place all omomyiforms in one family, even after the removal of the Microchoeridae there remain two clearly distinct groups, for which I propose reviving a formerly adopted family division: Anaptomorphidae and Omomyidae. The most obvious distinction is the reduced third molars of the former versus their large size in the latter; but details of molar crown pattern distinguish them as well.

The earliest anaptomorphids are placed in the subfamily Teilhardininae. *Teilhardina*, from the early Eocene of Belgium, is remarkable in providing an example of evolution in progress: specimens are known both with and without the anterior premolar (Quintet 1966). All later omomyiforms have only three premolars (sometimes, only two), parallelling the two cases of independent P1 loss we have seen in the adapiforms. In later anapto-morphids, indeed, if P2 occurs at all it is reduced in size; in *Teilhardina* and the North American lower Eocene representaive, *Chlororhysis*, it is not reduced, and reduction of canines and enlargement of incisors, conspicuous characters of later anaptomorphids, are not developed. An omomyiform resembling *Teilhardina* has recently been noted from the Oligocene of the Fayûm, Egypt (Simons *et al.* 1986).

The Omomyidae are entirely North American in distribution; throughout the Eocene and into the Oligocene they are common, and there is even an early Miocene representative. In some of the subfamily Omomyinae, such as the well preserved *Ourayia*, the upper incisors are clearly not separated in the midline; and in *Washakius*, of the subfamily Washakiinae, they are probably not. The totality of evidence is poor, however.

The best known omomyid is the Oligocene washakiine genus *Rooneyia*, represented by a nearly complete skull. The auditory tube and reduction of the premolars to two led early commentators such as Wilson (1966) to speak of it as a potential catarrhine ancestor; but in essentials, including derived features, it is an omomyid. The presence in North America of *Ekgmowe-chashala* in the Miocene recalls that of the Sivaladapidae in India; but in this case the fossil is early Miocene, not late, and the lineage can be traced back continuously to Eocene forebears.

Three genera–*Donrussellia, Copelemur*, and *Kohatius*—must for the moment remain *incertae sedis*. The first of these is, however, a potential omomyiform plesion, being of late Palaeocene age and intermediate between Omomyidae and Anaptomorphidae.

### 5.3.2 Tarsiiformes

Various authors (notably Cartmill 1981) have argued that the Omomy-iformes are the sister group of the other haplorhines. While they possess such haplorhine synapomorphies as the reduced stapedial artery, they lack the full suite of features, especially the presence of the posterior bony orbital wall, and the characters of the ear. These character states are on the contrary possessed by the tarsiers; it has been argued that the development of the post-orbital septum must be convergent because the pattern of bones is different, but Cartmill shows that the differences are purely a consequence of the enormous size, in *Tarsius*, of the orbits and hence of the septum itself, despite the presence (as in some platyrrhines) of a relatively large remnant fissure.

This line of argument is convincing: the Omomyiformes are truly the sister group of the other haplorhines, not—as long supposed—Eocene tarsioids. In Chapter 4, section 4.1 I proposed a principle: that stem-groups should not be ranked at a higher taxonomic level than branches of the crown-group. So we do not create a new taxonomic level, uniting the Tarsiiformes and Simiiformes and opposing them to the Omomyiformes: instead, we rank the Omomyiformes equally (as an infraorder) to the other two.

Hardly had the Omomyidae and Microchoeridae been ejected from the Tarsiiformes than their place was taken by an Oligocene form, *Afrotarsius chatrathi* Simons and Bown, 1985. This would seem to share certain derived features with *Tarsius* alone: the postero-labial inflection of the $M_3$ entocristid, the very tall hypoconid, and other distinctive features. With both *Tarsius* and the microchoerid *Pseudoloris* it shares a broad paraconid, separated from the other cusps by a deep valley, on all molars; but the possession of a large paraconid is plausibly a primitive character, and in this case the placement of the paraconid is more lingual in both *Tarsius* and *Afrotarsius*, again linking them to the exclusion of other fossils.

*Afrotarsius* is from the Oligocene of the Fayum, Egypt, where it occurs alongside the remains of early catarrhines (see subsection 5.3.4 below). Its occurrence, unaccompanied by adapiform or omomyiform primates, in a Gondwanaland site along with catarrhines—also missing from the European Eocene—enhances its potential importance, and suggests—or rather, reinforces the conclusion based on other mammals–that the Tethys Sea, in separating Gondwanaland continents to the south from Laurasian ones to the north, was a major barrier to migration, and that it is to the south of it that we should look for future revelations about the early history of extant groups of Old World primates.

### 5.3.3 Platyrrhini

The earliest primate in South America is *Branisella boliviana* from the early Oligocene of La Salla, Bolivia (see Hoffstetter 1969; Wolff 1984). Szalay and Delson (1979) place this in a monotypic subfamily Branisellinae of the Cebidae (*sensu* Rosenberger), but in the terms of this thesis it occupies the position of plesion to all other platyrrhines.

It is probably true to say that, had it been discovered in North America, *Branisella* would have been dismissed as an omomyid and no more would have been heard about it. But because it is where it is, there has been careful analysis of it to determine whether it really is a platyrrhine. Szalay and Delson (1979) are quite clear that it differs from any omomyiform in its lack of a buccal cingulum and paraconule; its small, single-rooted $P^2$, which resembles such omomyids as *Tetonius, Omomys*, and *Ourayia*, but the root is very wide buccolingually, resembling only *Washakius* among omomyids; its short $M_3$, with the talonid constricted; and its evidently U-shaped mandibular symphysis. Wolff (1984) follows Szalay and Delson (1979) in drawing attention to its close similarity in many respects to *Saimiri*, in what are presumably common retentions of primitive characters.

The question of where the platyrrhines came from has been endlessly discussed. Supposed similarities of particular North American omomyids to platyrrhines have been dismissed by Hoffstetter (1974) in favour of a direct South Atlantic derivation; this would of course reinstate, palaeontologically speaking, the neontologically unassailable monophyly of the Simiiformes, since there are Oligocene catarrhines in Africa, approximately contemporary with *Branisella*. In terms of dispersal routes, however, it is little help, since the Oligocene South Atlantic would have been much the same width as the seaway between North and South America at the same time. The idea of proto-platyrrhines, even aestivating in a cheirogaleid manner, rafting somehow, even were there islands as stepping-stones, across a South Atlantic already half as wide as today's, somehow carries little conviction. Lavocat (1974) points out that the caviomorph rodents also appear in South America at the same time (in the same deposits, in fact), and also look as if they have African affinities; so we are not talking about chance events, but about genuine dispersal routes. It seems to me that in all this North–South versus East–West argument it has been forgotten that the earliest catarrhines of all are from the Eocene of Burma, not the Oligocene of Africa (see subsection 5.3.4 below). Perhaps it is time to look carefully at still other routes: maybe the notorious continent of Pacifica, beloved of many vicariance bio-geographers (see papers in Nelson and Rosen 1981; and a more realistic assessment of trans-Pacific contacts in Craw and Weston 1984).

The platyrrhine fossil record supports the observations of Rosenberger and others, and the molecular evidence, in revealing very long separate lineages for the major groups of New World monkeys. The evidence has recently been reviewed by Rosenberger (1977, 1980) and by Rose and Fleagle (1980) and Setoguchi and Rosenberger (1985).

Callitrichidae: after the removal from marmoset affinities of *Dolichocebus* (Rosenberger 1979), this family was left without fossil representation for a while; now there is once again a fossil (Miocene) marmoset, *Micodon* Setoguchi and Rosenberger, 1985.

Cebidae: the presence of an interorbital fenestra at once identifies skulls of extant *Saimiri*, and its presence in the Oligocene *Dolichocebus* equally indicates the affinities of this genus (and is supported by other features too). The Miocene *Neosaimiri* is indistinguishable in its known parts from the modern squirrel monkey. Rimoli (1977) has also described a very large subfossil squirrel monkey, *Saimiri bernensis*, from the Dominican Republic, dated at 3850 BP.

Aotidae: the Miocene *Tremacebus* has very large orbits and (probably correlated) a large post-orbital fissure, as in the modern *Aotus*, to which it is probably related. The post-orbital septum is however damaged, not naturally incomplete as was originally claimed for it by its describer (Hershkovitz 1974).

Callicebidae: the only possible 'fossil' representative of this family is the peculiar *Xenothrix*, from the early Holocene of Jamaica, to which Rosenberger (1977) has devoted a special discussion.

Pitheciidae: the Miocene genus *Cebupithecia* has long been associated with this family. Hershkovitz (1970*b*) cast doubt on this affinity, but it has been supported by all recent workers. The long enigmatic early Miocene *Homunculus* may also pertain, but more distantly, to this family (Rosenberger 1978).

Atelidae: Hershkovitz (1970) described the Miocene genus *Stirtonia*, which has since been accepted as related to the extant *Alouatta*. The genus *Kondous* Setoguchi, 1985, described as a fossil spider monkey, has more recently (Kay *et al.* 1981) been shown to be a synonym of *Stirtonia*. There is, however, a 'fossil' *Ateles*: this is *A. anthropomorphus*, originally described as a separate genus *Montaneia*, from 2000-year-old deposits in Cuba. The correct allocation of this form has recently been determined by Arredondo and Varona (1983), who published in addition an aboriginal rock painting from Cuba of what can only be this animal.

The presence up to the Holocene of monkeys on the islands of Jamaica, Hispaniola, and Cuba is very interesting. Two of the three have been referred to extant genera [although Rosenberger (1978) considers that *Saimiri bernensis* should probably be referred to a new genus]. It is not profitable at the moment to speculate on possible dispersal routes; the existence of an endemic mammalian fauna (recently extinct insectivores such as *Nesophontes*, and the still extant insectivore *Solenodon* and the capromyine rodents) is probably not relevant, but mammals of detectable neo-tropical affinity, such as *Mylodon*, may be suitable analogues.

### 5.3.4 Catarrhini

Catarrhines above all should be identifiable from the fossil record, as the extant forms have so many synapomorphous features: premolars reduced to two, distinctive canine/sectorial complex, deflated bulla and tubular ectotympanic, and so on. It is to be expected, of course, that some of these features will be shown emerging through the fossil record; but as both living superfamilies share these features they must have developed (unless extreme homoplasy is envisaged) quite early on. It is therefore slightly surprising to find that there is some controversy over the catarrhine placement of two genera: *Amphipithecus* and *Oligopithecus*.

#### 5.3.4(a) Eocene catarrhines of Burma

Recent findings (Maw *et al.* 1979; Ciochon *et al.* 1985) have added to knowledge of the primates of the Pondaung Formation, Burma, and have dated them to approximately 40 to 44 MA.* *Pondaungia cotteri* is now known from several specimens—all jaws and teeth—and it is quite clear that, despite earlier misgivings, it is a catarrhine. The upper molars are bunodont, four-cusped, the lowers six-cusped, with a weak cingulum on both uppers and lowers; the mandibular body is both deep and thick. The lower molars are strongly corrugated, with many accessory cuspules and low ridges. The describers note a resemblance in the upper molars to the notharctid *Pelycodus*, and a general (but not detailed) likeness to the Oligocene catarrhine *Aegyptopithecus*.

The other Pondaung primate is *Amphipithecus mogaungensis*. Szalay and Delson (1979) claim this as an adapiform; previously it had generally been considered a catarrhine. *Amphipithecus* has a thick, deep mandible like *Pondaungia* (and most catarrhines), but differs in the cheek teeth having smooth, uncorrugated occlusal surfaces. The talonid cusps are poorly differentiated, being essentially excrescences on a continuous crest, and there is no hypoconulid (Ciochon *et al.* 1985). Paraconids are present as in *Pondaungia*. Unlike the latter, which is known only from molars, the

*\*Millioni Anni* = millions of years.

premolars of *Amphipithecus* are known: three are present (though only $P_{12}$ are represented by their crowns). Ciochon *et al.* stress the great jaw depth and the presence of a superior transverse torus in *Amphipithecus* as diagnostic catarrhine features.

I would rank both the Burmese genera as plesia of other catarrhines.

### 5.3.4(b) Oligocene catarrhines of Egypt

The Jebel al-Qatrani Formation of the Fayum, Egypt has been known since the turn of the century; its richness in fossil primates has gradually become clearer since the reopening of excavations in the 1960s. The African tarsiid, omomyid and lorisid, have already been mentioned; catarrhines fall into two families, Parapithecidae and Propliopithecidae. Temporally, the deposits fall into three levels: the lower and upper fossil wood zones (FWZ), and the intervening Yale Quarry I.

The Parapithecidae are represented by two genera in the middle and upper levels; *Parapithecus* and *Apidium*. Each shows evolution over these two time levels. *Parapithecus fraasi* from the middle level has a single pair of (lower) incisors; *P. grangeri* from the upper FWZ has lost even these (Simons 1986). *Apidium moustafai* from the middle levels retains the primitive two pairs of incisors, but has distinctive cheek-tooth characters; *A. phiomense* from the upper FWZ has complex molars with a centroconid on the lowers. To these has recently been added the primitive *Qatrania wingi* from the lower FWZ; smaller in size than other parapithecids, it among other distinctive features retains a paracristid, supporting a large paraconid (usually absent in other genera), on $M_1$.

The Propliopithecidae in the middle and upper levels are generally (Fleagle and Kay 1983) assigned to two genera, *Propliopithecus* and *Aegyptopithecus*. The four species involved—*P. haeckeli, P.* (formerly *Aeolopithecus*) *chirobates, P.* (formerly *Moeripithecus*) *markgrafi*, and *A. zeuxis*—form a morphologically graded series: thus $M_1$ is not shorter than $M_2$ in *P. haeckeli*, is 5 per cent shorter in *P. chirobates* and *markgrafi*, and is 17 per cent shorter in *A. zeuxis*. Molar crowns are broad and steep-sided in the first two, compressed with lateral flare in the second two. The molar cingula are well-developed in *P. haeckeli*, moderately in *P. chirobates*, and weakly in *P. markgrafi* and *A. zeuxis*. In other characters the poorly represented *P. markgrafi* is unknown, but the graded series runs in the same direction through the other three species.

It would be too easy to see this as a primitive-to-derived series; closer inspection suggests, instead, that *P. haeckeli* is derived in some features. At this point we must bring in the lower FWZ representative, *Oligopithecus savagei*.

*Oligopithecus*, known only from a single mandible, was ranked by its describer (Simons 1962) as a probable catarrhine with primitive dental features, but Szalay and Delson (1979) have argued that it is an adapid. For this assessment they cite especially its strong molar cristid development, with emphasis on crest rather than cusp function. Kay (1977) has concurred in its expulsion from the catarrhines: as he shows, all other catarrhines have a distinctive facet X, a wear facet on the lower molars, on the posterior slope of the protocristid, formed by wear from the enlarged protone in the upper molars. Parapithecids all have a big facet X, but *Oligopithecus* lacks it. This would seem on the face of it to rule out *Oligopithecus* as a catarrhine, but the evidence presented by Kay in fact better supports the possession of facet X as a simiiform rather than a strictly catarrhine synapomorphy: not only is it seen in *Pondaungia*, but it occurs also in the majority of New World monkeys (exceptions being marmosets, *Saimiri*, and *Alouatta*). It seems much more parsimonious to suppose that the ancestral simiiform possessed this functional system, and that a few—*Oligopithecus* among them—have since lost it. In other respects, *Oligopithecus* makes excellent sense as a catarrhine; in particular the canine–sectorial system seems at least incipiently developed (long canine, unicuspid $P_3$).

If there is a primitive precursor, the species of *Proliopithecus* and *Aegyptopithecus* should obviously be compared to it. $P_3$ in *P. haeckeli* is simply rounded; in *P. chirobates* and *A. zeuxis* it is long oval, with strong mesio-buccal flare; in *Oligopithecus* it is long oval and mesio-buccally extended, making *P. haeckeli* derived in this character. In both well-known species of *Propliopithecus* $P_4$ is broad, with a variably developed lingual cingulum; in *Aegyptopithecus*, compressed, with no cingulum; the latter, by reference to *Oligopithecus*, is derived. $M_1$ is much smaller than $M_2$ in *Aegyptopithecus*: a derived condition. The mandibular corpus is very deep in *P. haeckeli*, shallow in *P. chirobates* and *A. zeuxis*; here, as with the anterior premolar, *P. haeckeli* is derived. Molar crowns are flared, with no cingulum, in *Oligopithecus*, as in *P. markgrafi* and *A. zeuxis*. In all these respects, *P. markgrafi* (where its characters are known) is primitive; *P. chirobates* and *P. haeckeli* successively derived in one direction; *Aegyptopithecus* derived in an opposite direction. Unfortunately the stratigraphic position of *P. haeckeli* and *P. markgrafi* is unknown; the other two are both from the upper FWZ. The above interpretation predicts that, when specimens of *P. haeckeli* and *P. markgrafi* are eventually found *in situ*, the former will be from the upper FWZ, the latter from middle levels (e.g. Quarry I).

From a lower level than any previously known *Propliopithecus* comes the newly described *P. ankeli* (correctly, *ankelae*) (Simons *et al.* 1987). This form, as its describers note, is probably not ancestral to later species.

Crania of both *Aegyptopithecus* and *Apidium* are clearly catarrhine, with complete post-orbital closure and no metopic suture. Those of *Aegyptopithecus* are the more complete, and show that, unlike the canine/sectorial complex, the catarrhine tubular ectotympanic was not yet developed, there being a simple tympanic ring. Post-cranial material is also known; whether ascribed to Propliopithecidae or to Parapithecidae, all post-crania show primitive features such as an entepicondylar foramen on the humerus. The general primitiveness and lack of autapomorphic features of the Propliopithecidae suggest that they could be representative of a stage when the ancestors of the Hominoidea and Cercopithecoidea were not yet separate; and the condition of the tympanic region and post-cranial bones imply strongly that indeed they were well anterior to the splitting of the two extant lineages.

And what of the Parapithecidae? Hoffstetter (1982) discussed the question of whether they truly are catarrhines at all. Most commentators have accepted that they are, even though more primitive than any other (the Burmese fossils excepted); Delson (1985$b$) calls them 'paracatarrhines', and considers them the sister group of all other catarrhines. Andrews (1985) lists the following derived character states which they share with the catarrhines *sensu stricto*: prominent glabella, broad interorbital pillar, premolars 3–4 with heteromorphic cusps, lower molars five-cusped, talus longer than broad with long neck. They have also some autapomorphic characters, of which Andrews cites the shape of the capitulum; they have also characteristically block-like molars with a tendency to be medially constricted, mandibles that shallow anteriorly, and $P_2$ is the tallest premolar. On the other hand, much evidence now seems to suggest that the Parapithecidae might even be the sister-group of catarrhines plus platyrrhines (Fleagle and Kay 1987).

Simons and Kay (1983) suggest that the lower FWZ is closer to 40 than to 30 MA, the middle quarry and upper FWZ much later, perhaps 28–30: the Jebel Qatrani Formation is underlain by an Eocene formation, and capped (but unconformably) by a basalt dated at 25–27 MA. Recently, however, Fleagle *et al.* (1986) have produced a more accurate data for the overlying basalt of $31.0 \pm 1.0$ MA. Consequently at this period (Lower Oligocene) we have evidence of primitive catarrhines which could well be (and there is no contrary evidence) ancestors of all living members, as well as a primitive sideline.

### 5.3.4(c) Fossil Cercopithecoidea

The first evidence of the separation of the Old World monkey lineage is *Prohylobates* from the early Miocene of Wadi Moghara, Egypt, and Gebel Zelten, Libya (Simons 1969; Delson 1978). Cercopithecoid molars can at once be recognized by their bilophodonty (see Chapter 4, subsection 4.3.2(b) ); *Prohylobates* has this character developed incompletely, and additionally has remnants of a cingulum (which later cercopithecoids have lost), relatively broad $M_2$, and a more robust mandibular corpus than later cercopithecoids, to which on present evidence it is the plesion. Both sites where it occurs are placed by Delson at about 18 MA, at which time we may therefore hypothesize that proto-cercopithecoids were living.

The later history of the Old World monkeys has been reviewed by Delson (1975; and in Szalay and Delson 1979). The earliest complete evidence for the separation of Colobidae and Cercopithecidae is the middle/late Miocene occurrence of the undoubted colobid [even colobine, in the sense of this thesis, and probably related to Asian langurs (Dostal and Zapfe 1986)] *Mesopithecus*; there is some suggestion that the fossils from Maboko (*c.* 14 MA) described as *Victoriapithecus* may include very early stages of both families, but Benefit and Pickford (1986) regard this as highly improbable, as *Victoriapithecus* was still not fully bilophodont. The Pliocene European *Dolichopithecus* was a colobid superficially narrow-faced like *Nasalis*, but Delson has shown that this appearance is illusory, and it is evidently a direct descendant of *Mesopithecus*. There are a variety of African Miocene and Plio-Pleistocene genera, all detectably allied to *Colobus* and/or *Procolobus*, some being of giant size (Leakey 1982; Benefit and Pickford 1986).

The early history of the Cercopithecidae includes a North African species referred to *Macaca* from the late Miocene, and a Pliocene African genus, *Parapapio*, probably morphologically indistinguishable from *Macaca* as that genus is presently constituted, from which *Papio* and *Cercocebus* can be seen emerging at some sites. The commonest monkey at most Plio-Pleistocene sites in Africa is *Theropithecus*, nowadays restricted to the Ethiopian highlands; at least 300 specimens from this period are assigned to it, compared to only 21 of *Papio* (Eck and Jablonski 1984).

### 5.3.4(d) Fossil Hominoidea—and others!

Because the molar teeth of Old World monkeys are highly derived in shape but those of the Hominoidea remain relatively primitive to this day, the monkeys have proved fairly easy to separate out in the fossil record, leaving a residue of 'dental apes'. It has, in the past, been all too easy to fall into the trap of equating 'dental ape' with simply 'ape', and so of ascribing every fossil catarrhine that is not detectably a monkey to the Hominoidea. Groves (1972*b*, 1974*b*) pointed out that some of these 'dental apes', including those from the Fayum which had been so designated, fit on the cladogram well below the cercopithecoid–hominoid split, and it has gradually become accepted that they ought not to be included in the Hominoidea at all. The larger fossil members of the Hominoidea (strictly defined) have recently been reviewed very thoroughly by Kelley and Pilbeam (1986), a paper which appeared too late to be taken fully into consideration in this section.

*5.3.4(d)(i) The Pliopithecidae*   Andrews (1985) places the Fayum non-parapithecids in a superfamily Propliopithecoidea, and certain Miocene genera in another superfamily, Pliopithecoidea. I have argued above that the Fayum propliopithecids (including *Oligopithecus*) are a virtually autapomorph-less sequence of stem forms. I earlier argued (Groves 1972*b*) that the Pliopithecidae are very little advanced beyond the Propliopithecidae, and I included them together in the same family. Andrews (1985) shows, however, that there are some features which they do share with hominoids and cercopithecoids which the propliopithecids do not have: prominent glabella (separate from the supraorbital ridges), bicuspid and slightly elongated $P_4$, somewhat reduced post-orbital constriction, and in particular partial elongation of the tympanic ring into a tube, although it is by no means as complete a tube as in extant catarrhines. Cladistically speaking, it is evident that the Pliopithecidae are the sister group of all extant catarrhines, and the propliopithecidae are the sister group of extant catarrhines plus pliopithecids. Under these circumstances Andrews recognizes both at super-family level: Pliopithecoidea and Propliopithecoidea, and formally this is correct. But in the Fayum I believe, and have argued above, that we see a stepwise anagenetic(?) series. Unless we recognize (at least?) three plesia— plesion *Oligopithecus*, plesion *Propliopithecus*, plesion *Aegypto- pithecus*—we will simply have to maintain a fiction that a taxonomic group is represented in the Fayum, while recognizing tacitly that this is not the sort of situation that the Linnaean system was designed to cope with. Further discoveries may clarify matters and ease the problem; meanwhile Andrews's scheme should be maintained.

The Pliopithecidae have been revised by Ginsburg and Mein (1980). These authors pay particular attention to distinguishing them from the Hyloba-tidae, with which they are constantly confused: indeed, prior to Remane (1965) the pliopithecids had been universally accepted as ancestral gibbons, a scheme which, now has become impossible to maintain (Fig. 5.1). Ginsburg and Mein show that the Pliopithecidae can be distinguished by their possession of the pliopithecine triangle, a mandibular molar occlusal pattern of one crest running disto-lingually from the protoconid, meeting another running mesio-lingually from the hypoconid: although the nominotypical genus, *Pliopithecus*, is known fully enough to show that it had a number of autapomorphic features (especially of the facial skeleton), the other included genera are less well-known so that the pliopithecine triangle remains for the moment the family's only confirmed synapomorphy. Andrews (1985) lists the characters which the other catarrhines (non-Fayûm) share but which are not found in the Pliopithecidae: narrow bicipital groove, flat deltoid plane, and loss of entepicondylar foramen—all characters of the humerus—and complete ectotympanic tube.

The Pliopithecidae seem to have achieved some diversity in the middle Miocene of Europe. Ginsburg and Mein (1980) recognize two subfamilies: Pliopithecinae (with a single genus, *Pliopithecus*, with three known species) and Crouzeliinae (with three genera), with molar patterns emphasizing cusp and crest function respectively. To these has recently been added *Laccopithe- cus* from western China (Wu and Pan 1984). Though described as a hylobatid, it clearly resembles *Pliopithecus* in such features as its tall narrow incisors, robust low-crowned canines, short broad $P_3$, molarized $P_4$ with long broad talonid, molars with bulky cusps and lacking the broad hylobatid central basin, and large superior and inferior transverse tori on the symphysis. The pliopithecine triangle is present. The facial skeleton

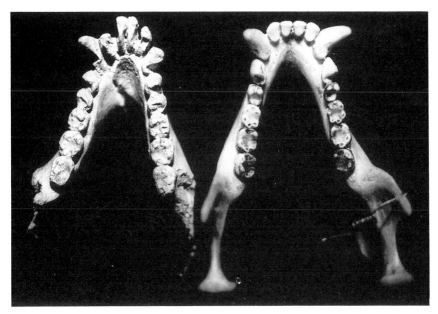

**Fig. 5.1.** Mandibles of the type of *Pliopithecus antiquus* (left) and of a gibbon, *Hylobates lar* (right). Similarities in molar structure, amounting mainly to the peripheralization of the major cusps and expansion of the central basins, are of the sort to be expected in a small-sized (non-Cercopithecoid) catarrhine. Pliopithecids remain much more primitive in symphyseal morphology, M₃ shape, incisor form, and the functional morphology of the canine/sectorial system.

resembles *Pliopithecus*, which means that there are some features (homoplastic, I think, rather than primitive retentions) also recalling *Hylobates* and, indeed, *Cercopithecus*. It has the characters of subfamily Pliopithecinae, not those of the Crouzeliinae.

*5.3.4.(d)(ii) The Proconsulidae* A number of genera from the early Miocene of East Africa—sites such as Rusinga, Songhor, Napak, Koru, all dated *c*. 18 MA—have been varyingly placed by Andrews (1985). Broadly, they may be divided into a small, lightly built group (*Dendropithecus, Micropithecus, Nyanzapithecus* (see Harrison 1986) ) and a larger, though variable, usually more robustly built group (*Proconsul, Rangwapithecus, Limnopithecus*). Andrews unites the latter group into a family, Proconsulidae, and leaves the former *incertae sedis*. This is an enormous advance over the classification of Simons and Pilbeam (1965), which lumped *Proconsul* into the catch-all genus *Dryopithecus*, along with all other Miocene hominoids (except *Ramapithecus*), although at the time their system did indeed 'clear the air', brushing away innumerable, genuinely useless, described genera. The taxonomy within the two larger-sized genera,

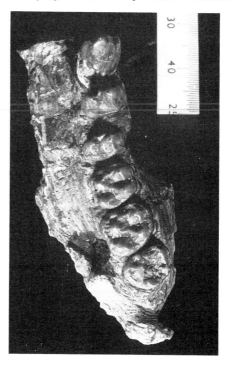

**Fig. 5.2.**    Holotype of *Proconsul africanus*. Photo courtesy of Dr P. Andrews.

*Proconsul* and *Rangwapithecus*, has recently reviewed thoroughly by Kelley (1986); sexual dimorphism may have been greater than in living hominoids.

The Proconsulidae (Fig. 5.2) share various derived character states with both Hominoidea and Cercopithecoidea: the tubular entotympanic is complete, the bicipital groove is wide, and the entepicondylar foramen is absent. They share characters with the Hominoidea that are almost certainly derived, and are not shared with the Cercopithecoidea: frontal bone wide at bregma, narrowing anteriorly; low-crowned $P_3$; reduction of cusp heteromorphy on upper premolars; development of a maxillary jugum; scapula with robust acromion and elongated vertebral border; humeral head rounded, larger than femoral head, and medially oriented; and strong medial and lateral keels on the trochlea of the humerus. The forelimb and scapular features suggest some special forelimb use in locomotion, feeding, or other activity: not necessarily brachiation, but a whole spectrum of alternative activities as described by Andrews and Groves (1976).

One character shared by *Proconsul* and related genera has always seemed to need special explanation: the enormously developed superior transverse torus on the mandibular symphysis, and total lack of an inferior torus. In that some inferior torus development, at least, is seen in all other catarrhines, and none has such a remarkably hypertrophied superior torus, there would seem to be a derived character state of the Proconsulidae here; though what might be its functional significance—if any?—is obscure. The occlusal basins of the molars are also enlarged. I conclude that the family is a genuine lineage, not just a collection of common ancestors for the other hominoids.

Recently two further genera have been described from the early Pliocene of East Africa: the remarkably long-faced *Afropithecus* and the small *Turkanapithecus* (R. and M. Leakey 1986). Lacking symphyseal transverse tori and other features, they are evidently not proconsulids; indeed, they both (*Turkanapithecus* in particular) have a remarkably primitive look about them. They must remain *incertae sedis* until further comparisons are made.

*5.3.4(d)(iii) The Dendropithecus group* *Dendropithecus macinnesi*, formerly included in the proconsulid genus *Limnopithecus*, is of less certain affinity. The maxillary jugum is undeveloped, but its small size and lack of buttressing could be the explanation; a similar explanation might serve for the smaller humeral trochlear keels. $P_3$ is rather higher crowned than unquestioned hominoids or proconsulids; $P^3$ has a projecting buccal cusp. The superior transverse torus is very large, but an inferior torus is also present. Other characters differentiating the Proconsulidae from later hominoids are unknown for *Dendropithecus*.

Andrews (1985) tends to view *Dendropithecus* as a remnant of a pre-proconsulid/pre-hominoid stock. This is plausible; or else it could be a member of the Proconsulidae, but a sister group to the other genera. There are few characters available for analysis, and they are not entirely consistent. At any rate, it is a far cry from the former interpretation of it as an early Miocene gibbon: Le Gros Clark and Leakey (1952) described it (as *Limnopithecus macinnesi*) as related to *Pliopithecus*, which in those days was regarded as a hylobatid. Throughout their monograph these authors seemed to take it for granted that it was a hylobatid, describing totally un-gibbon-like post-cranial anatomy, but ending up with it brachiating in the approved hylobatid fashion.

The only authors specifically to defend its allocation to the Hylobatidae were Ginsburg and Mein (1980). As these authors were concerned to differentiate pliopithecids from hylobatids, their arguments deserve special attention. They note on the upper molars of the Hylobatidae a tendency for a semilunar rounding of the lingual borders, and see the beginnings of this shape in *Dendropithecus*. A glance at the upper molars of any modern gibbon confirms that this shape is diagnostic; but I cannot see it in *Dendropithecus*. Ginsburg & Mein state that in this latter genus the lingual border of $M^1$ is 'encore rectiligne dans sa portion moyenne' [remains straight in its middle part]. What appears to describe the shape is that the lingual margin (of all three molars, in fact) is nearly straight, or slightly rounded by virtue of the presence of a large lingual cingulum, with bevelled mesial and distal corners.

It is not fundamentally gibbon-like; nor does it have the characteristically broadened central basins, with marginalized cusps, of gibbon molars. Much the same morphology can be seen in another, still smaller, East African Miocene genus, *Micropithecus* (see Fleagle 1975, 1978).

If the Proconsulidae, then (provisionally including *Dendropithecus* and *Micropithecus*), share derived features with the Hominoidea which the Cercopithecoidea do not have, they ought clearly to be placed in the Hominoidea. Andrews (1985) agrees, and includes them in the Hominoidea as one of four families.

*5.3.4(d)(iv) The Oreopithecidae*   A genus of arguable affinities is the late Miocene *Oreopithecus* from Italy. Straus (1963) assigned it to its own family, Oreopithecidae, within the Hominoidea. Szalay and Delson (1979) saw it as a cercopithecoid. Recently Harrison (1986) has once again urged its hominoid affinities, and suggested its origin from the East African *Nyanzapithecus*. The problem is that it has such curiously specialized teeth that the overwhelmingly hominoid nature of its post-cranial skeleton has been forgotten. The molars are crested and in some respects almost bilophodont, but the lowers have prominent centroconids; other dental characters are quite different, notably the short, though pointed, canines and the homomorphic lower premolars parallelling the human condition. When its characters are compared to Andrews's (1985) list, it is fundamentally hominoid: $P_3$ is low-crowned; upper premolar cusps are not hetromorphic; the trochlear keels are strongly developed, the olecranon is short, the sigmoid notch broad, the ulnar shaft is bowed, the iliac blade broadened, the talus neck is short and broad. All the character states are found in the extant Hominoidea, but many of them are not seen in the Proconsulidae: that is to say, it evidently shared a common ancestor with the Hominidae and Hylobatidae subsequent to the splitting off of the Proconsulidae. Overall, as Straus discovered more than 20 years ago, *Oreopithecus* had what at that time was thought of as a brachiating morphology; and, indeed, considering its very long arms (intermembral index about 120) it may even have been a genuine brachiator.

The Oreopithecidae would seem, thus, to be part of radition which gave rise also to the Hylobatidae and Hominidae. It is not possible at the moment to decide whether this three-way split can be resolved; possibly in fact the Hylobatidae and Hominidae are sister groups, and the Oreopithecidae the sister group to them, as the first two share a shortening of the lower molars (they are long in the Oreopithecidae), and the femoral condyles are asymmetrical (they are curiously symmetrical in *Oreopithecus* a condition that has been reverted to in *Homo*). But there are not many characters on which to base a decision.

*5.3.4(d)(v) The Dryopithecidae*   The other family which may be part of the same radiation is the Dryopithecidae of Andrews (1985). This family contains a single genus, *Dryopithecus*, which, shorn of its spurious cadres (*Proconsul, Sivapithecus*) is a monophyletic, fairly distinctive middle to upper Miocene ape. Not much is known of its post-cranial skeleton, but what is known resembles acknowledged hominoids, as do the characters of the skull (known from Hungarian remains assigned to a separate genus, *Rudapithecus*). The mandibles are rather slender and comparatively deep, and the superior transverse torus is fairly well developed, although there is a large inferior torus as well. The robust, stout canines are shared with the Hominidae but not with the Hylobatidae; $P_3$ is not bilaterally compressed; the tooth-rows are nearly parallel, and spaced more widely apart; and there is a large maxillary sinus. The precise evolutionary position of the genus is disputed by Andrews (1985) and Delson (1985 *a*), but it is evidently not far from the root of the Hominidae.

*5.3.4(d)(vi) The fossil Hylobatidae*   And what of the Hylobatidae? When the bogus 'proto-gibbons', *Pliopithecus* and so on, have been stripped away, what remains of the gibbons fossil record? Ginsburg and Mein (1980) have argued, convincingly in my opinion, that there are a few, but very few, fossil hylobatids, and that they may be recognized by their semilunar lingual maxillary molar margins; I would add that in the fossil hylobatids there is at least some detectable tendency for the molar cusps to become peripheral, creating broad central basins. *Krishnapithecus*, known only from single $M^3$s (one from the Siwaliks of northern India, one from Mongolia: both apparently late Miocene in age), has a relatively large metacone; *Dionysopithecus*, from the middle or upper Miocene of China, has more reduced third molars with no metacone. As the third molars of modern gibbons tend to be very reduced with only protocone and paracone well-developed, *Dionysopithecus* would seem to be the more derived of the two fossil genera. An upper molar ($M^1$ or $M^2$) from the Kamlial Formation of the Siwaliks (16.1 MA) seems also to be hylobatid (Barry *et al.* 1986).

*5.3.4(d)(vii) The Kenyapithecus question*   Before turning to the Hominidae there is a further genus of uncertain affinity, the middle Miocene *Kenyapithecus*, which has to be considered. Andrews (1985) considers this genus to be closer to the Hominidae than the Dryopithecidae, because of its elongated lower premolars and its thick-enamelled molar teeth. Delson (1985a) disagrees: the acknowledged hominid genus *Sivapithecus* (see-below) does not have elongated premolars, and the polarity of enamel thickness is arguable. On the other hand the Dryopithecidae have spatulate $I^2$, enlarged maxillary sinuses, and a keeled humeral trochlea, which *Kenyapithecus* does not. So few character differences are involved that the relationships are hard to disentangle; but recently Pickford (1985) has considerably expanded our knowledge of *Kenyapithecus*, and a few extra characters have become available. In particular the maxillary sinuses of this genus turn out to be really quite respectable; and the large canine fossa is a character which is found in at least some hominids (not all), but not outside the family. (Pickford gives the canine fossa as a reason for placing *Kenyapithecus* on the orang lineage, but the presence of such a structure is not a specifically pongine trait). In the mandibular symphysis there is an extremely large inferior transverse torus, far more developed than in any dryopithecid, but only a small superior one. I consider it more likely that *Kenyapithecus* is a hominid, and rank it as the hominid plesion. The remains come from the late middle Miocene sites of Maboko and Fort Ternan (the latter dated at somewhat under 14 MA), and perhaps (Delson 1985*a*) from Buluk (a minimum of 17 MA) making it in fact of the right time slot to have been precisely the hominid common ancestor.

*5.3.4(d)(viii) Sivapithecus, Gigantopithecus, and the Ponginae*     Within the Hominidae we have to try to distinguish the Ponginae, the lineage leading to the orang utan, from the Homininae, the common gorilla–chimpanzee–human lineage. In a classic paper, Andrews and Cronin (1982) demonstrated that three species of the middle/late Miocene genus *Sivapithecus* show derived features of the Ponginae: naso-alveolar clivus smoothly sloping into the floor of the nasal cavity; incisive foramina very small; orbits high oval in shape; interorbital pillar very narrow; zygomatic bone flattened, facing anteriorly; large zygomatic foramina; glabella undeveloped; great size discrepancy between upper central and lateral incisors. The best-known species of *Sivapithecus, S. sivalensis* (Fig. 5.3), from the late Miocene of the Nagri zone, Siwaliks, India/Pakistan, has every one of these characters, giving it a thoroughly orang-like facial skeleton (Preuss 1982); it lacks mainly the refinements of the orang's multiple zygomatic foramina, situated above the plane of the lower orbital margin. Another species, *S. meteai* from contemporaneous deposits at Sinap, Turkey, shares the same states in those characters that can be seen on the less-complete material. Finally, a third species, *S. punjabicus* from the Siwaliks, though the material is less-complete still, shows the same conditions in those parts which we have.

The importance of this is not only that the Ponginae can be traced back into the late Miocene—about the time predicted by the molecular clock, according to Andrews and Cronin (1982)—but also that it disposes of the notorious *Ramapithecus*. The story of this genus, and its promotion as a very early human ancestor, has become widely known. The characters ascribed to it are in part those of female apes, but in the main just those to be expected of a small species of *Sivapithecus*: relatively short, thick jaws, on which in fact the whole case for its anthropocentrically significant position rested. That its characters, cladistically analysed, turn out to be those of the Ponginae reveals it to be not a proto-human but a red herring. It may now happily take its place as a perfectly valid, but humble and unassuming, species of the genus *Sivapithecus*.

**Fig. 5.3.** Holotype of *Palaeopithecus* ( = *Sivapithecus*) *sivalensis*.

There are two earlier species assigned to *Sivapithecus* by Martin (1983): *S. darwini* (Austria and Turkey) and *S. alpani* (Turkey). Both are known entirely by teeth and jaws, and so cannot show the specific features of either hominid lineage; clearly, their assignment to *Sivapithecus* is provisional only (Martin 1983). The site of Pasalar, in Turkey, where both species occur, is some 14 million years old. They could profitably be compared to *Kenyapithecus*. What seems to be a similar type of hominoid has recently been described by Gu and Lin (1983) as *Platodontopithecus jhanghuamaiensis*, from the Middle Miocene of China.

Fossils referred to *Sivapithecus* and *Ramapithecus* have turned up in large numbers (including some very fine cranial specimens) at Lufeng in Yunnan, China. These have been described in a number of places [see especially Wu Rukang (1984), who assigns them an age of approximately 8 MA]. The original discoveries, well-preserved mandibles, were assigned to two genera, each wtih a new species: *Ramapithecus lufengensis* and *Sivapithecus yunnanensis* (Xu Qingha and Lu Qingwu 1979). The question naturally arose whether these two taxa, a large and a small species at the same site (and the type mandibles so clearly female and male respectively), did not in fact represent the two sexes of one species. Wu Rukang and Oxnard (1983) showed that this was not the case: dividing the large available sample into two size groups, they were able to demonstrate that there was (presumably sexual) dimorphism within each, although it was less within the small-sized form (ascribed to *Ramapithecus*) than within the large one (ascribed to *Sivapithecus*).

If, then, there are two different species at Lufeng, what are they? The best-preserved specimens (shown to me courtesy of Professor Wu Rukang) show almost all the features ascribed by Andrews and Cronin (1982) to the *Sivapithecus—Pongo* lineage: orbits tall oval, smooth subnasal plane, very flat zygomatic bone, incisor heteromorphy, very large zygomatic foramina (doubled in each), no glabella, small incisive foramina, and slit-like palatine foramina. But they both have extremely broad interorbital pillars, quite un-pongine; and very prominent, raised, superior temporal lines, meeting just behind the supraorbital region in the large cranium but not meeting in the small one. The broad interorbital space is flattened between the raised orbital rims, which are continuous around the whole superior and medial margins of either orbit instead of being superior only and merging into the interorbital pillar (perhaps, perforce, due to the interorbital narrowing) as in the orang utan and the Siwalik skull of *Sivapithecus indicus*.

This remarkable interorbital broadening could be seen as a primitive feature, since a narrow interorbital space is seen as derived for the Ponginae. However, the morphology is so idiosyncratic in the Lufeng skulls that it might be better viewed as a re-broadening; the orang-style orbital rimming, originally developed as a functional concomitant of the narrowing, would now be thrown into relief as the orbits had moved apart again. If this last interpretation is correct, then we are in the presence of a new (as yet undescribed) pongine genus, with two (already described) species. [Since this was written, Prof. Wu has drawn to my attention that he has indeed referred the Lufeng fossils to a new genus, *Lufengpithecus* (Wu 1987)].

Another possible synapomorphy of the Lufeng genus would be reduction in premolar size. In *Pongo* the length of the premolar row (in the upper jaw: figures for the lower teeth are similar) is some 55 per cent the length of the molar row, and this is about standard for catarrhines; i.e. the primitive condition. In the Homininae the premolar row measures under 50 per cent of the molar row (*Pan* and *Gorilla*, 45–46 per cent; *Homo* only 35 per cent). In both Lufeng forms, according to figures given by Wu Rukang and Oxnard (1983), the percentage is 37–39. Unless the polarity is wrong (which is possible), or the Lufeng apes are really proto-humans with a lot of parallelisms to the Ponginae (which is unlikely), then they are derived in this feature and parallel to the human line; if the latter, then they share this feature with *S. meteai*, where the percentage is 42: perhaps this figure, as in quite another context, is the answer to life, the universe, and everything (Adams 1982).

The genus *Gigantopithecus* has been endlessly discussed. The jaws are the largest known for a primate; whether this necessarily means that the animal was the largest primate that ever lived is arguable [see, for two different viewpoints on body-size extrapolation from skeletal parts, Smith (1985) and

Steudel (1985)]. Robinson (1972) associated it in the same subfamily as *Paranthropus*; but as long ago as 1962, Woo Ju-Kang (now known as Wu Rukang) demonstrated its affinity to 'pongids', and most authors have supported this assessment, most recently Gelvin (1980). The presumably Pliocene species, *G. giganteus* (synonym *G. bilaspurensis*), from the Haritalyangar region of the Siwaliks (in the upper levels assigned to the Dhok Pathan Formation), is really very like an enlarged *Sivapithecus indicus*. The Chinese species, *G. blacki*, is early Pleistocene or even late Pliocene in age (Eckhardt 1974). Or it may even linger on in the snowy wastes of the Himalayas (Heuvelmans 1955). . . .

And what of enamel thickness? *Sivapithecus* has long been known (Greenfield 1980) to have had much thicker enamel, at least on the molars, than living or fossil apes, and this similarity to fossil hominins was a major reason why—in its previous reincarnation, as *Ramapithecus*—it was so constantly cited as a human ancestor (Simons 1972, for example). The whole question has been thoroughly investigated by Martin (1983), who finds that, when overall tooth size is properly controlled, *Pongo* does in fact have rather thick enamel, though not as thick as *Sivapithecus* or *Homo*. Viewed microscopically, enamel can be seen to be composed of tubules of different patterns, two of which—fast-forming pattern 3, and slow-forming pattern 1—are present together in the molar enamel of the Hominoidea. the differences are simply a function of different rates of enamel formation: the Hylobatidae, *Sivapithecus*, and *Homo* have almost entirely the fast-forming pattern 3, from the enamel–dentine junction to the tooth surface, with only a thin outermost layer of pattern 1; but in the Hylobatidae the enamel layer as a whole is very thin (i.e. its formation is completed very rapidly in ontogeny). In *Pongo* over 80 per cent of the enamel thickness is pattern 3, with a relatively thin outer layer of pattern 1; so the difference from the first group is not great. In *Pan* and *Gorilla* the pattern 3 deep layer is only somewhat over half the enamel thickness: the outer layer, of pattern 1, forms over 40 per cent of the whole. The identity between *Homo* and *Sivapithecus*, and their great similarity to the Hylobatidae despite the superficial difference, leads Martin to propose that this is the primitive state, and that the three extant apes have undergone parallel evolution towards thinner enamel, more marked between the two African genera than between them and the orang-utan. Thus for a hominoid a great thickness of pattern 3 is primitive; for a hominid, and for each of the subfamilies within the Hominidae, the macroscopic condition of 'thick enamel' is primitive. This reconstruction, made on the basis of living forms and a single fossil genus, therefore agrees very well indeed with the evidence of other, less well-known, fossil taxa.

*5.3.4(d)(ix) The search for the earliest Homininae*  So we have early representatives of the Ponginae. If we are correct that they really are Ponginae, that the two subfamilies had now separated, then we would be justified in seeking for contemporaneous stem-group representatives of the other subfamily, Homininae: fossils belonging to the period before the divergence of the human, chimpanzee, and gorilla lineages.

Apart from a couple of isolated teeth—Ngorora, Lukeino—which are virtually indecipherable [although Corruccini and McHenry (1981) have proposed a placement for the latter], we have two candidates for the common hominine stem: the Samburu hominid and *Ouranopithecus*.

The maxilla from the Namurungule Formation, Samburu Hills, Kenya (Fig. 5.4) is dated at about 9 MA (Ishida *et al.* 1984). It is described as having a low-placed zygomatic root, but otherwise is gorilla-like, with a pneumatized anterior zygomatic root, sharp-edged anterior margin to the nasal aperture, arched palate, considerable prognathism, and well-curved alveolar process: all these features are those to be expected in a large-sized (prognathism, pneumatization) hominine (nasal aperture form), and I would doubt that they are specifically gorilla-like. The molars however, are, not gorilla-like: they are bunodont, with rather thick enamel, the cusps being low and rounded (they are noticeably high and crystalline in gorillas), the trigon basin is not enlarged as in a gorilla, there is a large protocone cingulum; and the protocone is the largest cusp, with the other three cusps being more or less equal in size [in gorillas the hypocone is smaller than the other cusps, and on $M^1$ (but not on $M^2$ or $M^3$) the paracone is noticeably smaller than the metacone]. Some of these characters approach the specimen to the human and chimpanzee conditions, while others are more primitive than either. Thus molar bunodonty is a primitive feature from which the gorilla has departed more than the chimpanzee, while humans have retained the primitive state; thick enamel, as has been seen above, is a primitive state

**Fig. 5.4.**   The Samburu palate. Photo courtesy of Dr H. Ishida. The differences between this specimen and other hominoids (see Figs. 5.3 and 5.4) are described in the text; the Samburu specimen corresponds well to the reconstructed morphotype of the Hominidae.

retained, among the Homininae, by the human line alone; the trigon basin is opened out, by pushing the cusps towards the tooth margins, in the gorilla and somewhat in the chimpanzee, but not at all in humans; the protocone cingulum is reduced to a (variably developed) carabelli complex in all three hominines; in all homines the protocone is the largest cusp, but in chimpanzees and humans the hypocone and (to a lesser degree) the metacone are progressively reduced in size from $M^1$ to $M^3$. In the Samburu maxilla the molar size sequence is $3 > 2 > 1$, the sequence that is commonest in the gorilla; in the chimpanzee the usual sequence is $3 < 2 > 1$; in humans it is usually $1 > 2 > 3$; the Samburu/gorilla sequence is probably primitive. Finally, in the Samburu specimen the premolars are said to be 'enlarged', but here I disagree. The premolar row is 40 per cent of the length of the molar row; it was shown above that in the Ponginae this proportion is above 50 per cent, in the Homininae considerably below 50 per cent, so the Samburu maxilla agrees here with the Homininae. (Expressed as a percentage of $M^1$ alone, the figures for premolar row length are: Ponginae, over 150; *Pan* and *Gorilla* 147–149; *Homo* 132; Samburu maxilla 146).

There is thus no evidence of any pongine derived features in the Samburu maxilla; some evidence of hominine derived conditions; and much evidence that, if it is indeed a hominine, it is not on any of the three surviving lineages.

The other candidate for a place in the pre-split Homininae is *Ouranopithecus macedoniensis*, from the late Miocene (*c.* 10 MA) of Rain Ravine, Greek Macedonia. Continuing excavations have turned up an ever larger sample of ape remains; de Bonis and Melentis (1984) give a fairly full summary of the morphology as known up to then, including arguments against the suggestion of some other authors that more than one species is represented. Martin and Andrews (1984) draw attention to the poorly preserved mandible known as *Graecopithecus freybergi* from Pyrgos, further south in Greece, and argue that the two sets of remains represent the same taxon. If this argument is accepted, *Graecopithecus freybergi* von Koeniswald, 1972 would have priority over *Ouranopithecus macedoniensis* de Bonis *et al.* 1974; Martin and Andrews make a good case for the synonymy, and further argue that both are synonyms of *Sivapithecus meteai* (Ozansoy 1957) (originally ascribed to a separate genus, *Ankarapithecus*) (for full synonymy see Martin and Andrews 1984). De Bonis and Melentis (1984) seem to agree that synonymy of the two Greek forms is likely, but because of the excellence of the Macedonian material compared to the single rather nondescript Pyrgos specimen, suggest that they continue to be kept apart for the moment; but they argue against the synonymy of *O. macedoniensis* with *S. meteai*, maintaining that at the Turkish site of Sinap both are present, and are clearly separate. The Sinap facial skeleton, the basis for the association of *S. meteai* with *S. indicus* and the orang lineage, is one of those Sinap specimens which they wish to refer to *Ouranopithecus*.

The most recent specimen to come from Rain Ravine is a maxilla, described by de Bonis and Melentis (1985), and this specimen seems to offer a most interesting resolution of the problem. A short digression on the incisive fossa is first in order.

As shown by Ward *et al.* (1983) and Ward and Pilbeam (1983), the primitive form of the fossa takes the form of a direct communication between the incisive foramina and the floor of the nasal cavity. This is the condition in most primates, from lemurs through to Old World monkeys, Proconsulidae and Hylobatidae; only in the Dryopithecidae is there some backward growth of the naso-alveolar clivus to cover the fossa. In *Pongo* and *Sivapithecus* there is a true naso-palatine canal running from the tiny incisive foramina through the thickened palate to the incisive fossa in the nasal cavity floor; the canal is very reduced in size, and curves backward, as does the clivus, which enters the nasal cavity smoothly and at the same level as the floor of the cavity. Finally, in the Homininae, while the canal remains primitively broad, the clivus ends abruptly, overhanging the fossa, so that there is a sudden drop to the level of the nasal cavity floor. In the new *Ouranopithecus* maxilla the incisive fossa opens broadly behind and below the posterior margin of the clivus, and there is a steep drop from the clivus to the floor of the nasal cavity. This is precisely the hominine pattern.

The discovery that *Ouranopithecus* has an important derived condition of the Homininae at once banishes all thoughts of synonymy with *Sivapithecus*. Indeed, Martin and Andrews (1984), while arguing for this synonymy, did admit that there are some differences in the parts then known (mainly mandibular): the mandible is more slender, less heavily buttressed, with enlargement of $M_3$ (as judged by the roots) and large, broad $M_2$. In noting that, where known, *S. meteai* tends to resemble *Ouranopithecus* rather than its Indo-Pakistani congeners, they considered it more probable that the differences were either of an individual nature, or differentiated a single western (Greek/Turkish) species from the Indian ones. It now seems time for another look at all the material. The questions that need to be asked are: (1) Given that the Sinap face and the Rain Ravine maxilla simply cannot, if the accepted polarities of the character states are correct, belong to the same species, how else do the two taxa they represent differ? (2) Are there one or two species at Sinap, and at Rain Ravine? (3) Are, therefore, all the Macedonian specimens hominine and all the Turkish ones pongine, or were the two subfamilies, in the early days after their separation, still sympatric? The answers to these questions will perhaps help to decide, once and for all, the position of the wretched Pyrgos mandible, and so allocate the name *Graecopithecus freybergi* to one or other taxon.

Although the new maxilla has not been fully described, except for its incisive fossa, some comparisons can be made between previously described Rain Ravine specimens and the Samburu maxilla. First is that the molar size progression distally is much more marked in the Samburu specimen; the Samburu molars are also very much narrower. In their general morphology they are very much alike, but this need not indicate anything other than primitive features. It has already been stated (above) that *S. meteai* has small premolars, as in the Samburu specimen, and so does the previously known maxillary specimen of *Ouranopithecus*.

*5.3.4(d)(x) The fossil record and the molecular clock*   Now that we have laboured along the paths of primate evolution from then to now, what have we learned? The first thing, I suppose, is how fickle the fossil record really is. Reasonable supposition there often is; hypothesis there is in plenty; but what does it tell us for sure? One thing, evidently, is minimum dates for splits, and this is bound to be valuable for the calibration of molecular clocks. Separation dates are always being enquired after; the answers given are usually of the kind: 'The oldest fossil gibbon (lemur, ape, whatever) is $X$ million years old; another $Y$ million years are needed for it to have differentiated from its common stem with its nearest relatives; so the date of the split is $X$ plus $Y$'. In this sort of reasoning, $Y$ is little or no better than speculation, or personal bias; for a Darwinist, it will be a long time during which the characters of the lineage in question will have slowly accumulated, but for a punctuationist it can be a very short period indeed. So let us henceforth eschew estimates of the time needed for a given amount of evolution, and concentrate on first appearance dates, which are knowable.

I have tried to show that some of the often-quoted first apperance dates are dubious if not spurious, but others are much better, and yet others have emerged quite recently to replace the more dubious if long-accepted ones. The following minimum-separation dates I would regard as quite well-authenticated, and acceptable calibrations for a molecular clock:

1. The earliest Lemuriformes (in my usage, lemurs including lorises, probably excluding the aye-aye): 45–49 MA, the date span of the Messel Formation where occurs a lemur with a toilet claw (von Koenigswald 1979).

2. The earliest Lorisidae: 18–19 MA, the dates of the Rusinga–Songhor–Napak group where undoubted lorisids occur (Walker 1974).

3. The earliest catarrhine: 31 MA, the date of the basalt overlying the middle and upper levels of the Jebel Qatrani Formation where *Propliopithecus* and *Aegyptopithecus* are found (Fleagle *et al.* (1986)). I personally would accept *Oligopithecus* as an even earlier catarrhine, but this is controversial. The Parapithecidae and the Burmese fossils are perhaps catarrhines, but may not be. *Branisella*, earliest Oligocene in date (*c.* 37 MA), is suitable to be a platyrrhine (Hoffstetter 1982), but again definitive evidence is lacking.

4. The earliest Cercopithecoidea: *c.* 18 MA, the age of *Prohylobates* spp. (Delson 1978). This of course marks the cercopithecoid–hominoid split.

5. The earliest gibbon is *Krishnapithecus krishnaii* from the Nagri zone of the Siwaliks (Chopra 1978). The Nagri is dated between 10 and 8.5 MA by Pilbeam *et al.* (1979), but Thomas (1984) puts its lower limit at 9.5 MA, and according to Barry *et al.* (1980) the lithofacies that is identified as Nagri has a fluctuating lower limit that may dip down in places to 10 million year levels. (If, as I incline to think, the new Kamlial molar (Barry *et al.* 1986) is hylobatid, then the earliest date will be as much as 16.1 MA).

6. The earliest dates for the *Theropithecus* and *Pongo* lineages have already been satisfactorily matched up with molecular dates (Cronin and Meikle 1979; Andrews and Cronin 1982).

I have tried to survey 65 million years of primate evolution in a few pages, giving no more than a respectful nod at the precursors of lemurs, tarsiers, and sundry monkeys, but going in more detail into the ancestry of the Hominoidea. There is, as always, much to be resolved; new specimens are always needed, and we must continue to devise new ways of looking at known ones. We are now at the brink of the separation of the human lineage, and this will be the topic of consideration from now on.

# 6
# Background to human evolution

Before examining in detail the remains of fossil hominins, there are a few topics that must be got out of the way. First, there is the question of an interim nomenclature; although I will attempt to establish a new taxonomy later on, we need some shorthand means of reference for different fossil types. Next, there will be a brief overview of the sites at which the fossils have been found, and their dating. Finally, a consideration of the principle of variation: if we are going to erect a taxonomy, we need to know how much variation to expect within a species (or subspecies), and where to draw the line.

## 6.1 THE STAGES OF HUMAN EVOLUTION

In the days when it was widely believed that human evolution was unilineal, it was possible for Brace (1967) to write a book with the above title. If there had been only one proto-human at any one time, all that needed to be done was to chop up that single lineage into arbitrary segments: below a certain point we will agree to call them australopithecines, between here and here we will call it *Homo erectus*, and so on. While the Single-Species Hypothesis is now well and truly dead (Leakey and Walker 1976), what we may call residual uni-linealism is alive and well, indeed tends to be an unspoken assumption in many analyses. This work intends to test that assumption.

So it is with no intention of prejudging issues that I want here to establish an informal nomenclature for proto-human (fossil hominin) reference. If the story of human evolution is, crudely speaking, a story of increasing brains, reducing jaws, and development and refinement of bipedal locomotion, then it should be possible to segregate grades of fossils along the spectrum. The clearest criterion is cranial capacity, simply because we are here dealing with numbers so that borderlines can be drawn. But, especially at the more recent end of the scale, where samples become quite respectable, this criterion breaks down and it is necessary to use additional, perhaps less objective, criteria. So I will speak of the following stages, or phases:

1. Australopithecines. This term will be used here in much the same way as it is used by most other authors: hominins with a cranial capacity of 530 cc or less (associated with large cheek teeth). Australopithecines are customarily spoken of as being gracile or robust, qualifications which have meaning in

most cases but become more dubious when the Hadar and Laetoli samples are being considered.

2. Habilines. Hominins with cranial capacities of more than about 530 but less than 800 cc, associated with large or medium-sized cheek teeth. This grade and the last together comprise what Groves and Mazak (1975) referred to as the 'early phase'. Like the australopithecines, the braincase when viewed from the back is bell-shaped.

3. Erectus phase. Hominins with cranial capacities above 800 cc, but retaining primitive features such as relatively prognathous faces, low-crowned vaults, and large, often torus-like, brow ridges. The braincase, viewed from the back, is arch-shaped.

4. Sapiens phase. Hominins with cranial capacities averaging larger than the Erectus phase but more orthognathous, higher vaulted, smaller browed. We can subdivide this type into modern types, with chins and short, rounded braincases (slab-sided); and Neandertalers, with mandibular symphyses that are at most vertical (not with a chin) and longer, lower, but still very large braincases ('baggy', i.e. with convex lateral walls).

These terms imply grades in a general way only, and do not imply a unilineal evolutionary scheme; but they facilitate reference, especially when saying what sort of hominin occurs at a particular site.

## 6.2 WHERE AND WHEN?

A survey of the main sites is now necessary. All the early sites must be mentioned, but in the late Pleistocene there are simply so many fossil hominin sites that only the main ones can be mentioned.

### 6.2.1 Plio-Pleistocene African sites

These sites contain in the main australopithecines and/or habilines, and a few Erectus-phase specimens.

#### 6.2.1(a) Earliest sites

A mandibular fragment with $M_1$, from Lothagam (southwest of Lake Turkana, Kenya), is the earliest fossil considered to be a hominin (i.e. a member of the Hominini), as opposed to hominine (i.e. a member of the Homininae), with a date of 5.0 to 5.5 MA (Patterson *et al.* 1970). A slightly more informative mandibular fragment, with $M_{12}$ and the root of $P_4$, has recently been reported from Tabarin, Lake Baringo, Kenya, with boundary dates of 4.96 $\pm$ 0.3 and 5.25 $\pm$ 0.04 MA (Hill 1985).

The Kanapoi Formation (also southwest of Lake Turkana), whence comes the lower end of a humerus, is overlain by a basalt dated at 4.01 $\pm$ 0.01 MA

(Patterson *et al.* 1970). The hominin fossil is only a little below the basalt.

There are two hominin finds from the sites of Maka and Belohdelie, in the Middle Awash region, Ethiopia. Between the levels where the fossils were found lies the Cindery Tuff, dated at 4.0 $\pm$ 0.2 MA (White 1984). The Belohdelie frontal bone lies 11m below the level of the Cindery Tuff, the Maka femoral fragment 7m above it.

### 6.2.1(b) Laetoli

A rich series of mostly somewhat fragmentary fossils—26 with LH ( = Laetoli hominid) numbers, plus two 'Garusi hominids' found in the 1930s—comes from Laetoli in Tanzania, dated at 3.5 to 3.8 MA. (Harris 1985). As well as the fossils there are series of bipedal footprints. One of the Garusi specimens is the type of an unattached (hence unavailable) generic name *Praanthropus* Hennig, 1948; and of *Meganthropus africanus* Weinert, 1950, separated generically as *Praeanthropus* Senyurek, 1956. LH4 is the type of *Australopithecus afarensis* Johanson, 1978 (the original authorship and citation have been in dispute, but are correctly cited by Day *et al.* 1980). The name *Australopithecus africanus tanzaniensis* Tobias, 1979, applied to Laetoli hominids, lacks a type designation though would not be for that reason unavailable.

### 6.2.1(c) Hadar

A large number of australopithecine fossils have been found at Hadar, in the Afar Triangle, northern Ethiopia. The dates of the fossiliferous deposits are disputed; though they all come from levels overlain by a tuff, BKT-2, dated at 2.88 $\pm$ 0.08 MA by K/Ar dating and 2.7 $\pm$ 0.2 MA by fission track (Walter and Aronson 1982). Below this are three hominin-bearing Members, in increasing order of age the Kada Hadar, Denen Dora, and Sidi Hakoma Members. Towards the top of the Sidi Hakoma Member lies the KMB Basalt, originally given a minimal date of about 3.0 MA (Aronson *et al.* 1977) but later revised to 3.6 $\pm$ 0.15 MA (Walter and Aronson 1982). At the base of the Sidi Hakoma Member is the SHT Tuff; geochemical analysis (Brown 1982) matches that of the Tulu Bor Tuff from Koobi Fora (see subsection 6.2.1(f) below) and the B and U-10 Tuffs from Omo (Subsection 6.2.1(d) below), suggesting that the four Tuffs represent the same volcanic eruption and that SHT is therefore about 3.3 million years old, i.e. *later* than KMB.

White *et al.* (1984) agree that the four Tuffs must have come from a common source, but cite evidence that different eruptions from the same source may also have the same chemical composition. They point out also that the dating of the Tulu Bor and the two Omo Tuffs is less than secure, while the Sidi Hakoma and Denen Dora faunas contain more elements in common with Laetoli and even Kanapoi than with younger deposits.

The palaeomagnetism of Hadar is not a great help. BKT tuffs are normal; the middle Kada Hadar is reversed; the lower Kada Hadar is normal. This could represent the end of the Cochitti subchron at 3.7 MA followed by the Gilbert/Gauss transition at 3.4 MA supporting the earlier date for the SH tuffs, or the beginning and end of the Mammoth subchron at approximately 3.0 and 2.9 MA, supporting the later date.

On the whole, White *et al.*'s (1984) arguments seem cogent, and earlier dates will be supported here; but, as they point out, it in fact makes relatively little difference in palaeoanthropological terms.

Tobias (1980) gave the name *Australopithecus africanus aethiopicus* to the Hadar hominid taxon, but without designating a type; this failure does not render the name unavailable, but a reading of his paper shows that the name was proposed conditionally, and is therefore unavailable under Art. 15 of the International Code of Zoological Nomenclature (1985 edn.). Olson (1981*b*) considered that two hominin species are represented at Hadar; one identical to the Laetoli form, the other a smaller species referred to as *Homo sp. indet.* Later (1985*b*) the same author resurrected Tobias's name as *Homo aethiopicus*, with lectotype AL 288-1 (1985*b*) (the specimen known as 'Lucy'); this is therefore the first available usage of the name *aethiopicus*. In the meantime, however, another name had been proposed, *Homo antiquus* Ferguson, 1984, with 'Lucy' as holotype. [A name *Homo antiquus* Adloff, 1908, quoted by Campbell (1965), would seem to preoccupy Ferguson's usuage, but a study of Adloff's paper has failed to confirm the reference.]

### 6.2.1(d) Omo

The Shungura Formation on the Omo River, Ethiopia, has a long fossiliferous sequence, punctuated by dated tuffs, giving it the status of, as Johanson and Edey (1981) put it, a 'magic ruler', a yardstick against which all other sites can be measured. The dates and palaeomagnetism are most recently summarized by Brown *et al.* (1985*b*), while Delson (1984) discusses the significance of the faunal turnover within the sequence for the dating of sites where radiometric dating is impossible, or disputed. Key tuffs of the sequence are:

B-10: $2.98 \pm 0.03$ and $2.93 + 0.03$;
C: *c.* 2.6;
D: $2.58 \pm 0.03$ to $2.42 \pm 0.03$ (8 determinations);
F: $2.39 \pm 0.04$ and $2.32 \pm 0.04$;
G: $2.34 \pm 0.03$ to $2.32 \pm 0.03$ (3 determinations);
H-2: $1.85 \pm 0.02$ to $1.83 \pm 0.02$ (3 determinations);
H-6/7 boundary: *c.* 1.67 ( = end of Olduvai subchron);
J: *c.* 1.55–1.50.

There are three other formations of the Omo, of which only the Usno Formation, with its White and Brown Sands localities, is of importance for palaeoanthropology. The WS-1 bassalt of White Sands has dates of $4.11 \pm 0.06$ and $4.08 \pm 0.06$ MA.

The hominin fossils from the Omo deposits are unfortunately rather disappointing. All those so far discovered are of australopithecine grade. There are a few teeth from the Usno Formation, including an enormous $DM^1$ from White Sands and a large $DM_2$ from Brown Sands, both of which fit better in the 'robust' australopithecine category, although some permenent molars and premolars from Usno fit better with 'graciles' (Howell 1969). In the Shungura Formation most of the remains are isolated teeth, but a few others are more satisfactory and have attracted special attention: a mandible from Member C (below Tuff D) was made the type of a new genus and species *Paraustralopithecus aethiopicus* Arambourg and Coppens, 1968 (the same name was proposed by these authors a year earlier as well, but was stated to be provisional, hence unavailable according to Art. 15 of the Code; so the name dates from the 1968 description). From Member G, above Tuff G and so some 2.2 million years old, come two mandibles, both ascribed to 'robust' australopithecines by Howell (1969), but differing in size and morphology: L-7, the largest hominin mandible yet found, and the much smaller hemi-mandibular fragment L-74. Finally, from Member E, and so some 2.4 MA, comes a juvenile partial braincase, L-338y-6, considered 'robust' by Rak and Howell (1978) but 'gracile' by Holloway (1981).

### 6.2.1(e) Olduvai

Several important fossils come from Olduvai Gorge, including three types. The chronology is summarized by Leakey and Hay (1982): Bed I dates from 1.85 to 1.71 MA, with the key Tuff IB being dated at $1.84 \pm 0.03$ MA; Bed II goes up to about 1.2 million (extrapolated from calculated deposition rates), with an important marker Tuff IIA in the Lemuta Member of Lower Bed II dated at about 1.65 MA; Bed III goes up to about 0.8 MA; Bed IV, up to about 0.62 MA, with Tuff IVB being apparently at or just below the Brunhes/Matuyama boundary; then come the Masek Beds, and above them the Lower Ndutu Beds, the boundary being at about 0.4 MA.

From Bed I come OH5, type of *Zinjanthropus boisei* Leakey, 1959 (later transferred to the genus *Australopithecus* by Tobias 1968), and OH7, type of *Homo habilis* Leakey, Tobias and Napier, 1964. The former is an australopithecine of 'robust' type; the latter, of habiline grade. From the upper part of Bed II, above Tuff IID, comes OH9, type of *Homo leakeyi* Heberer, 1963, a name proposed conditionally and never, as far as I am aware, made available; it is also the type of *Homo erectus olduvaiensis* Tobias, 1968, which is available.

Specimens ascribed to *Homo habilis* come from Bed I and Bed II up to about the middle. Large molar teeth, OH3 and OH38, from the upper part of Bed II, have been referred to the robust australopithecine. A fragmentary cranium, OH12, and pelvic and femoral specimens, OH28, from bed IV have been considered of Erectus grade. Various mandibles and other specimens of varying states of completeness have been referred at times to habilines and Erectus-grade hominins.

### 6.2.1(f) Koobi Fora and West Turkana

Perhaps the most important fossil hominin site in East Africa, this lies to the east of Lake Turkana (formerly Lake Rudolf), and was formerly referred to as the East Rudolf site. The stratigraphy of the Koobi Fora Formation was long misunderstood, and a spurious date of 2.6 MA for the key KBS Tuff also retarded the understanding of the site. The latest information on the Tuff sequence and dating, and the Omo Tuffs to which they correspond, is given by Brown *et al.* (1985*b*) and McDougall (1985):

Moiti: $4.10 \pm 0.07$ ( = WS Basalt at Usno);
Lokochot: ( = Shungura A)
Tulu Bor: *c.* 3.35 ( = Shungura B);
Toroto: $3.32 \pm 0.02$;
Ninikaa, Hasuma: ( = Shungura C);
Ingumwai: ( = Shungura C4);
Burgi, Lokeridede: ( = Shungura D);
--------------------- disconformity
Lorenyang, KBS: $1.88 \pm 0.02$ ( = Shungura H2);
Malbe: ( = Shungura H4);
Okote, Lower, Middle, BBS: *c.* 1.60 – 1.48 ( = Shungura J7);
Koobi Fora, Chari: $1.39 \pm 0.01$ ( = Shungura L);
'Tuff above Chari': $1.25 \pm 0.02$;
Silbo: $0.74 \pm 0.01$.

In addition, a tuff at Nariokotome, west of Lake Turkana, stratigraphically above the Chari Tuff, was dated at $1.33 \pm 0.02$ MA.

The hominin fossils from the Koobi Fora Formation are found almost throughout the sequence. ER-7727 is a tooth from the Hasuma tuff level, and ER-5431 consists of some isolated teeth from the Tulu Bor and Burgi tuff levels. From below the KBS Tuff, but above the disconformity, come a series of crania (ER-1470, 1590, 3732) and mandibles (ER-1482, 3734); the crania have been variously referred to as gracile australopithecines (Walker 1975) or habilines (Pilbeam and Gould 1974), while mandible 1482 has been likened to *Paraustralopithecus aethiopicus* (Leakey 1974), assigned to a robust austra- lopithecine (Chamberlain and Wood, 1985), or, usually, not mentioned at all. [Recently ER 1470 has been made the type of *Pithecanthropus rudolfensis* Alexeev 1986.]

From the level of the KBS Tuff, whether originating above or below it being not quite clear, comes a well-preserved cranium ER-1813. This is called a gracile australopithecine by Walker and Leakey (1978) and by Walker (1981*b*), but a habiline by Wood (1985). Above the KBS Tuff, but below the Okote/Koobi Fora tuff complex, are known specimens referred to the Erectus grade (ER-3733, 730), along with an associated partial cranium and mandible, ER-1805, assigned to Erectus by Wolpoff (1980*b*) but, apparently with a query, to the same taxon as 1813 by Walker (1981*b*). It is a at this level, too, that robust australopithecines definitively appear (ER-406, 407, 732), although one specimen (ER-1500) from below the KBS Tuff is also generally referred to this grade.

Above the Okote complex have been found further 'robusts' (ER-733) and Erectus-grade fossils (ER-3883). There is also a well-preserved mandible, ER-992, from this level, the type of *Homo ergaster* Groves and Mazak, 1975; this taxon was inadvertently described under the impression that it was in the public domain, which it was probably not (an unintended breach of ethics); unfortunately, too, it was named before the description of such fossils as 1813 and 3733, with either of which the jaw 992 could perhaps be conspecific.

Sites west of Lake Turkana have begun to be prospected more recently. So far, they have yielded some magnificent specimens: a subadult skeleton of Erectus grade (WT 15000) from just above a tuff representing the Okote Tuff (Brown *et al.*, 1985*b*), and an early robust australopithecine cranium (WT 17000) and mandible (WT 16005) from, respectively, somewhat below and above a tuff representing the Burgi Tuff (Walker *et al.* 1986).

### 6.2.1(g)  Other East African sites

The Chemeron (Lake Baringo, Kenya) deposits are capped by a tuff dated at 2 million years (Martyn 1967). Below the Upper Tuff are the Upper Fish Beds, within which was found a temporal fragment, including the mandibular fossa and tympanic. Tobias (1967*b*) likens the specimen to a gracile australopithecine or habiline.

Leakey and Leakey (1964) announced the discovery of the mandible of a robust australopithecine from Peninj, by Lake Natron, on the Kenya–Tanzania border. The mandible was found just below the Wa-Mbugu Basalt, a normally magnetized layer in the reversed magnetized Humbu Formation. Several dates gave a wide spread of results; two of the samples were less altered than the rest, and gave results of 2.27 and 1.55 MA (Isaac and Curtis 1974). The basalt is presumably, therefore, from one of the normal sub-chrons in the early half of the Matuyama.

The main hominin from Chesowanja, Kenya, is the partial cranium of a robust australopithecine (Carney *et al.* 1971). A second specimen consists only of two fragmentary molars. Both are from the Chemoigut Formation, which contains the giant elephant-like *Deinotherium bozasi*, elsewhere not known from a later time than 1.5 MA (below Tuff J at Omo). The magnetism was reversed throughout, i.e. Matuyama.

### 6.2.1(h)  South African sites

Five sites in South Africa contain Plio-Pleistocene hominin remains: Taung, Sterkfontein, Kromdraai, Swartkrans, and Makapansgat. They are caves/solution cavities in limestone or dolomite, and contain no materials known to be suitable for radiometric dating. Attempts to analyse remnant magnetism were successful only at Makapansgat, so the sites have been placed chronologically by fauna (Vrba 1975; Delson 1984) or by geomorphology (calculation of rates of nickpoint recession: Partridge 1973). The results of various dating attempts are summarized by Partridge (1982).

The Taung site, long since destroyed by lime quarrying, contains an upper 'hominid' deposit and Lower 'Baboon' Sands. All attempts to place either level have arrived at surprisingly young results: Partridge (1973) puts it between about 1.1 and 0.8 MA.

Kromdraai also has two separate deposits, A (faunal) and B (hominid). The fossil hominins come from Kromdraai B, Member 3, which Partridge (1973) places around 1.2 to 1.0 MA.

There are two fossiliferous Members at Swartkrans, with almost certainly a considerable hiatus in deposition between them. Mbr 1 is placed by Partridge (1973) between 1.4 and 1.2 MA, and Mbr 2 is much later, about 0.5 to 0.3 MA.

At Sterkfontein there are five Members, of which only the upper two contain fossil hominins. Mbr 4, the so-called 'type site', is suggested to be between 3.0 and 2.2 MA: it seems not possible to place the deposit within closer limits than that. Mbr 5, formerly known as the 'extension site', is put at some 1.7 to 1.5 MA.

Makapansgat is the oldest of the five sites. There are four Members, with hominin fossils occurring in the upper two. Mbr 1 has reversed palaeomagnetism; Mbr 2 begins reversed then changes to normal; Mbr 3 is unfortunately unknown; Mbr 4 begins reversed, switches to normal, and at the top is reversed again (MacFadden *et al.* 1979). The two reversed zones in Mbr 4 are identified as probably the Mammoth and Kaena subchrons; while it is uncertain how much is missing in Mbr 3, it cannot be too much, so its base is put by Partridge (1973) at about 3 MA. White *et al.* (1981) suggested an alternative, putting Mbr 3 at around 2.6 MA, but this does, as Partridge notes, fit less well with the palaeomagnetic data.

The hominins at these sites are mostly australopithecines: gracile at Taung, Sterkfontein, and Makapansgat, and robust at Kromdraai and Swartkrans. In addition there is a habiline at Sterkfontein (Mbr 5, whereas all the australopithecines are from Mbr 4), and what is considered an Erectus-grade form at Swartkrans (Mbr 1 and Mbr 2, the australopithecines being from Mbr 1 only).

These sites were all discovered during the 1920s, 1930s, and 1940s, earlier than any site in East Africa. It is hardly surprising, therefore, that the sample from each includes at least one type.

The only specimen from Taung, a juvenile skull, is the type of *Australopithecus africanus* Dart, 1924. This is the earliest name ever given to a gracile australopithecine. A specimen from Sterkfontein was made type of *Australopithecus transvaalensis* Broom, 1936. The species was later separated generically as *Plesianthropus* Broom, 1938. The earliest name ever given to a robust australopithecine is *Paranthropus robustus* Broom, 1938, from Kromdraai. The Makapansgat sample includes the type of *Australopithecus prometheus* Dart, 1948. Finally specimens from Swartkrans were made the types of two different taxa: *Paranthropus crassidens* Broom, 1949 (a robust australopithecine) and *Telanthropus capensis* Broom and Robinson, 1949 (an Erectus-grade form).

## 6.2.2 Sites of the Asian lower and middle Pleistocene

### 6.2.2(a) Java

Six sites in Java have yielded hominin fossils that are thought to be early or middle Pleistocene in age: Trinil, Kedungbrubus, Sangiran, Perning (Mojokerto), Ngandong, and Sambungmacan. Of these, Sangiran is by far the most important, both in the number of fossils it has yielded and in the time span it covers.

After a long period of at times acrimonious debate, the dating of Sangiran seems at last to have been settled (Itihara *et al.* 1985). The site of the Sangiran Dome is divided into four formations, previously called Kalibeng, Pucangan (formerly spelt Putjangan), Kabuh and Notopuro; the faunas were known as Ci Julang/Kali Glagah, Jetis (formerly Djetis), Trinil and Ngandong. However, as none of these names derives from the site itself, being proposed correlations of formations at other sites with those of Sangiran, Itihara *et al.* propose a new set of names for the four formations: from bottom to top these are Puren, Sangiran, Bapang, and Pohjajar. The faunas, on the other hand, seem to differ very little from one another (and are in any case controversial: Sondaar 1984), and will not be further discussed here.

The Lower Tuff in the Puren Formation is dated at 2.99 ± 0.47 MA. No other dates are available from this formation, which in any case has no hominins, and few land mammals of any kind. The Lower Lahar of the Sangiran Formation dates to 2.0 ± 0.6 MA (by K/Ar), which is similar to the 1.9 MA date quoted by Curtis (1981) from what is supposed to be an equivalent level at Perning. Eleven tuffs succeed one another through the Sangiran Formation. Above Tuff 10, dated at 1.16 ± 0.24 MA (by fission track), fossil hominins appear; but about this level there is a reversed-to-normal magnetic transition, which does not correspond to any well-authenticated palaeomagnetic event. Hominin fossils continue into the lower half of the Bapang Formation, through the transitional zone (the so-called Grenzbank), up to the Lower Tuff; the Middle Tuff, not far above the Lower, is dated at 0.78 ± 0.15 MA (fission track). The date of the top of the Bapang is not known; but between the Middle and Upper Tuffs occurs a layer of tektites, supposed to be extraterrestrial bodies which fall in showers over restricted periods; those occurring at Sangiran are called Javites. The Javites have been dated to 0.71 $\pm$ 0.10 and 0.71 $\pm$ 0.09 MA (two runs). At the Javite level occurs a reversed-to-normal transition, presumably the Brunhes/Matuyama boundary. One of the hominins, Sangiran 10 (so-called Pithecanthropus VI), was apparently found at the Javite level.

There must still be a few queries about this generally satisfactory scheme. Thus Semah (1982), on the basis of an independent palaeomagnetic study, would place the upper (hominin-bearing) levels of the Sangiran Formation later than would Itihara *et al.* (1985). Curtis (1981) regards a K/Ar date of 1.2 MA, stated to be from hominin levels of the Bapang Formation, as reliable. Siesser and Orchiston (1978) analysed Foraminifera from claystones adhering to the Sangiran 9 ( = Pithecanthropus C) mandible, finding no species present which is known to occur later than 1.6 MA.

Nearly all of the hominin fossils from Sangiran were collected by people from nearby villages, not by trained fossil hunters, and the professionals had to rely on their memories to settle their provenance. Their exact stratigraphic origin was finally settled by Matsu'ura (1982, 1985), who analysed the fluorine content of mammal bones from excavations specially undertaken, finding consistent differences between bones from the different levels. He then compared the fluorine content of all the important hominin fossils with those of known origin, and found that all must have come from a relatively restricted section above and below the Grenzbank, with the exception of Sangiran 10. It may be noted that, if the datings proposed by Itihara *et al.* (1985) are correct, there would be contradiction here as far as Sangiran 9 is concerned: from Matsu'ura's (1982) placement it would have to be less than 1.16 million years old, from Siesser and Orchiston's (1978), more than 1.6 MA.

The site of Trinil, where the earliest Javan fossil hominins were discovered, is presently the subject of a controversy as to whether Dubois, the excavator, dug through one or more levels, how old the main level is, and what sort of fauna is represented (de Vos and Sondaar 1982; Bartstra (1982). All one seems permitted to say at the moment is that the single known calvaria fits easily within the range of those from the Bapang Formation, Sangiran.

Kedungbrubus, the other site excavated by Dubois in the 1890s, is similarly controversial. A mandibular fragment, the only hominin fossil found at the site, is so nondescript as to be virtually indeterminable.

Perning, near Mojokerto (Modjokerto), in East Java, has recently been re-investigated by Sartono *et al.* (1981). These authors reject the 1.9 MA date, commonly quoted from this site, as being stratigraphically indeterminable with respect to the juvenile calvaria (the 'Mojokerto child') that is the only hominin from the site, and propose that the Lidah Formation, from which the specimen came, is equivalent to the Bapang Formation at Sangiran (not the Sangiran Formation as usually supposed).

Sambungmacan, a site which has yielded a single skull-cap, was considered by Jacob (1976) to be middle Pleistocene in age, equivalent to the Bapang Formation; Sartono (1979), however, examined the stratigraphy and considered it much later in age, perhaps upper Pleistocene.

The other hominin site in Java is Ngandong, on the Solo River, site of the dozen or so calvariae referred to as 'Solo man'. Opinions on the age of this site have varied widely, as well as on its stratigraphy (Sartono 1976*b*; Bartstra *et al.* 1976). The only thing that seems to be clear is that the river terraces overlie the Notopuro/Pohjajar Formation, so must be later in time; as there is a fission-track date of 0.25 $\pm$ 0.07 MA for the upper part of the Pohjajar, this means the Ngandong terraces must be quite late in age, perhaps even upper Pleistocene (Itihara *et al.* 1985). It is interesting that this conclusion implies a return to the views of Weidenreich (1951).

The Trinil material is the type of *Anthropopithecus erectus* Dubois, 1892; the generic name, preoccupied by de Blainville (1838) for the chimpanzee, was soon altered to *Pithecanthropus* Dubois, 1894. The controversy that followed the discovery, and Dubois's interpretation of its significance, resulted in the creation of seven further specific names, and one new generic name, between 1893 and 1932, by commentators under the misapprehension that a scientific name has to be 'appropriate' (see Chapter 1, subsection 1.1.2). These are listed by Campbell (1965), who selects the calotte as lectotype; it is still unclear whether the femora are or are not genuinely associated with the taxon (Day and Molleson 1973).

The Mojokerto child is the type of *Homo modjokertensis* von Koenigswald, 1936. Sangiran fossils have been made types of three taxa: *Pithecanthropus robustus* Weidenreich, 1944 (Sangiran 4 = Pithecanthropus IV; the specimen is a maxilla and a calvarial fragment, of which the calvaria is here selected as lectotype); *Meganthropus palaeojavanicus* Weidenreich, 1944 (Sangiran 6 = Meganthropus A); and *Pithecanthropus dubius* von Koenigswald, 1949 (Sangiran 5). All these three taxa are from the Sangiran Formation. For the Ngandong fossils the available name is *Homo (Javanthropus) soloensis* Oppenoorth, 1932; Ngandong 1 was chosen as lectotype by Campbell (1965).

Sartono (1976*a*) has proposed a reclassification of the Java fossils, and in the process has unfortunately proposed new names. For Trinil and some of the (later) Sangiran specimens he uses the name *Homo erectus trinilensis*, and for Ngandong and at least one of the Sangiran specimens (Sangiran 17 = Pithecanthropus VIII), *Homo erectus ngandongensis*. For Meganthropus A he employs the name *Homo palaeojavanicus sangiranensis* (and for other early Sangiran specimens, *Homo palaeojavanicus modjokertensis*, which is a new combination but not a new name as such). [Later (1985), following on the conclusion that the Mojokerto child is late in age, and the discovery of an apparently sagittally crested calotte ascribed to Meganthropus, he abandoned the species *Homo palaeojavanicus*, dividing it between *Homo robustus* and *Australopithecus palaeojavanicus*. Robinson (1954) had earlier also considered Meganthropus an australopithecine, calling it *Paranthropus palaeojavanicus*.]

It should be mentioned, finally, that Krantz (1975) considered the Sangiran 4 maxilla separate from the calavarial specimen usually thought to be the same individual, and conspecific with the mandible Sangiran 5, and that both of them represent not a hominin at all but an archaic orang-utan, to which he gave the name *Pongo brevirostris*. This is of course a junior subjective synonym of *Pithecanthropus dubius* (if Krantz's association of S5 with the S4 maxilla is accepted); and would become a junior objective synonym of *Pithecanthropus robustus* were it ever to be shown beyond any doubt that the maxilla and calvaria of S4 really do represent the same specimen.

### 6.2.2(b) China

The site of Zhoukoudian (Choukoutien) is the richest and best known in China. Layer 14 shows reversed polarity, but all layers above this are normal (Qian 1980). Layer 10, the lowest level from which a hominin fossil has come, is dated by fission track at $0.462 \pm 0.045$ MA. (Guo *et al.* 1980). U/Th dating of layers 8–9 gives a date of $0.420^{+ >0.180}_{-0.100}$; MA of layers 1–3 $0.230^{+0.030}_{-0.023}$, (animal teeth), and $0.256^{+0.062}_{-0.040}$ MA (animal bone). These give the outside limits for the Erectus-level fossils found there.

At Lantian, a cranium from the Gongwangling Formation is in reversed levels, and a mandible from Chenjiawo is in normal levels. These are interpreted as being either side of the Matuyama/Brunhes boundary, and from sedimentation rates the cranium has been provisionally placed at 0.80 to 0.75 MA, the mandible at 0.65 MA (Ma Xinghua *et al.* 1978).

Another Erectus-level fossil, a calvaria (also a few pieces of a second individual) from Hexian, Anhui, is correlated on the basis of associated fauna with $O^{18}$ stage 8, i.e. 0.28 to 0.24 MA (Xu Quingha 1984).

Two incisors from Yuanmou, Yunnan, were reported by Cheng *et al.* (1977) to be dated to the Olduvai subchron, but new studies of the palaeomagnetism and fauna of the site have now concluded that it is considerably later than this, perhaps 0.73 MA or later (Liu and Ding 1983). [But see, more recently, Qing 1985].

Two specimens considered of 'intermediate' type, or early Sapiens grade, are from Maba (Ma-pa), Guangdong (Woo Ju-Kang and Pang 1959), and from Dali, Shaanxi (Wu Xinzhi 1981). Both are considered, on faunal correlation grounds, to be late middle or early upper Pleistocene.

Fossils from each of these sites have been awarded special scientific names. From Zhoukoudian the name is *Sinanthropus pekinensis* Black, 1927 (renamed *Praehomo asiaticus sinensis* von Eickstedt, 1932). From Lantian, *Sinanthropus lantianensis* Wu Rukang, 1964 (type, the Chenjiawo mandible); but this name was specifically stated to be 'provisional', hence is unavailable, its first available usage being Wu Rukang, 1965, when it is used for both cranium and mandible (the latter here selected as lectotype). From

Hexian, *Homo erectus hexianensis* Huang, Fang, and Ye, 1982. From Yuanmou, *Homo (Sinanthropus) erectus yuanmouensis* Hu, 1973. From Mapa, *Homo erectus mapaensis* Kurth, 1965. From Dali, *Homo sapiens daliensis* Wu Xinzhi, 1981.

There is also the name *Sinanthropus officinalis* von Koenigswald, 1952, given to isolated teeth of Chinese Erectus type, but larger than those from Zhoukoudian, purchased from pharmacists in Hong Kong and hence of unknown origin. The yellow earth adhering to some of them suggested to von Koenigswald that they originated from early or early middle Pleistocene deposits in southern China. A supposed hominin taxon, *Hemianthropus peii* von Koenigswald, 1957*a* (the generic name being preoccupied was altered to *Hemanthropus* von Koenigswald, 1957*b*), appears to me from the photographs of the teeth to belong to *Pongo*.

### 6.2.2(c) India

A calvaria recently discovered in the upper part of unit I at Hathnora, near Hoshangabad in the Narmada Valley, India, is considered by its discoverer (Sonakia 1985) to be 'fully middle Pleistocene' in age, although previous authors had suggested an middle upper Pleistocene boundary for the unit. The specimen was named *Homo erectus narmadensis* by Sonakia (1985).

### 6.2.3 African middle Pleistocene

#### 6.2.3(a) Ndutu

The fragmented cranium of an Erectus- or Sapiens-grade hominin was discovered at Lake Ndutu, Tanzania, in 1976 (Mturi 1976; Clarke 1976). It is considered middle-to-upper middle Pleistocene in age (Leakey and Hay 1982).

#### 6.2.3(b) Bodo

A cranial specimen, consisting largely of the facial skeleton, discovered at Bodo d'Ar in the Middle Awash region of Ethiopia, is again considered middle-to-upper middle Pleistocene (Conroy *et al.* 1978).

#### 6.2.3(c) Yayo

The Yayo (Koro Toro) levels in Chad which yielded the facial skeleton of a fossil hominin are also middle Pleistocene in age (Maglio 1973). The specimen was named *Tchadanthropus uxoris* Coppens, 1965; but the name was given provisionally and so is unavailable.

#### 6.2.3(d) Kabwe

Formerly known as Broken Hill, the site of Kabwe, Zambia, has yielded a well-known nearly complete, well-preserved cranium and some less well-known post-cranial fragments, as well as skull fragments of a second individual. The site is probably middle-to-upper middle Pleistocene (Partridge 1982). The names *Homo rhodesiensis* Woodward, 1921 and *Homo primigenius africanus* Weidenreich, 1928, and the generic name *Cyphanthropus* Pycraft, 1928, have been awarded to Kabwe 1.

### 6.2.3(e) Ngaloba

A cranium from the Ngaloba Beds at Laetoli, Tanzania (LH18) was described by Day *et al.* (1980). The deposits contain a tuff thought to be correlated with one in the Ndutu Beds at Olduvai, dated to 120 000 $\pm$ 30 000 BP.

### 6.2.3(f) Omo

The Kibish Formation at Omo has yielded two well-preserved calvariae, one with facial (including mandibular) remains as well, and a third less-complete. They are assigned to Erectus-Sapiens transitional forms by Day (1971). Omo I was found partly *in situ* within Member 1 of the Kibish; Omo II, 'from a clay residual next to a hill consisting of the upper part of Member 1', reputedly at the same level as Omo I. Member I is dated by U/Th at 130 000 BP.

### 6.2.3(g) North African sites

The sites of Sidi Abderrahman, Sale, Rabat, and Thomas 1 in Morocco, and Tighenif (formerly Ternifine) in Algeria, have produced important remains of hominin fossils. The sites and the fossils were reviewed by Jaeger (1981), who dates them by geomorphological means according to sea-levels. Tighenif would be slightly under 500 000 years old; the Thomas Cave slightly over 300 000; Sidi Abderrahman and Sale slightly over 200 000; and Rabat about 150 000. More recently Hublin (1985) placed Tighenif at 0.6 to 1.0 MA, the Sale group of sites (including Sidi Abderrahman and Thomas) at about 400 000, and Rabat more recent. It is especially interesting that, over this wide time–span (on either model), Jaeger could detect no evolutionary change.

The Ternifine I mandible is the type of *Atlanthropus mauritanicus* Arambourg, 1954; the three somewhat diverse mandibular specimens from Ternifine were split among two separate species, *Homo (pithecanthropus) atlanticus* and *ternifinus*, by Dolinar-Osole (1956).

### 6.2.3(h) Other African sites

A mandible from Kapthurin, Lake Baringo, Kenya, was described by Leakey *et al.* (1969). It was associated with an Acheulian industry, and so presumably late lower or, more likely, middle Pleistocene.

A calvaria and mandibular fragment from the Hopefield/Elandsfontein district, Saldanha Bay, South Africa, is considered by Partridge (1982) to be probably middle-to-upper middle Pleistocene. It is the type of *Homo saldanhensis* Drennan, 1955.

The facial skeleton from Florisbad, Orange Free State, South Africa, has often been considered late Pleistocene but is almost certainly middle Pleistocene ('well in excess of 100 000 years'—Clarke 1985*b*, p. 304). It is the type of *Homo (Africanthropus) helmei* Dreyer, 1935.

Fragments from Eyasi are considered to be middle Pleistocene by Mehlman (1984). The most complete skull, Eyasi I, is the type of *Palaeoanthropus njarasensis* Reck and Kohl-Larsen, 1936, later (Weinert 1938) referred to the genus *Africanthropus*.

### 6.2.4 European sites

The dates of many European sites have recently been reviewed by Stringer (1981).

#### 6.2.4(a) Mauer

The famous mandible discovered in a gravel-pit at Mauer, near Heidelberg, West Germany, in 1908, is associated with a late Biharian fauna; the site is of normal polarity. Stringer (1981) suggests, on faunal grounds, a date of over 350 000 BP, and probably (1985) over 450 000. The specimen is the type of *Homo heidelbergensis* Schoetensack, 1908 (later referred to a new genus *Palaeanthropus* Bonarelli, 1909) and of *Praehomo europaeus* von Eickstedt, 1934.

#### 6.2.4(b) Vertesszollos

An ionium date from travertines gives a span of 475 000–225 000 BP. The occipital bone, Vertesszollos 2 (specimen 1 being a couple of broken and eroded teeth), is the type of *Homo erectus (seu sapiens) palaeohungaricus* Thoma, 1965.

#### 6.2.4(c) Arago

The continuing discoveries at Arago thus far include a partial cranium (facial skeleton plus parietal), two mandibles, an innominate and other pieces. An amino-acid racemization date of 450 000 is reported by de Lumley *et al.* (1977); whatever ones opinions of this method, a middle Pleistocene age seems on faunal grounds about right.

#### 6.2.4(d) Petralona

Stringer's (1981) review of the dating of the skull from this site concludes in favour of a date of somewhat over 350 000 BP. The specimen is the type of *Homo erectus petralonensis* Murrill, 1981.

#### 6.2.4(e) Bilzingsleben

The extremely fragmentary skull remains from this site have been dated by uranium series to *c*. 228 000 years (Harmon *et al.* 1980). The remains, supposedly of a single individual, are the type of *Homo erectus bilzing-slebenensis* Vlcek, 1978.

### 6.2.4(f)  Other middle Pleistocene sites

Swanscombe, Steinheim, and Biache are considered by Stringer (1981) to be most likely close in age to Bilzingsleben. To the Swanscombe remains (an occipital and both parietals, all belonging to the same individual) have been given the names *Homo marstoni* Paterson, 1940 (*nomen nudum*), *Homo swanscombensis* Kennard, 1942 (*nomen nudum*), and, by lectotype selection (Campbell 1965), *Homo sapiens proto-sapiens* Montandon, 1943 (available, though it must have its hyphen removed). The nearly complete but damaged Steinheim cranium is the type of *Homo steinheimensis* Berckheimer, 1936 and *Homo murrensis* Weinert, 1936.

From Ehringsdorf come fragments of a calvaria (supposedly all from one individual), and a mandible with a trace of a chin. The lower travertines at the site have been dated by uranium series to 200 000–220 000 BP. The cranium is the type of *Homo heringsdorfensis* Moller, 1928 and *Homo ehringsdorfensis* Paterson, 1940 (both *nomina nuda*).

All these specimens have been considered to belong either to the archaic end of the Sapiens grade, or Erectus–Sapiens transition, or Sapiens of Neandertal type.

### 6.2.4(g)  Neandertal sites

The dates of sites containing undoubted specimens of Neandertalers have been reviewed by Stringer and Burleigh (1981). Western European sites appear to range in age from about 60 000 (Saccopastore, Italy) to 35 250 $\pm$ 530 (La Quina); those in Southwestern Asia from 50 600 $\pm$ 3000 (Shanidar, Iraq) to 28 000 (Amud, Israel); in North Africa from 47 000 $^{+3200}_{-2300}$, (Haua Fteah, Libya) to '>32 000' (Jebel Irhoud); and in Southeastern Europe there is a date of 30 700 $\pm$ 750 for Krapina (Yugoslavia), though there is a doubt whether the date is associated with the Neandertal fossils, and at Vindija, also in Yugoslavia, there is a claimed association between a Neandertal fossil and an Aurignacian bone point, both dated at '>27 000' (by $C^{14}$), though there seems to be some doubt about the associations and the dating.

It is of interest of compare these dates with those of Sapiens of modern type. At Velika Pecina, Yugoslavia, a frontal bone of modern type is dated at 33 850 $\pm$ 520; if the Krapina date is valid, this would mean that the two taxa overlapped, with no signs of interbreeding. Similarly, a frontal bone from Hahnofersand, near Humburg, has been dated at 36 000 (Brauer 1981); it is said to be intermediate in form though to my mind it is a Sapiens of modern form.

In the Middle East the fully modern Sapiens remains from Skhul and Qafzeh, Israel, are of uncertain date; if, as argued by Jelinek (1982), they are over 30 000 BP, then the possibility is opened up that they may overlap in time with Amud, and possibly with Tabun (dated at 40 900 ± 1000). [Since this was written the remains of Qafzeh have been dated at 92 000 ± 5 000 B.P. (Vallados *et al.*, 1988).]

Finally, there is the fully Neandertal skull from St Cesaire, in France, associated with a Chatelperronian industry (a supposedly upper Palaeolithic technology) and dated at some 31 000.

The possibility that there was some chronological overlap between Neandertals and Sapiens of modern form must therefore be kept open.

## 6.3 THE QUESTION OF VARIATION

As we are concerned with detecting taxonomic differences in fossil hominids, the question must be asked: how do we recognize such differences, especially when more than one species has been claimed at a single site? There have been various answers proposed to this problem.

Blumenberg (1985) works with the 'coefficient of variation' (CV.) This is calculated as the mean (of a given sample) divided by the standard deviation; it is a way of utilizing the normality expected (!) of the variation in a population, while taking into account the influence of absolute size. His method is to compare the CVs of suspected fossil populations with those of known living ones, and to see if there are any taxonomic differences between the CVs of the groups compared. The particular comparisons he makes are in dental measurements. He finds that in the three living great apes the CV values for most teeth are between 5 and 11, mostly being around 7; but they are a little higher for $M_3$, higher again for $P_3$, and very high for canines (both upper and lower).

Martin (1983) performed the same calculations, but on a different set of modern ape specimens; it is therefore extremely interesting and significant that his results are nearly identical to Blumenberg's (1985). The comparisons are described in Table 6.1. The teeth showing the highest CVs are those where sexual dimorphism may be expected to be greatest: the canines, $P_3$ which would be expected to vary along with C', and $M_3$ which expands the jaw at the back and so would not disrupt occlusal patterns if expanded or contracted. (Certainly, third molars are noticeably sexually dimorphic in many other mammals: in baboons among the primates, and in Suidae elsewhere.)

Martin (1983) however, does issue a caution on the subject of sexual dimorphism. He calculated CVs for single-sex samples of apes, as well as for mixed-sex samples. In chimpanzees, the CVS for the mixed-sex samples rarely exceeded those for single-sex samples, except for the canines; gorillas and orang-utans did tend more often to show greater CVs in mixed-sex samples.

Different species do not necessarily show sexual dimorphism in the same ways. Martin (1983) finds that there is clear bimodality, with no overlap between the modes, in $M^1$ b–1 in *Pongo*, but in no other cheek teeth—even other first molar measurements. Such sexual dimorphism is not found in

**Table 6.1**　Coefficients of variation for dental metrics in Great Apes.

| Dimensions | After Blumenberg(1985) | After Martin(1983) |
|---|---|---|
| $C'$ m–d | 13–20 | 13–20 |
| b–l | 14–19 | 14–19 |
| $C_1$ m–d | 12–18 | 12–18 |
| b–l | 12–16 | 12–19 |
| $P_3$ m–d | 5–13 | 5–13 |
| b–l | 8–12 | 9–11 |
| $P_4$ m–d | 7–10 | 7–11 |
| b–l | 5–15 | 5–11 |
| $M_1$ m–d | 5–8 | 5–6 |
| b–l | 5–8 | 6–7 |
| $M_2$ m–d | 5–11 | 6–8 |
| b–l | 6–9 | 6–8 |
| $M_3$ m–d | 6–17 | 7–10 |
| b–l | 5–11 | 6–9 |

gorillas and chimpanzees. Blumenberg (1985) stresses that different levels of sexual dimorphism, in different teeth, are found in different living and fossil hominids. Oxnard (1984*a*) treats the question in multivariate terms, and confirms that, while there may be generalizations one can make (I would doubt, for example, that canines would ever be actually smaller in males than in females), the patterns of sexual dimorphism vary a good deal. Both he and Blumenberg stress this in relation to what I have called *Sivapithecus*, particularly the Chinese taxa, and particularly the differences between the large and small Chinese species.

Martin (1983) also looks beyond CV to other measures: small samples will tend to vary more in CV, especially towards high values. He suggests that in some instances it may be preferable to use observed range as a percentage of mean, though this would have the opposite effect of CV: it would give spuriously low values in small samples. Using the two measures together would probably be more valid than using CV alone.

To test this proposition, and to get some idea of the variation within fossil hominin samples, I calculated range in per cent of mean (RM) in the fossil hominin taxa as recognized by Blumenberg (1985), using the measurements usefully set out in Blumenberg and Lloyd (1983), and compared the results with the CV values as calculated by Blumenberg. The results are shown in Table 6.2. There are several interesting things about this picture. First, is that the samples of all these quasi-taxa are large enough so that the RM values are standardized: none is outrageously large, though some would appear to be

**Table 6.2** Coefficients of variation for dental metrics in the fossil biominin species recognised by Bluemenberg (1985), compared to those of living Great Apes.

| Species | Canines | | $P_3$ | | $P_4$ to $M_2$ | |
|---|---|---|---|---|---|---|
| | CV | RM | CV | RM | CV | RM |
| Living apes | 12–20 | 1.5–1.9 | 5–13 | 1.1–1.9 | 5–11 | 1.6–1.6 |
| A. afarensis | 13–17 | 1.3–1.6 | 8–12 | 1.3–1.6 | 8–13 | 1.3–1.5 |
| A. africanus | 8–10 | 1.1–1.4 | 8–8 | 1.4–1.4 | 5–10 | 1.2–1.4 |
| A. robustus | 8–13 | 1.1–1.4 | 5–8 | 1.4–1.5 | 5–6 | 1.2–1.3 |
| A. boisei | 6–12 | 1.1–1.4 | 13–20 | 1.5–1.7 | 6–12 | 1.2–1.6 |
| H. habilis | 14–16 | 1.3–1.5 | 6–9 | 1.3–1.5 | 7–11 | 1.2–1.5 |
| H. erectus | 8–9 | 1.3–1.4 | 7–12 | 1.3–1.4 | 7–10 | 1.3–1.5 |

remarkably small. The value of the CV is enhanced by this, and we can probably treat it more confidently because of it.

Looking first at the values for $P_4$ to $M_2$, we can see that in two cases (*A. afarensis* and *A. boisei*) hominin samples have higher CVs than any of the ape ones. The figures are, in fact, $P_4$ length in the case of the former, $M_1$ breadth for the latter. [*A. boisei* also has a CV value of 12 for $P_4$ breadth, but this is covered by an anomalous value of 15 for an ape sample in Blumenberg's (1985) table, a value which has been omitted in the above table.] These two quasi-taxa must be, as it were, kept under observation as candidates for being heterogeneous in some way; as these teeth are not ones subject to sexual dimorphism in other primates, at least in hominids, while bearing in mind Oxnard's (1984*a*) warnings that sexual dimorphism can take unsuspected forms, the primary suspicion must fall on taxonomy as the source of heterogeneity.

In the $P_3$ comparison, the outlier is *A. boisei*. The highest CV value for apes' $P_3$ lengths, 13, is equalled by *A. boisei*; the highest value for apes' $P_3$ breadths, 12, is enormously exceeded by *A. boisei* (20). Again, there must be a suspicion of heterogeneity in some way in this quasi-taxon; unless it is the very considerable sexual dimorphism of the apes in this measurement that is exceeded by the East African robust australopithecines, then the category *boisei* seems likely to be taxonomically heterogeneous.

The other values that are interestingly high, though not excessive when compared to apes, are the upper canine CVs for *Homo habilis* (14 and 16). These are the only ones that are within the apes' range, and they are very much larger than any others among the hominins. This group, too, must evidently be watched for heterogeneity, though as it concerns canines the presence of sexual, rather than taxonomic, differences is not excluded.

In Table 6.2, values for $M_3$ have not been included. It has been seen that in living apes the CVs for this tooth are high; in hominins they are not, and in no case do they exceed the limits for the other cheek teeth. It must also be noted that both upper and lower canines are included under the one heading in the tables; in apes, they are about equally variable, or lower canines slightly less than upper, but in the australopithecines (NB *A. boisei* is not represented for upper canines) the lower canines are every case more variable than the upper, in *H. habilis* the uppers are far more variable than the lowers; only in *Homo erectus* are the upper and lower canines about equally variable. I cannot explain this curiosity as far as the australopithecines are concerned; I will note only that in *A. africanus* and *robustus* (the only two taxa which have not thus far been tainted with the suspicion of heterogeneity) the variability of the lower canines concerns breadth (buccolingual), not length, as is also the case for *A. boisei*; while in *A. afarensis* it concerns length even more than breadth.

What of *Homo sapiens*? Dental measurements are available from a number of sources; below are listed ranges for CV in four human populations: American Whites and Blacks (Moss *et al.* 1967), Nasioi (Bailit *et al.* 1968), and Yemenite Jews (Rosenzweig and Zilberman 1967).

Incisors: 5.9–13.6 (next highest value, 9.6);
C'lengths: 5-6–8.8;
$C_1$ lengths: 3.7–6.4;
C'breadths: 7.1–9.3;
$C_1$ breadths: 7.6–8.6;
$P_3$: 4.3–9.7;
$P^3/P_4$ to M2: 3.2–12.2 (next highest value, 9.9);
$M^3$: 8.8–12.6;
$M_3$: 6.0–8.7.

It should be noted that these are in two cases (Yemenites and Nasioi) separate-sex samples, in the other two combined-sex; there are no differences at all in the CVs.

These figures confirm the results from apes: CVs for dental metrics are normally in the 5–12 range, usually in fact less than 10; suggesting that when higher values are obtained, one should become suspicious—of something or other.

During my professional career I have measured some thousands of mammal skulls, and have calculated CVs for numerous popualtions. In building up a taxonomic picture, it is essential always to work at the lowest possible level: popualtion samples, in as far as these can be obtained, which can then be combined if appropriate into subspecies, and so on. Such a procedure is, of course, ideally suited for the fossil case: at a single site, one will (by definition) not find more than one subspecies within a species, and the range of variation will be that of a single deme (provided there is not too much time depth represented). So, do skull measurements vary more or less than, or the same as, tooth measurements?

Consider the information contained in Table 6.3. The values for non-

**Table 6.3** Coefficients of variation for cranial metrics in some living primates.

| Group | Cranial length | Biorb. breadth | Face height | Cranial breadth | Palate length | Tooth-row length |
|---|---|---|---|---|---|---|
| Gorillas | 2.6–9.8 | 2.7–7.3 | 4.1–9.0 | 2.5–9.0 | 2.4–6.9 | 2.6–7.8 |
| Gibbons | 1.7–4.9 | 1.8–6.7 | 1.6–9.5 | 1.8–8.4 | 2.5–7.8 | — |
| Langurs | 1.5–5.1 | 3.0–6.1 | 2.7–3.8 | 2.3–4.5 | 3.8–7.1 | 2.2–6.0 |
| Mangabeys | 1.1–7.6 | 0.7–4.1 | 2.8–7.0 | — | 1.0–7.8 | 2.9–5.0 |

*Source:* living primates.

primates vary along much the same lines. Moreover, those collected by Yablokov (1966) are of the same order or only slightly greater (Yablokov does not generally take his samples below the subspecific level). Skull measurements are, then, somewhat less variable than tooth measurements, for whatever reason. It looks very much as if the well-known principle of Simpson *et al.* (1960) is absolutely justified: that we should suspect heterogeneity, of some sort, if the CVs stray above 10.

And in *Homo sapiens*? Howells (1973) has published a full set of figures for various human populations. The CVs, again, rarely even approach 10. Skull samples in the hominin fossil record, unlike tooth samples, are generally not large enough to calculate CVs, but in some cases hypotheses can be tested in this way. For example, Thorne and Wolpoff (1981) propose to test the peripheral population model (that peripheral populations are less variable than central ones) by comparisons between samples of *Homo erectus* from the centre of its range (or rather, near the point of origin of the species) and from the periphery. They list a number of measurements and compare their CVs for East African, Indonesian, and Zhoukoudian samples. (In my analysis, I will exclude the bone and torus thicknesses from their Table 4, p. 347, and utilize only the figures for standard craniometric variables.) The CVs for the Java hominins vary from 4.6–10.0; for Zhoukoudian, 3.2–8.4; for East Africa, 5.7–14.8. Their point, that the 'central' sample varies more than the 'peripheral' ones, is certainly proved: but in fact the 'peripheral' ones vary about as much as one might expect for a standard population (or even, in the Java case, perhaps a little more), while the East African sample is far, far too variable to be a single population by any stretch of the imagination. This conclusion will become important in my next chapter.

As noted above, fossil skull samples are generally far short of adequate for comparisons like this. What, then, can we do? Wood (1985) approaches the problem as follows. It is desired to calculate the probability of two Koobi Fora crania, ER1470 and ER1813, being members of the same population. The cranial capacity of ER1470 is 770 cc; that of 1813, 510 cc. The evidence suggests that the CV for cranial capacity is generally around 10. Thus a population with a mean cranial capacity of 640 cc (mean of 770 and 510) would have a standard deviation of about 64; the two crania in question have capacities almost exactly two standard deviations from the mean. The chance of either of them being drawn from such a population is therefore 5 per cent; of both of them being drawn from such a population (i.e. that the two came from the same population) is the square of this, namely 0.25 per cent. This is really vanishingly small.

Finally, then, what about the peripheral population (centre vs. edge) model? I have calculated overall mean CVs for a number of species I have studied in the past. We must be careful what we mean, of course, by 'centre' and 'edge' populations, so I have distinguished two kinds of peripheral populations: those which are at or towards the edges of the species' range, and those which are actual outliers, more or less isolates. The results are tabulated in Table 6.4. In this table, all the values are from measurements taken by me, except *Homo sapiens*, which is from Howells (1973). The centre/edge model seems to be justified. There are exceptions, but on the whole variability does indeed diminish towards the periphery of a species' range. The implications for centrifugal speciation are clear. So are the implications for species recognition in fossil hominins.

It is interesting to note, finally, how in the table some of the peripheral *isolates* have higher CVs than the means for the central populations [which is hardly ever the case for 'edge' (non-isolated peripheral) populations]. It seems that sometimes an isolated population can act as its own diversity generator, though variation is reduced often enough that one can still visualize them as remnant 'edge' populations that have since divided off and started afresh.

**Table 6.4** Coefficients of variation of central, peripheral, and isolated populations of various mammalian species.

| Species | Central | Peripheral | Isolate |
| --- | --- | --- | --- |
| *Diceros bicornis* | 3.15 | 3.17–3.24 | 2.37–2.41 |
| *Babyrousa babyrussa* | 5.90 | 3.45 | 2.94–4.76 |
| *Sus scrofa* | 5.02 | 4.72–5.47 | 3.09–6.07 |
| *Gazella dorcas* | 3.27 | 2.97–3.08 | |
| *Gazella rufifrons* | 3.73 | 3.48–5.06 | |
| *Cervus nippon* | 2.72 | | 2.52–6.60 |
| *Trachypithecus spp.* | 4.80 | 3.69–4.02 | 4.36 |
| *Cercocebus spp.* | 2.53 | 2.30 | |
| *Lophocebus spp.* | 4.08 | 2.98–3.52 | |
| *Hylobates spp.* | 5.02 | 3.50–4.19 | 3.10–4.67 |
| *Gorilla gorilla* M | 5.52 | 3.96–4.92 | 5.23–5.99 |
| F | 4.91 | 3.99–4.41 | 3.55–4.66 |
| *Homo sapiens* | 4.12 | 3.82–4.94 | 3.87–4.76 |

*Source:* author's data.

# 7
# Species and clades in human evolution

The general assumption of most writers I mentioned earlier has been that human evolution was largely unilineal. This assumption is so deeply ingrained that, even among those who do not espouse this view, communication is most readily achieved by talking in grade terms; which is why I did the same in Chapter 6, section 6.1. But this may not be how it did actually happen: here, I would like to test the unilineal view by cladistic analysis assisted by a realistic (section 6.3) view of variation.

Cladistic analysis is relatively new in palaeoanthropological studies. So far, only three studies have appeared which satisfy the normal tenets of such work, namely that proper attention be paid to polarities, and that cladograms be constructed independently of stratigraphic assumptions. These are the analyses of Corruccini and McHenry (1981), Skelton *et al.* (1986), and Wood and Chamberlain (1986).

## 7.1 VIEWS OF THE AUSTRALOPITHECINES

Are the australopithecines just a grade, or are they a clade as well? Wood and Chamberlain (1986) ask this question, and find it extraordinarily hard to answer satisfactorily. The six cladograms they generate are almost equally parsimonious—or rather, almost equally un-parsimonious!—although one, in which the australopithecines are not a clade, but *A. africanus* is part of the *Homo* lineage. is consistent with more of their functional region analyses than the other five. In contrast, Skelton *et al.* (1986) find that their most parsimonious cladogram places *A. robustus* and *boisei* on the *Homo* lineage, and apart from the other australopithecines. The third solution (Corruccini and McHenry, 1981) agrees with the non-cladistic assessments of Rak (1983) and of White *et al.* (1981) in placing *A. africanus, robustus*, and *boisei* all on one lineage, with *Homo* spp. on another, with *A. afarensis* as sister group to all the rest.

Essentially, therefore, one can pick one's preferred cladogram and use it to support the conviction one started with. The only thing the three analyses seem to have in common is the probability that *Australopithecus* is not a clade, at least not if it includes the four species usually assigned to it. On the other hand *Homo* really does seem to be a clade; only Wood and Chamberlain (1986) really test this proposition, but it seems at least consistent with the results of the other two analyses.

So, what is wrong? And how else could it be done to enable firmer conclusions to be drawn?

Corruccini and McHenry (1981) measured teeth and it was these measurements that were used in cladogram construction. The measurements were not simply gross length/breadth/height measurements, but distances between cusp tips, cusp sizes, and the like. They also measured a series of apes, and these in theory determined the polarities. For them, the *A. africanus–robustus–boisei* clade was distinguished by sharing the derived conditions of shortened $M_1$ and reduced $M_1$ entoconid, reduced mandible depth under $M_2$ and $P_4$, broadened lower premolars, higher crowned $P_4$ and reduced length of $I_1$. But inspection shows that not all these characters are clear-cut. Thus *A. afarensis* (represented, in fact, only by Laetoli specimens) shared the short $M_1$ with its reduced entoconid, and the reduction of $I_1$ length; as Laetoli is generally placed as the sister group of the other hominins, this immediately halves the number of uniquely derived states of the *africanus–robustus–boisei* clade. *A. boisei* reverses its mandibular shallowing under $M_2$. The only derived features left to the clade, then, are broadened premolars and increased crown height of $P_4$; large premolars, that is.

Skeleton *et al.* (1986) find that *A. robustus* and *boisei* resemble the *Homo* lineage in: reduced prognathism, vertical mandibular symphysis, nearly vertical tympanic plate, strong basicranial flexion, small size of posterior relative to anterior temporalis fibres, extramolar sulcus broader at $M_3$ than at $M_2$, post-glenoid process converges towards tympanic plate superiorly, tympanic not tubular, a strong mandibular articular eminence, foramen magnum horizontal and forwards in position, petrous bones coronally oriented, $M^3$ close to temporo-mandibular joint, no lingual ridge on lower canine, lower incisor mammelons only three in number. In addition the robust clade approaches *Homo*, but is less-advanced, in its reduced canine jugum and fairly horizontal nuchal plane. The strength of these numerous characters is considered to place the robusts and *Homo* together in a clade defined by decreased prognathism and parabolic upper dental arcade.

The journal in which Skelton *et al.*'s (1986) paper was published is noted for its publication of commentaries by other professionals on the main papers. So it is not without interest that we read, immediately following on from Skelton *et al.*'s paper, the opinions of Bilsborough, Chamberlain and Wood, and Vancata. All three commentaries make the point that the characters used by Skelton *et al.* are hardly uncorrelated. In fact, one could claim that every one of the characters in the long list given above is part of a single functional complex: dental reduction. This leads automatically to a remodelling of the basicranium in the interests of head balance, and to reduced crown relief (to a limited extent) on the front teeth. Yet as long ago as 1954 Robinson pointed out that in the robusts it is the front teeth alone that are reduced—the cheek teeth are, if anything, enlarged—while in *Homo* the whole dental apparatus is reduced. That is to say, the characters listed by Skelton *et al.* are prevented from being autapomorphic by the fact of their dependence on different functional bases.

Another comment made in all three commentaries on the Skelton *et al.* paper was the way the OTUs (quasi-taxa) were chosen. If the controversy over how many species are involved in *Australopithecus afarensis* is skated over by lumping all Hadar and Laetoli specimens into this one category, then any chance of testing for diversity, let alone of whether some components belong on different clades, is closed off. Vancata (1986), in his commentary, makes this point, and Bilsborough (1986) makes it in regard to *Homo habilis*.

Wood and Chamberlain (1986), authors of an alternative set of cladograms, might have been expected to offer the most cogent criticism of Skelton *et al.* (1986), and so it is that they put their finger on the weakness: their preferred cladogram is supported by 45 characters, their second-best by 44! As usual (Chapter 1, section 1.5) the system is so replete with homoplasy that the conclusions return to the number you first thought of.

Wood and Chamberlain (1986) are well aware of the limitations of the sort of analysis they undertake. As far as the definitions of their OTUs go, they 'have adhered to conventional definitions and assumed that all our units are monophyletic ones' (p. 223), although they do not explore the consequences of such an assumption not being justified. They divide the skull, to which the analysis is restricted, into functional regions, and analyse them separately, and eliminate, as far as possible, traits which are likely to be correlated. (Not in every case, however; for example, 'alveolar toothrow divergence' appears under both 'palate' and 'mandible', and there are a few other, less-obvious cases.) They then code the character states numerically. The results have already been alluded to: overwhelming parallelism, so that it is almost impossible to choose one cladogram over another.

## 7.2 A NEW AUSTRALOPITHECINE CLADOGRAM

Certainly Wood and Chamberlain's (1986) cladistic analysis is the most realistic so far, and reasons for here offering a new analysis include some element of desperation: it ought to be possible, why has it not been up to now? I will also admit to wanting to cram in as much information as possible: cranial, dental, post-cranial. I plundered the literature for all these sorts of characters, and ended up with 87 that seemed usable. Some of Wood and Chamberlain's characters were used, others rejected for reasons either of redundancy (as explained above) or of overlap (i.e. the means of a given measurement or index were different, but the ranges overlapped so much that no real divisions could be drawn).

I first restricted the analysis to three groups, which I called genera: *Australopithecus* ( = graciles), *Paranthropus* ( = robusts), and *Homo*. These associations are about the only ones that seem generally agreed upon, and uncontroversial. I drew up three alternative cladograms, in the form of the three-taxon problem.

In Table 7.1, all the characters that would appear to differentiate the three putative genera are listed (together with those that differentiate *A. afarensis*: see below). In the two columns headed 'supports cladograms', the first column lists the cladogram, if any, supported by the data (the three cladograms are labelled A, B, and C in the series of diagrams that accompany Table 7.1). The primitive states of the characters are not specially listed, but will be obvious from the layout of the cladograms; a character state is here considered primitive if it occurs in both *Pongo* and one (or both) of *Pan* and *Gorilla*.

Cladogram A is supported by 18 characters, with two others being somewhat equivocal. This is the cladogram that associates *Australopithecus* and *Homo* as sister groups, with *Paranthropus* their sister group. Cladogram B, in which the australopithecines together form a clade, is supported by only a single character. Cladogram C, in which *Paranthropus* forms a clade with *Homo* (as favoured by Skelton *et al.* 1986), is supported by four characters,

CLADOGRAMS

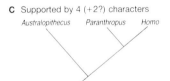

**Series A–C**

**A** Supported by 18 (+2?) characters

  *Paranthropus  Australopithecus  Homo*

**B** Supported by 1 character

  *Paranthropus  Australopithecus  Homo*

**C** Supported by 4 (+2?) characters

  *Australopithecus  Paranthropus  Homo*

**Series a–e** (*H*=*Homo*, *P*=*Paranthropus* )

**a** Supported by 3 characters

  *afarensis  H,P*

**b** Supported by 31 characters

  *H  afarensis  P  afarensi*

**c** Supported by 2 characters

  *H  afarensis  P*

**d** Supported by 1 character

  *H  afarensis  P*

**e** Supported by 26 characters

  (*afarensis* is a polyphyletic taxon)

**Table 7.1** Characters of the australopithecine taxa and Homo.

| Character | Australopithecus africanus | Paranthropus spp. | Homo s.s. | Australopithecus afarensis | Supports cladograms | Source |
|---|---|---|---|---|---|---|
| (1) DC$_1$ distal apical edge | oblique | oblique | oblique | elong., vertical | — a | Grine 1985 |
| (2) Crown area index: C$^1$ cf. P$^3$ | 78–88 | 46–78 | 76–108 | 83–121 | — a/b | Kimbel *et al.* 1985 |
| P$^4$ | 81–83 | 38–54 | 66–125 | 93–99 | | " |
| C$_1$ cf. P$_3$ | 72–99 | 41–72 | 69–104 | 70–120 | | " |
| P$_4$ | 69–85 | 27–59 | 70–104 | 68–94 | | " |
| (3) DM$^1$ paracone cf. | ≫ | = | > | ≫ | A — | Grine 1985 |
| (4) tuberculum molare | moderate/good | virtually abs. | moderate | moderate/good | A — | " |
| (5) mesial fissure on buccal face | very distinct | weak | distinct | weak | A b | |
| (6) DM$_1$ protoconid cf. metaconid | mesial | transverse | mesial | mesial | — | " |
| DM$_2$ ditto | fairly mesial | same plane | fairly mesial | very mesial | — | " |
| (7) DM$_1$ trigonid cf. talonid | > | =/< | > | > | — | " |
| (8) DM$_1$ and M$_1$ buccal groove | shallow; no pit | deep; ends in pit | shallow; usually no pit | shallow; pit varies | — e | " |
| M$^1$ buccal groove | poor | deep, broad fissure | shallow | varies | — | Robinson 1956 |
| (9) DM$_1$ fovea anterior | basin-like | fissure | basin | basin | — | Grine 1985 |
| (10) I$_1$ lingual marginal ridges | big | faint | big | vary | — b/e | Grine 1982 |
| (11) I$_2$ incisive edge | slopes distally | straight across | slopes | varies? | — e | Wood and Chamberlain 1986 |
| (12) I$^1$:I$^2$ crown area ratio | 169–182 | 155–165 | 114–167 | 153–179 | — | Grine 1982 |
| (13) C$_1$ lingual relief | strong | none | strong to fair | strong to fair | — b | Robinson 1956 |
| (14) C$_1$ lingual swelling | absent | present | slight or absent | slight or absent | — | " |
| mesial lingual ridge | present | absent | present | present | — b | " |
| (15) C$^1$ shape | pointed | rounded | pointed | pointed | — b | " |
| C$_1$ shape | asymmetrical | symmetrical | asymmetrical | asymmetrical | — | "  ; McHenry 1985 |
| canines projection | yes | no | yes | yes | | " |
| (16) Lower molars: shape | rectangular edges | oval | rectangular edges | rectangular edges | — b | Clarke 1985a |
| cusps position | | central | | | | |

**Table 7.1**  *contd*

| Character | Australopithecus africanus | Paranthropus spp. | Homo s.s. | Austalopithecus afarensis | Supports cladograms | | Source |
|---|---|---|---|---|---|---|---|
| (17) Ditto, protoconidal cingulum | strong | poor/none | usually poor | present | C | b | Robinson 1956 |
| (18) Upper molars: | | | | | | | |
| anterior fovea | varies | slight/absent | present/poor | present | — | b | " |
| proto-style | present | absent | present | present | | | " |
| Carabelli complex | present | absent/trace | present/trace | present | | | " |
| (19) $P^3$ crown tapers occlusally | yes | no | yes | yes | — | b | " |
| shape | asymmetrical | rounded | usually asymmetrical | asymmetrical | — | b | " |
| (20) $P_3$ buccal grooves, relative devt. | mesial > distal | < | > (or both poor) | vary | A | e | " |
| $P_4$ ditto | mesial = distal | none | = | < | — | e | |
| (21) $P^4$ shape index | < 75 | < 75 | > 75 | vary | — | e | Kimbel *et al.* 1985 |
| $P_4$ ditto | 70–88 | 77–96 | > 96 | vary | | | |
| (22) Premolar root type | A or C | A or B | C or D | A, C, or D | C | b | Abbott and Wood 1985 |
| (23) $M_1$ eruption cf. $I_1$ | earlier | same time | same time | earlier | — | b | Dean 1985 |
| (24) $M^3{:}M^1$ m–d length index | 102–115 | 107–110 | < 100 | 106–118 | | | Tobias 1985 |
| (25) Hypoconulid size as % crown area | 15 | 18–19 | 14 | ± 20 | A | — | Wood *et al.* 1983 |
| (26) $P_4$ cf. $P_3$ breadth | = or < | > | = or < | vary | A | e | orig. obs. |
| $P^4$ cf. $M^1$ breadth | < | > | = or < | < | | | " |
| (27) Premaxilla | rounded | straight | rounded | rounded | C | b | Clarke 1985*a* |
| Alveolar projection index | 128–134 | 122–126 | 98–128 | — | | | Wood and Chamberlain 1986 |
| Zygomatic root on tooth-row | $P^{34}$–$M^1$ | $P^3$–$P^4$ | $M^1$ + | $P^4M^1$–$M^1$ | | | " |
| (28) Maxillo-alveolar index | 91–99 | 90–99 | > 100 | 91–98 | — | b | Tobias 1985 |
| (29) Palate projn.: pre-masseter/total | 47–65 | 26–45 | > 50 | — | — | — | Rak 1983 |

| Character | | | | | | | Reference |
|---|---|---|---|---|---|---|---|
| Palate length cf. biorbital br. | 79 | 74–76 | 47–65 | — | — | — | " |
| (30) Subnasal projection index | 81–117 | 92–156 | 36–83 | vary | — | e | Wood and Chamberlain 1986 |
| (31) Prosthion angle | 47° | > 60° | 55–60° | vary | — | e | Stringer 1984 |
| (32) Po-zygomax./po-alveolare | 59–65 | 71–73 | 56–66 | — | — | — | Wood and Chamberlain 1986 |
| (33) Palate breadth:length | 86–100 | 86–105 | > 107 | 90 | — | b | " |
| (34) C'–C'/M²–M² | 66–98 | 65–91 | 50–68 | 80–87 | — | b | " |
| (35) Diastema I¹–C', C₁–P₃ | absent | absent | occasional | varies | — | e | Coppens 1982 |
| (36) Curve of Spee | present | extreme | present or absent | varies | — | e | Osborn 1983; orig. obs. |
| (37) Mandib. fossa, depth/length | 33–36 | 30–36 | > 40 | varies | — | e | Tobias 1969; Coppens 1982 |
|     depth/breadth | 24–27 | 24–27 | > 30 | varies | — | — | " |
| (38) Mandib. fossa br./biporionic br. | 26–32 | 27–31 | 18–26 | 27 | — | — | Wood and Chamberlain 1986 |
| (39) Glabella | triangular | rectangular | triangular | — | — | — | Rak 1983 |
| (40) Nasion cf. glabella | below | at | below | below | — | b | " |
| (41) Nasals (raised or not) | slightly | not | yes | not | A | b | Clarke 1985*a* |
| (42) Nasals, breadth:length | 38–56 | 36–67 | 46–100 | — | — | — | Wood and Chamberlain 1986 |
| (43) Nasals, position greatest br. | inferior | superior | inferior | inferior | — | b | Olson 1978; McHenry 1985 |
| (44) Superior end pyriform aperture, cf. orbits | at or above | well-below | at or above | slightly below | A | b | orig. obs. |
| (45) Vomer insertion cf. ant. nas. spine | immed. behind | continuous | well behind | vary | — | e | Ward and Kimbel 1983 |
| (46) Naso-alv. clivus cf. nasal floor | sill | smooth | sill | vary | A | e | Clarke 1985*a* |
| (47) Incisive canals, size | fairly big | small | fairly big | big | — | b | " |
| (48) Lateral nasal border | sharp | blunt | sharp | sharp | — | b | Rak 1985 |
| Anterior pillars | usually present | big | trace or absent | absent | A | b | " |
| (49) Transverse buttress | absent | present | rudim./absent | big | A | b | " |
| (50) Zygomatic prominence | bulbous | prominent | flat/present | absent | — | — | " |
| (51) Frontal process of malar | flares laterally | flares | vert./slight flare | — | — | — | " |

**Table 7.1**   *contd*

| Character | Australopithecus africanus | Paranthropus spp. | Homo s.s. | Austalopithecus afarensis | Supports cladograms | | Source |
|---|---|---|---|---|---|---|---|
| (52) Frontal process of maxilla faces | forward | forward | laterally | forward | — | b | McHenry 1985 |
| (53) Face ht./biorbital br. | 101–102 | 96–121 | 71–82 | — | — | — | Rak 1983 |
| /upper face br. | 77–81 | 75–97 | 66–79 | — | A | — | Wood and Chamberlain 1986 |
| (54) Infraorbital foramen position | high | low | high | high | — | b | Rak 983 |
| (55) Bizygomatic/biorbital br. | 134–137 | 135–159 | 114–129 | — | B | — | Wood and Chamberlain 1986 |
| (56) Foramen magnum slope | – 20 | – 7 to – 10 | + 3 to + 13 | – 6 | — | c | " |
| (57) Petrous axis/bicarotid line angle | > 60° | 42–49° | 46–55° | > 60 ± | C | b | Dean and Wood 1981, 1982 |
| (58) Tympanic plate (mm) | < 30 | > 30 | > 30 | ± 20 | — | — | " |
| (59) Petrotympanic angle | 51–58° | 17–34° | 29–41° | 35–45° | C? | b | Wood and Chamberlain 1986 |
| (60) Foramen ovale cf. pterygoid plate | behind | at base | behind | at base | A | b | orig. obs. |
| (61) Digastric scar | in notch | on mastoid | in notch | in depression | A | c | Olson 1985b; Kimbel et al. 1985 |
| (62) Occipito-mastoid crest | present | absent | present | present | — | — | Olson 1981b, 1985b |
| (63) Biporionic/biparietal br. | 100–110 | 89–92 | 94–128 | 80 | A | a | Wood and Chamberlain 1986 |
| (64) Temporal lines diverge | well-above lambda | well-above | well-above | vary | — | e | McHenry 1985 |
| (65) Occipital index | 72–80 | 64–67 | 60–69 | 64 | — | — | orig. obs. |
| (66) Occipital marginal sinus | no | yes | no | yes | — | d | Wood and Chamberlain 1986 |
| (67) Post-cranial elongation index | 29–33 | 27–29 | 37–48 | — | — | — | " |
| (68) Relative cranial capacity | 1.9 | 1.7–1.9 | > 2.0 | 1.6 | — | a | " |
| (69) Bone thickness: asterion | 9 | 5.5–7.3 | 9–15 | vary | — | e | Wood 1984 |
| bregma | 4.5–7.0 | 5.5–7.0 | 9–11 | vary | | | |
| (70) Supraorbital sulcus | absent | absent | usually present | absent | — | b | Rak 1983 |
| (71) Frontal squama angle | 45° | 20–34° | 35–90° | vary | A | e | Wood and Chamberlain 1986 |
| (72) Post-orb. constrn.: | | | | | | | |

| Feature | | | | | | | Reference |
|---|---|---|---|---|---|---|---|
| biorbital br. | 78–88 | 61–86 | 80–104 | 79 | | | " |
| bizyg. br. | 48–50 | 34–48 | 66–74 | — | — | b | Rak 1983 |
| (73) Humerus: | | | | | | | |
|   lateral anterior crest | — | poor | poor | vary | — | e | Senut and Tardieu 1985 |
|   lateral epicondyle | — | salient | weak | weak | | | " |
|   capitulum | — | strong | poor | vary | | | " |
|   shaft | — | flattened | rounded | vary | | | " |
| (74) Femur: | | | | | | | |
|   distal epiphysis index | > 75 | < 75 | > 75 | vary | A | e | " |
|   asymmetry | no | yes | no | vary | | | " |
|   Tibial spine index | > 50 | < 50 | > 50 | vary | | | " |
| (75) Femur, intercondylar notch index | = or < | < | > | vary | A | e | Kennedy 1983; orig. obs. |
| (76) Femur platymeria | 91 | 66–79 | 65–86 | vary | — | e | " |
| (77) Femur shaft, breadth: length | 65 | — | 33–55 | vary | — | e | Senut and Tardieu 1985 |
| (78) Meniscus insertions on tibia | — | one | two | one | — | b | Jungers and Stern 1983 |
| (79) Humero-femoral index | 50–60? | 76–90 | 72–74 | vary | A? | e | McHenry 1978 |
|   Elbow br.: femoral head | — | 112 | < 100 | [105] | | | McHenry 1974 |
|   Femur length: pelvis height | 162 or 179? | — | 216 | [177] | | | |
| (80) Lateral margin femoral head | — | ant.–med./ post.–lat. | ant.–lat./ post.–med. | vary | — | e | Stern and Susman 1985 |
| (81) Intertrochanteric line | present | absent | present | vary | A? | e | " |
| (82) Femoral neck index | 59 | 68 | 76–83 | vary | — | e | Kenney 1983 |
| (83) Tilt of distal tibial articuln. | — | posterior | anterior | vary | — | e | Stern and Susman 1985 |
| (84) Femur, shaft: head diameter | 58 | 63–69 | 53 | 63–64 | A | b | Robinson 1972 |
| (85) Ischium length/iliac crest br. | 21 | 25 | 21 | [18] | — | — | " |
| (86) Acetabulum ht./ilium ht. | 38 | 34–44 | 53–56 | [35] | — | — | " |
| (87) Pelvic torsion angle | 83–87 | 75(–87?) | 48–83 | — | C? | — | " |

with two others equivocal; the low support I take to reflect the attempted sweeping away of redundancies.

For once, the result is clear-cut: the view espoused, for example, by Robinson (1972) or Olson (1981*b*) is supported. It is also the view supported by Groves and Mazak (1975). I think I have a good enough record of backing down from previous positions, in the face of contrary evidence, to avoid charges of coming to a preconceived conclusion.

This means, clearly, that if there is to be more than one genus in the tribe Hominini (as most authorities would think reasonable), then the prime candidate for generic separation must be *Paranthropus*.

*Australopithecus afarensis* is also listed in the table, with characters ascribed to it. In view of the controversies that are still raging, it was thought best to restrict the recording of character states to just three groups: (1) AL288-1 (Lucy), plus the small hemipalate AL199-1, (2) the AL333 and 333-w series, whose taphonomic circumstances are so remarkable as to make it almost inconceivable that they are anything but a (conspecific) social group, and (3) the Laetoli sample, which have never been considered anything but homogeneous. The cladograms supported by their character states are listed in the table (in the 'supports cladograms' column) by small letters (a,b,c,d,e). (Note that these listings were made only after the three-taxon analysis above had been made.) In cladograms a,b,c, and d (see the table drawing), *A. afarensis* is a homogeneous taxon: in a, it is the sister group to both *Homo* and *Paranthropus*; in b, any one of three placements is possible from the evidence; in c, it is the sister group of *Homo* only; in d, of *Paranthropus* only. In alternative e, hardly a 'cladogram' at all, *A. afarensis* is not homogeneous: the 'taxon' is polyphyletic, as insisted by Tobias (1980), Olson (1981*b*), Ferguson (1983, 1984), and others.

Cladograms a, c and d are each supported by one to three characters only. The 'flabby' cladogram, b, is supported by 31 characters. Alternative e, the polyphyletic alternative, is supported by 26 characters. Again, a clear-cut result: the putative taxon *A. afarensis* is polyphyletic.

This being so, we must now take each of our three components of *afarensis* separately, and test them against the original three-taxon cladogram. The results are given in Table 7.2 and the series of cladograms that accompany Table 7.2. In cladogram A, the subgroup in question (Lucy, 333 or Laetoli) is a sister group of *Paranthropus*; in cladogram B, of *Homo*; in cladogram C, of a *Paranthropus/Homo* clade. D is a flabby cladogram, any of the above three being compatible. In E, no evidence is offered as to placement.

For Lucy, 'Hadar, small type', the evidence marginally favours cladogram C of the three definitive cladograms. If this is so, it favours White *et al.*'s (1981) placement of *A. afarensis* as the most primitive of the Hominini—for Lucy is certainly the most crucial specimen of this putative taxon in anyone's thinking. But the evidence is not, in the case, very clear-cut, though provisionally at least such a placement can be followed. It is interesting to note, in parentheses, that in all characters where the Hadar small form (i.e. AL199-1)

CLADOGRAMS

X=the *afarensis* group- under consideration: Laetol, Hadar small,Hadar –333

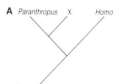

**A** *Paranthropus* X *Homo*

Supported by:

2 characters for Laetoli;
1 character for Hadar, small type;
2 characters for Hadar–333 group

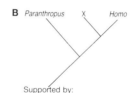

**B** *Paranthropus* X *Homo*

Supported by:

2 characters for Laetoli;
1 character for Hadar, small type;
15(+ 1?) characters for Hadar–333 group
of which 6 (+ 1?) characters support
cladogram *B1*, in which the branching
point is *ABOVE* that of *A. africanus*.

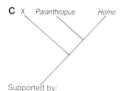

**C** X *Paranthropus* *Homo*

Supported by:

1? character for Laetoli;
4(+ 1?) characters for Hadar, small type;
0 characters for Hadar, 333 group

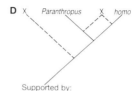

**D** X *Paranthropus* X *homo*

Supported by:

7characters for Laetoli;
18(+ 1?) characters for Hadar, small type;
9 characters for Hadar, 333 group

**E** No evidence offered as to placement.
Laetoli, 3 characters;
Hadar, small type, 6 characters;
Hadar–333 group, 5 characters,

differed from the 333 group, the larger palate A L200–1 resembled 199. So the 'Lucy' taxon is not just a small-sized one (Fig. 7.1 (A), (B)).

For the 333 group, matters are quite different. 15 characters support cladogram B, versus two for A, none at all for C. Moreover, there is a sub-ordinate cladogram, B1, in which the 333 taxon is placed above *A. africanus* on the *Homo* lineage, i.e. *A. africanus* is the sister group of 333 plus *Homo*. B1 is supported by six (and equivocally by one other) of the 15 characters; moreover, none of the remaining nine characters specifically oppose it. This unlooked-for result supports the view of Coppens (1983*a*), that the 333 specimens represent a taxon that is 'étonnement moderne'.

For Laetoli, the evidence is again not clear-cut: in two characters ($DM_1/M_1$ buccal groove ends in a pit; $P_4$ broader than $P_3$) it seems to go with *Paran-thropus*; in two ($P_3$ mesio-buccal groove more strongly developed than mesiodistal; well-developed intertrochanteric line on femur) it aligns with the

**Table 7.2**  Characters of A. afarensis under cladogram 'e'.

| Character | Laetoli | Cladogram | Hadar, small type | Cladogram | Hadar-333 | Cladogram |
|---|---|---|---|---|---|---|
| DM₁/M₁ buccal groove | ends in pit | A | no pit | E | ends in pit | A |
| M¹ buccal groove | weak | E | weak | E | deep, broad | A |
| I₂ incisive edge | slopes | E | slopes | E | straight | E |
| P⁴ shape | < 75 | D | < 75 | D | usually > 75 | B1 |
| P₄ shape | > 90 | D | < 80 | C | > 90 | D |
| I¹² lingual tubercle | weak | D | strong | C | weak | D |
| P₃ buccal groove | mes. > dist. | B | mes. < dist. | D | mes. < dist. | D |
| P₄ cf. P₃ breadth | > | A | > | A | = or < | D |
| Prosthion angle | | | 25° | C | 35–45° | E |
| Diastemata | small | C? | pres. or abs. | C? | abs. | E |
| Curve of Spee | pres. | D | pres. | D | abs. | B1 |
| Mandib. fossa depth | | | shallow | D | deeper | B1? |
| Alveolar projection | | | 97–100 | D | 73–92 | D |
| Vomer insertion | | | slightly behind nas. spine | D | well-behind | B1 |
| Clivus/nasal floor | | | sill (sometimes) | B | no sill | D |
| Temp. lines diverge | | | well above lambda | E | just above | B? |
| Vault bones, bregma | thin (4.7) | D | thin (4.5) | D | thicker (6.0) | B1 |
| Frontal angle | (low) | D | > 30°? | D? | c.30° | D |
| Humerus: lat. ant. crest | | | strong | C | poor | D |
| capitulum | | | strong | D | poor | B |
| shaft | | | flattened | D | rounded | B |
| Femur: distal index | | | < 75 | D | > 75 | B |
| notch ht./br. | | | < | D | > | B1 |
| asymmetry | | | yes | D | no | B |

| | | | | | | |
|---|---|---|---|---|---|---|
| platymeria | 79 | E | 68 | E | 78 | E |
| shaft index | | | 65–72 | E | 87–91 | E |
| Tibial spine index | | | < 50 | D | > 50 | B |
| Arms cf. legs | | | long | D | less so | D |
| Femoral head | | | am–pl* | D | al–pm† | B |
| Intertroch. line | good | B | poor | D | good | B |
| Femoral neck index | 'flat' | D | 69 | D | 79 | B1 |
| Dist. tibial tilt | | | post. | D | ant. | B |

\* anteromedial-posterolateral
† anterolateral-posteromedial

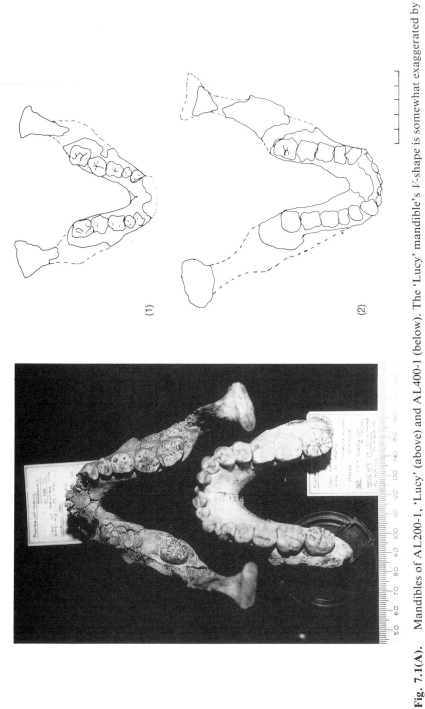

**Fig. 7.1(A).** Mandibles of AL200-1, 'Lucy' (above) and AL400-1 (below). The 'Lucy' mandible's *V*-shape is somewhat exaggerated by the loss of external symphyseal bone bilaterally, but this does not account for the whole of the difference.

**Fig. 7.1(B).** Hadar mandibles. (1) Gen. indet. *antiquus*. AL288-1. (2) *Homo* sp. AL333w-60. Both reconstructed following White *et al.*

*Australopithecus/Homo* lineage; in the presence of diastemata it may be more primitive than either (although a diastema can occasionally occur in *Homo*: viz: the much-discused occurrence in Sangiran 4). So many of the characters are simply unknown or indeterminate in the Laetoli sample that it is not possible to make a definitive placement on a cladistic basis.

If we simply compare Laetoli phenetically with other fossil samples, there seems little doubt but that it is to *Australopithecus africanus* and to the AL333 series that it bears most resemblance. Compared to the former, it differs in the larger size, on average, of the canines, the occurrence of premolar root type D, the two *Paranthropus*-like features, the presence in some specimens of a diastema, and the femoral platymeria. The presence of the pit on $DM_1$ and $M_1$ in the 333 group, which is so definitively *Homo*-like in other characters as we have seen, indicates that this character could be subject to homoplasy. Compared with 333, Laetoli differs in the weak expression of the buccal groove on $M^1$, the broad $P^4$ the predominance of mesial over distal buccal grooves on $P_3$, the broad $P_4$, the potential presence of diastemata, the development of a curve of Spee, the thinner vault bones, and the reportedly (White 1980) flat femoral neck.

On balance, therefore, the Laetoli sample seems most like *A. africanus*. Until such a decision can be corroborated or refuted cladistically, this must remain provisional; but in the meantime there is no denying the general likeness. I propose to leave the Laetoli fossils in the genus *Australopithecus*.

## 7.3 THE SPECIES OF *HOMO*

It is preferable wherever feasible to cut up samples as finely as possible, otherwise something may be missed. Possible over-lumping may obscure potentially interesting differentiation, as noted by Bilsborough (1986). For this reason I have not asumed the orthodox boundaries of either *Homo habilis* or *Homo erectus*, but have tried to separate out candidates for special taxonomic status wherever possible, using geographic criteria or, where there may be diversity at a single site, variability criteria in the manner discussed earlier (Chapter 6 section 6.3). The candidates for special consideration are, therefore:

—*Homo sapiens*: while originally keeping 'anatomically modern' representatives separate from the archaic forms referred to this species by Stringer (1984) and others, it became very clear that the former are simply more derived versions of the latter. They are therefore lumped in Table 7.3.

—*Homo erectus*: the characters of the Zhoukoudian and Java fossils were combined here, as the differences are in no case of cladistic importance.

—Olduvai Hominid 9: though treated as *H. erectus* even under such carefully analysed systems as that of Bilsborough and Wood (1986), this fossil ought clearly to be kept separate on geographic grounds alone. That there is no other specimen like it from any locality, and that it consists of a calvaria only, are unfortunate, so that it cannot be a primary basis for a cladistic decision, but must be added on the basis of available evidence to a cladogram built up on the basis of more complete material.

—Koobi Fora *Homo* cf. *erectus* fossils (Upper and Ileret Members). The cranium of this type is well-known from the beautifully preserved ER3733, and some less-complete examples (especially ER3883). The discovery (Leakey and Walker 1985) of similar cranial fragments associated with the earlier described mandible ER730 enables the characters of this mandible to be added to the potential taxon.

—Koobi Fora small *Homo*. The cranium ER1813, from about the level of the KBS Tuff at Koobi Fora, was referred to *Australopithecus africanus* by Walker and Leakey (1978), but Wood (1985) has shown that this is very far from appropriate. It has been taken separately here. The skull ER1805 from the Koobi Fora Upper Member was also, at first, considered separately, but its character states were in every case those of 1813, so in the end othe two were treated together. ER1805 has a poorly preserved mandible as well as cranial fragments, thus enabling comparisons between this potential taxon and the 3733 type to be extended to the mandible.

—*Homo habilis*: the sample of this species was restricted to the Olduvai specimens, primarily OH7, 13, 16, and 24. The case made by Leakey *et al.* (1971) for the inclusion of all these four specimens in the same taxon seems convincing, and where two or more of them could be assessed for present purposes their characters were invariably the same.

—Koobi Fora Lower Member taxon: the crania ER1470, 1590 and 3732 from below the KBS Tuff are almost certainly representatives of the same taxon. Wood (1978) argues that mandibles like ER1802 represent this form, and notes that ER1482, also from below the KBS Tuff, differs in a number of respects (overall V-shaped, with long post-incisive planum and developed transverse tori) from the more usual U-shaped, relatively poorly buttressed mandibles. One mandible from Omo can be associated with the V-shaped type, and four with the V-shaped 1482. The difference between the two forms at Koobi Fora is impressive, and the Omo specimens fall into the same groups, as Wood shows. Curiously, however, the measurements of the two types are the same (Fig. 7.2). Whether they are treated together or not makes no difference as far as the results of the analysis are concerned.

These, then, are the basic OTUs which will be used for the analysis. Other, less-complete, specimens, such as the Member 5 specimen from Sterkfontein, the *Homo* from Swartkrans, mandible ER992 from Koobi Fora, and the enigmatic Olduvai hominid 12, will be brought in afterwards.

The 'cladograms supported' section of Table 7.3 has five separate columns. The first (cladogram series A–C that accompanies Table 7.3) examines the relative positions of *H. sapiens*, *H. erectus*, and the ER3733 taxon. The second (cladogram series a–e in the drawing) determines the

# CLADOGRAMS

## Series A–C
Relative positions of *H. sapiens*, *H. erectus*, and 3733

**A** Supported by 0 characters

**B** Supported by 6 characters

**C** Supported by 2 characters

## Series a–e
Position of OH9

**a** Supported by 1 character

**b** Supported by 11 characters

**c** Supported by 0 characters

**d** Supported by 5 characters (compatible with either b or c)

**e** Supported by 2 characters

## Series 1–3
Relative position of *H. habilis*, ER1813, and the ER3733 *H. erectus / H. sapiens* clade

**1** Supported by 6 characters

**2** Supported by 3 characters

**3** Supported by 2 characters

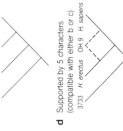

## Series I–II
Relative positions of ER1813, ER3733, and the *H erectus/H. sapiens* clade

**I** Supported by 12 characters

**II** Supported by 6 characters

## Series i–iii
Relative position of *H. habilis*, ER1470, and the ER1813/ *H. erectus/H. sapiens* clade

**i** Supported by 8 characters

**ii** Supported by 3 characters

**iii** Supported by 4 characters

**Table 7.3** Characters of the Taxa of Homo (s.s.).

| Character | ER1470/1802 | H.habilis | ER1813/1805 | ER3733/730 | OH9 | H.erectus | H.sapiens | Cladograms supported | Source |
|---|---|---|---|---|---|---|---|---|---|
| (1) Cheeks cf. pyriform aperture | slightly anterior | slightly anterior | post. | post. | — | post. | post. | 1 | Clarke 1985a |
| (2) Anterior pillars | trace | trace | abs. | abs. | abs. | abs. | abs. | B d 1 | Rak 1983 |
| (3) Nasal br./ht. index | 46 | 60 | 50–64 | 52 | 50 + + | 90 | 65–100 | d | Wood and Chamberlain 1986 |
| (4) Nasal skeleton | rounded | ridged | ridged | ridged | arched | arched | arched | d i | Clark 1985 |
| (5) Clivus cf. nasal floor | sill | sill | sill | smooth | — | smooth | sill | C | " |
| (6) Alveolar projection | 128? | 121 | 125 | 108 | — | c.108 | 99–100 | I | Wood and Chamberlain 1986 |
| (7) Upper face cf. midface br. | < | > | > | = or > | — | = | > | i | Wood 1985 |
| (8) Interorb. cf. bifrontal br. | < 25% | < 25% | < 25% | < 25% | c.25% | > 25% | c.25% | e | Stringer 1984 |
| (9) Supraorbital bar shape | straight | arched | arched | arched | straight | straight | arched | b i | Wood 1985 |
| (10) Supraorb. bar lateral wing | no | no | no | no | prominent | prominent | no | b | Macintosh and Larnach 1972 |
| (11) Supraorb. bar thickness, mm | 7 | 6–9 | 8–9.5 | 8–14 | 19 | 12–19 | 0–22 | I | Rightmire 1984 |
| (12) Frontal angle | 60° | 47–60° | 60° | 50–70° | 45° | 35–43° | 70–90° | b | Wood and Chamberlain 1986 |
| (13) Supraorbital notch | abs./slight | present | abs./sl. | slight | big | present | deep | 2 | orig. obs. |
| (14) Glabellar flatness | 33 | 40–46 | 25 | 43 | 49 | 15–40 | 10–38 | a | Spitery 1980 |
| (15) Face height cf. cranial length | 59 | 41–51 | 39–42 | 48 | — | 36–40 | 34–47 | i | orig. obs. |
| (16) Postorb. cf. biorbital breadth | 70 | 80–90 | (73–)81 | 73–76 | 72 | 82–89 | 94–104 | 3 II i | Rightmire 1984 |
| (17) Cranial height: length | 74 | 61 | 56–60 | 55–59 | 56 | 53–59 | 60–77 | i | Wood 1984 |
| (18) Biparietal: biporionic breadth | 98 | 98 | 94–96 | 95 | 103 | 108–117 | 112–128 | B e | Wood and Chamberlain 1986 |

| Character | | | | | | | | Code | | Reference |
|---|---|---|---|---|---|---|---|---|---|---|
| (19) Vault shape | rounded | rounded | flattened | rounded | rounded | flattened | rounded | d | 1 | Wood 1985 |
| (20) Braincase 1.: Biporionic br. | 149 | 131 | 140–156 | 136–138 | 153 | 153–160 | 151–171 | | II | iii Wood and Chamberlain 1986 |
| (21) Occipital index | 82 | 71–72 | 78–88 | 65–77 | (65–73) | 60–76 | 84–99 | | II | orig. obs. |
| (22) Occipital cf. nuchal scale | > | > | > | 93–106 | > 94 | < | > 67 | | I | Wood 1985 |
| (23) Occipital angulation | no | no | no | no | yes | yes | no | b | | Wood 1984 |
| (24) Occipital torus | no | no | weak | weak | strong | strong | weak | b | 1 | ″ |
| (25) Angular torus | no | no | no | no | yes | strong | sometimes | b | | Wood 1984 |
| (26) Supramastoid crest | small | absent | small | small | large | large | small/absent | b | | iii Rightmire 1984 |
| (27) Supernumerary lambda bones | no | no | may occur | may occur | (not seen) | may occur | may occur | | 1 | Wood 1984 |
| (28) Thickness at bregma, mm | 6 | 4–7 | 5–6 | 4–10 | (thick) | 8–16 | 6–13 | | 1 | orig. obs. |
| (29) Inion cf. opisthocranion | lower | lower | lower | at | at? | at | lower | C b | | ″ |
| (30) Endinion cf. ectinion | coincident | slightly lower | ? | lower | lower | lower | usually lower | | | i Wood 1984 |
| (31) Parietal keeling | no | no | no | no | no? | yes | no | | | Rightmire 1984 |
| (32) Frontal keeling | no | no | slight | slight | no | yes | sometimes | | 1 | Wood 1984 |
| (33) Coronal ridge | no | no | no | no | no | yes | no | | | ″ |
| (34) Mastoid length, mm | — | < 10 | 10–12 | 25–30 | (12 +) | 12–27 | (long) | | I | ″ |
| (35) Sup. surface post. zygomatic | flat | weak gutter | weak g. | weak g. | strong g. | strong g. | weak | b | | Clarke 1985 |
| (36) Occipitomastoid crest | small | small | small | small | present | big | small/absent | b | | orig. obs. |
| (37) Vaginal crest | — | high | low | high | high | absent | high | | | Wood 1984 |
| (38) Tympanic plate | — | horiz. | horiz. | horiz. | vert. | vert. | vert. | B d | | Wood and Chamberlain 1986 |
| (39) Tympanic 1.: bitympanic br. | 23 | 28 | 27–27 | 22–24 | 24 | 21–22 | 22 | | 1 | Wood and Chamberlain 1986 |
| (40) Basion cf. bitympanic line | level | anterior | ± level | anterior | level? | ± anterior | ± posterior | | | iii Dean and Wood 1981 1982 |

**Table 7.3**  *contd.*

| Character | ER1470/1802 | H.habilis | ER1813/1805 | ER3733/730 | OH9 | H.erectus | H.sapiens | Cladograms supported | Source |
|---|---|---|---|---|---|---|---|---|---|
| (41) Petrotympanic angle | — | 30–41 | 45–45 | 46 | c.33–35 | 2c.30 | 29 | d 2 | Wood and Chamberlain 1986 |
| (42) Mandibular fossa, depth:br. | 27 | 29–43 | 46 | 58 | — | 56–64 | 58–63 | I i | Tobias 1969 |
| (43) Mastoid–petrosal fissure | no | no | no | no | yes | yes | no | b | Stringer 1984 |
| (44) Palate length: biorbital br. | — | 64 | 65 | 58–62 | — | 47 | 47 | B I | Rak 1983 |
| (45) Palate, ant. cf. post. depth | < | < | = | < | — | = | = | II | Clark 1985a |
| (46) C'–C': M²–M² | 65 | — | 69–72 | 68 | — | 63–76 | 50–68 | I | Wood and Chamberlain 1986 |
| C₁–C₁:M₂–M₂ | 46 | 53–54 | 50 | 50 | — | 53–63? | 35–51 | | " |
| (47) Planum alveolare | slight | usually big | big | slight | — | present | slight/abs. | I | orig. obs. |
| (48) Sup. and inf. transverse tori | small | us. small | fairly big | small | — | present | small/abs. | 2 | " |
| (49) Mandible robusticity index | 59–67 | 61–74 | ?62–70 | 64–69 | — | 54–60 | 33–59 | B | Wood and Chamberlain 1985 |
| (50) Mandible marginal torus | no | no | no | no | — | no | yes | | orig. obs. |
| (51) Curve of Spee | slight | slight | good | present | — | absent | present | C I | Coppens 1982 |
| (52) I¹:I² crown area | — | 157–167 | 150 | 114 | — | 106 | 135 | | Grine 1982, 1984 |
| (53) C₁ lingual relief | strong | little | — | — | — | strong | fair | iii | Tobias 1968 |
| (54) C'–P₄ area | 90–104 | 73–84 | — | 87–94 | — | 83–96 | 69–100 | | |
| C'–p⁴ area | 98–99 | 68 | 66–87 | — | — | 92–108 | 107–128 | | |
| C'–p³ area | 102–104 | 65 | 76–91 | 87 | — | 91–108 | 93–138 | | |
| (55) P₃ m–d:b–1 | 101–102 | 93–109 | — | 95 | — | 75–97 | 76–92 | | Tobias 1966 |
| P³ m–d:b–1 | 75–76 | 73–75 | 59–72 | 61–73 | — | 68–86 | 71–80 | ii | " |

|  | C1 | C2 | C3 | C4 | C5 | C6 | C7 |  | Reference |
|---|---|---|---|---|---|---|---|---|---|
| (56) P₄ m-d:b-1 | 61–76 | 91–106 | — | 95 | — | 75–96 | 74–95 |  | " |
| P⁴ m-d:b-1 | 72–77 | 64–76 | 68–80 | 68–69 | — | 68–81 | 64–73 |  | " |
| (57) P₃ subsidiary cusps | present | virt.abs. | — | present | — | present | present |  | ii Vandebroek, 1969 |
| fovea anterior | small | small | — | present | — | pres./big | pres./big |  | " |
| (58) P₄ talonid | large | very large | — | large | — | small/abs. | absent | B 3 | " |
| P⁴ talon | poor | large | fair | large | — | poor | poor |  | orig. obs. |
| (59) P³ buccal grooves | fair | good | shallow | fair | — | poor | vary |  | " |
| (60) P₄:P₃ breadth | = | = | < | < | — | = | = | II | " |
| P⁴:P³ breadth | = | — | = or < | = | — | = or < | = |  | " |
| (61) Upper molars, ant. fovea | present | usually big | small | big | — | poor/small | poor/absent | II ii | " |
| Lower molars, tuberculum sext. | present? | present | — | — | — | rare | rare |  | " |
| (62) Upper molars, Carabelli | present | trace | present | trace | — | trace | trace (usually) | 2 | " |
| (63) Mesiobuccal groove end | no pit | pit | no pit | no pit | — | no pit | no pit |  | Grine 1982 |
| (64) M³:M¹ m-d length | 90–97 | 95–96 | 95–105 | (< 90?) | — | 88–89 | 88–89 | I | Tobias 1985 |

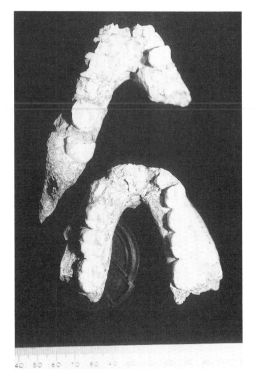

**Fig. 7.2.** The two *Homo* (or cf. *Homo*) mandibular types from the Lower Member, Koobi Fora: [cast of] (above) ER1482, here seen as representing *Homo aethiopicus* and (below) ER1802, representing *H. rudolfensis*.

placement of OH9 when it is added to the cladogram determined by the A–C analysis. The third (series 1–3 in the drawing) takes a *sapiens/erectus/*3733 clade for granted, and analyses the dispositions of this cluster, *H. habilis*, and the ER1813 form. The fourth (series I–II in the drawing) is really a check on series 1–3: it takes 3733 out of the *sapiens/erectus* cluster and tests its placement relative to them and 1813. The fifth (series i–iii in the drawing) tests the relative positions of *H. habilis* and the 1470 group.

### 7.3.1 *Homo sapiens, Homo erectus*, and the 3733 group

Three cladograms are, of course, possible in this three-taxon problem. Cladogram A, in which *H. erectus* is the sister group to a 3733/*sapiens* combination, is not supported by any characters. Choice B, in which 3733 is the sister group to the other two, is supported by six characters. Choice C, in which *H. sapiens* is the sister group of the other two, is supported by two characters (absence of a sill between clivus and nasal cavity floor, and inion level with opistho-cranion); possibly also the relative lack of size disparity between $I^1$ and $I^2$ (but factors of wear renders this uncertain). Even one of the features associating 3733 with *H. erectus*, the absence of the sill, loses much of its significance because of the frequent lack of much, or any, sill in some populations of *H. sapiens*.

There seems little doubt, therefore, that the 3733 taxon is the sister group of a *sapiens/erectus* combination. It cannot, therefore, be included in the species *Homo erectus*; a conclusion reached previously by such authors as Wood (1984).

### 7.3.2 The position of OH9

Given the correctness of cladogram B, as argued above, there are five possible placements for Olduvai hominid 9: sister group of 3733, of *H. erectus*, of *H. sapiens*, of the *H. erectus/sapiens* combination, or of all three. The last placement was not supported by any characters, so is not even figured; while the third, that it is the sister group of *H. sapiens*, is also unsupported (but is figured for compatibility's sake). As can be seen from Table 7.3 and Fig. 7.3, the evidence is overwhelmingly in favour of cladogram b—that it is a sister group of *Homo erectus*. (There are also five equivocal characters, which might support this placement but are not definitive.)

This result was unexpected, but it conforms with the opinion expressed by Bilsborough and Wood (1986). It may be taken as definite: although, unlike the 3733 result above, there is only a single specimen, the sheer weight of its *H. erectus* synapomorphies implies that the discovery of further specimens would not affect the essential result even though the mean resemblance might shift away somewhat. It also means that there is, after all, evidence for *Homo erectus* outside Asia.

### 7.3.3 *Homo habilis* and 1813

Assuming for the moment the monophyly of all the OTUs dealt with up to now, where do the smaller brained forms, contemporaneous with at least 3733, fit in? The series of diagrams following Table 7.3 gives the cladograms (series 1–3).

The evidence here is not so very clear-cut, but on balance *H. habilis* is placed as a sister group to a grouping of 1813 and the 3733/*erectus/sapiens* clade. Inspection of characters supporting cladogram 2 shows that they tend to be subject to variability: thus a supraorbital notch is slightly developed in 1813, and the petrotympanic angle of 3733 is in fact as great as in 1813, making the polarity uncertain. Probably, but not definitely, the 1813/1805 species is closer cladistically to the *erectus/sapiens* clade than is *habilis*.

### 7.3.4 1813 and 3733

The reality of the assumed monophyly of the 3733/*erectus/sapiens* clade needs, however, to be tested. It will be seen, from the drawings of series I-II in Table 7.3, that there are six characters in which 1813 is closer to the *erectus/sapiens* clade, but 12 in which 3733 is closer. Thus, while homoplasy

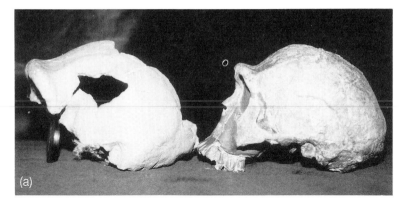

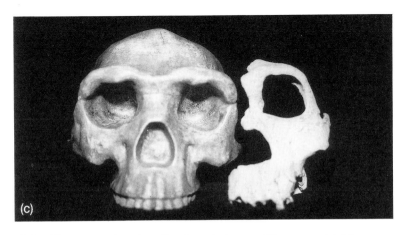

**Fig. 7.3.** *Homo* sp. (species 4f of text) is not *H.erectus.* (a) *Homo erectus olduvaiensis* (OH9) (left) with ER3733 (right), side view. (b) The same, front view. (c) *H.e.pekinensis* (reconstruction, 'Nellie') (left) with SK-847 (right).

as usual is rampant, one's subjective judgement that 3733 is 'more advanced' is supported.

### 7.3.5 The position of 1470

Bilsborough (1986) objected to Skelton *et al.* (1986) referring 1470 without demur to *Homo habilis*. That this placement of the 1470 taxon has a venerable history (back at least to Pilbeam and Gould 1974) does not by any means guarantee that it is correct. The drawings of series i–iii in Table 7.3 give a cladistic assessment of its status with respect to *Homo habilis* and the 1813/3733/etc. clade.

The cladogram (ii) associating 1470 with *habilis* is the least supported of the three. Best supported is cladogram i, which places 1470 as the sister group of *habilis* plus the rest. Once again there is plenty of homoplasy. Whether this phenomenon represents parallel development, or polymorphism in a common ancestor, it is a fact of life in the analysis of phylogeny in the genus *Homo*.

### 7.3.6 Are there ancestors?

While there is no reason to think that evolution travels in straight lines, parsimony clearly dictates that the most likely candidates for the role of ancestors are, as argued in Chapter 1, section 1.5.2, those forms with no demonstrable autapomorphies. The question is, then, whether there are such OTUs in fossil *Homo*. Table 7.4 shows that indeed there are.

**Table 7.4** Autapomorphic character states of the taxa of Homo (s.s).

| | |
|---|---|
| *1470 group* | *H. habilis* |
| Superior surface of posterior zygoma flat | $P_3$ subsidiary cusps virtually absent |
| Lower canine crown area > 90% (usually > 100%) of $P_4$ crown area | Supramastoid crest absent |
| | Basion well in front of bitympanic line |
| | Upper molar mesiobuccal grooves end in a pit |
| *1813 group* | Lower canine lingual relief reduced |
| Vault flattened | Upper canine crown area < 70% of $P_3$ area |
| $P^3$ mesiodistal length < 72% of buccolingual breadth | $P_4$ mesiodistal length > 90% (usually > 95%) of buccolingual breadth |
| *3733 group* | *H.erectus (without OH9)* |
| No autapomorphic features | Interorbital breadth > 25% of bifrontal |
| | Parietal keeling |
| *OH9* | Coronal ridge |
| No autapomorphic features | Vaginal crest absent |
| | Very flattened fault |
| | ? Curve of Spee absent |
| | ? Face height < 70% of bifrontal breadth |

Exactly as was argued by Andrews (1984), *Homo erectus* (excluding OH9 for the moment) is not such a one. It has several autapomorphic states: conditions not found in *Homo sapiens*. While theoretically a series of evolutionary reversals are certainly imaginable, the probability would seem to be that *Homo erectus* is not the ancestor of *Homo sapiens*. As far as the Chinese fossils are concerned, the two species are in any case contemporary; but the early dates now apparently confirmed [Chapter 6, subsection 6.2.2 (a) above] for the Java fossils, and the newly supported inclusion in *H. erectus* of OH9, might still enable an ancestor-descendant relationship between *H. erectus* and *H. sapiens* to be considered. This possibility will be raised again later (Section 7.3), but in view of the evolutionary reversals required it must be regarded as a 'poor relation'.

Olduvai hominid 9 has no autapomorphic states. Its whole morphology, it seems, is exactly that to be expected in a lineal ancestor of *Homo erectus*, albeit presumably at the rugged end of the range of its population.

The ER3733 group equally has no autapomorphic states. The group is fingered as an ideal common ancestor for the later representatives of the genus.

The 1813 group have a few uniquely derived conditions: two, as listed in Table 7.4. On the sort of criteria proposed above, the group is probably not precisely ancestral to the 3733-and-above clade, but is not far off it. Some members of the 1813 and 3733 groups overlap in time, although cranium ER1813 itself is apparently earlier than any of the 3733 group.

*Homo habilis* is quite evidently not an ancestor of anything else, unless quite a number of evolutionary reversals have occurred. Although it has been almost universally accepted, without any questioning, as the stock from which the *erectus/sapiens* clade was derived, as long ago as 1969 Vandebroek did in fact question this status, and some of the characters he mentioned (pointed out by him for OH7, but applicable to other specimens as well) have been used here.

ER1470 represents a form much closer to the lineage of later forms than *H. habilis*. Its primitive morphology was pointed out in 1975 by Walker, and in the present study it is only the large canines that strongly push it off the ancestral line, and the large cranial capacity—but more of that later.

### 7.3.7  The others

#### 7.3.7(a)  Sterkfontein

The Sterkfontein Member 5 fossil is, as Clarke (1985*a*) has shown, very like *Homo habilis*. Not much detail has been published on it, but what measurements have been made available, and what characters can be seen on the illustrations, suggest that it could indeed be a member of that species. One of the characteristic features of *H. habilis*, premolar narrowing, seems actually to be taken further in Stw 53: $P^3$ mesiodistal length is more than 90 per cent of its breadth, whereas in *H. habilis* from Olduvai this figure is 73–75 per cent.

### 7.3.7(b) Swartkrans

The Swartkrans form, as represented by SK15 and SK847, seems indistinguishable from the 3733 taxon. None of the measurements published so far seem to differentiate it, and the juxtaposition of SK847 and ER3733 by Walker (1981*b*) confirms not only a general similarity but the presence in the former of all the putatively 'ancestral' features of the latter, without any indication of the derived features of either *Homo erectus* or *H. sapiens*, except perhaps for the arched nasal bones of the latter. As, however, most of the derived states of at least *Homo sapiens* are in the vault, it cannot be excluded for the moment that the Swartkrans form might be more 'sapient' than the 3733 taxon.

### 7.3.7(c) Homo ergaster

The Koobi Fora mandible ER992 is important because, for better or worse, it is the type of *Homo ergaster* Groves and Mazak, 1975. Its position must therefore be settled in order to determine the nomenclature of these lower Pleistocene taxa. Coming as it does from the upper levels (Ileret Member) of Koobi Fora, it could conceivably pertain to either the 3733 or the 1813 taxon. Groves and Mazak (1975) also referred ER730 to *H. ergaster*, and this is now known (Leakey and Walker 1985) to be associated with a 3733-like cranium; but at the time of publication of the name there was no reason to question the existence of more than one 'late' gracile taxon at Koobi Fora—and indeed, did it depend on the mandibles alone, there would still not be. The question therefore largely depends on resolving the question: Which is 992 more like? 730 or 1805 (Fig. 7.4)?

The 992 mandible has teeth intermediate in size between the larger ones of 1805 and the smaller ones of 730. If 1805 has been correctly associated in a common taxon with 1813—and this now seems certain—then, as 1813 has smaller maxillary teeth than 1805, the 1805 mandibular teeth are not the minimum size possible for the taxon. This consideration therefore marginally favours the allocation of 992 to the 1813 taxon.

Against this might be the relatively large post-incisive planum of 1805, and its reduction in 992 as in 730. It can be argued, however, that this might be subject to individual variations within a taxon. The symphyseal tori of 992 are better developed than in 730, and approach 1805 in this respect. All three mandibles are heavily developed, with a robusticity index of over 60.

Until further mandibles are found of one type or the other—especially of the 1813 type, for the 1805 mandible is a decidedly scrappy piece—there seems little prospect of further comparisons. There is however, another approach: fitting mandibular morphology to cranial morphology. The

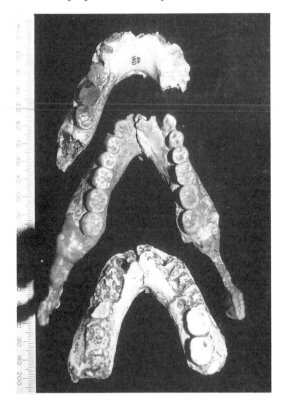

**Fig. 7.4.** Three mandibles from Koobi Fora (Upper Member): (top to bottom) ER730, ER992 (type of *H.ergaster*), ER1805. ER992 resembles ER1805 rather than ER730 (see text), permitting the name to be associated, with greater plausibility, with the small-brained 1805/1813 species rather than the large-brained 730/3733 species.

length of the 992 mandible is greater than its bicondylar breadth, differing from all *erectus/sapiens* grade mandibles but resembling those australopithecines in which these measurements can be taken. The correlation is of course with cranial broadening and, inferentially, cranial capacity enlargement. While these two measurements cannot be taken on 730 or 1805, both of which lack ascending rami, the body of the jaw clearly 'opens out' more in 730, while the horizontal rami of 992 are, if anything, more nearly parallel even than in 1805. Similarly, as pointed out by Robinson (1972), the lingual contour of 992 is V-shaped like a gracile australopithecine (and, in fact, even more markedly than in 1805), not U-shaped like a 'higher' form or like 730.

While the evidence, therefore, is not all it might be, it does on the whole support the view that 992 belongs to the 1813/1805 taxon, which therefore takes the name *Homo ergaster*.

## 7.3.7(d)  Olduvai hominid 12

Although discovered as early as 1964, it was not until 15 years later that OH12 was described in any detail (Rightmire 1979). It is from Olduvai Bed IV, hence about 750 000 years old. Rightmire (1980) has also described a well-preserved hemimandible, OH22, from Bed IV, and it will here be taken as a working hypothesis that they represent the same taxon.

The Bed IV hominid is not *Homo habilis*: although the cranial capacity is only 727 cc, and so (just) within the range of the latter, there is a fairly well-developed supramastoid crest, the lower premolars are not excessively narrowed ($P_4$ index in OH22 is 90), and the vault bones are extremely thick.

Nor is it *Homo erectus*. The supraorbital ridges are not developed into a true torus; occipital angulation, while appreciable, is not as marked as, for example, in OH9, and is not accompanied by a very strong occipital torus. An angular torus is present but not strong, and the supramastoid crest is very far from being like that of *H. erectus*. The vault is not flattened. The tympanic plate is not thickened. OH22 does however resemble *Homo erectus* in its lack of a curve of Spee.

While it does not have the derived states of *Homo erectus*, the Bed IV hominid does have those common to the *erectus/sapiens/*3733 clade: the supraorbital bar is 10 mm thick, the vault is thick (10 mm), the mastoid is long (25 mm), there is no $M_1$ tuberculum sextum, and the talonid of $P_4$ is reduced. For a member of this clade, the alveolar planum and symphyseal tori are perhaps unexpectedly large, and the mandible is very robust (robusticity index = 72), though none of these features is unknown within the clade.

Finally, the relatively rounded vault with its implied relatively great biparietal breadth, low position of inion, and reduction of occipital angulation all suggest that the Bed IV hominid is a member of the *Homo sapiens* clade. Certain features—especially the low cranial capacity, mandibular robusticity, and planum alveolare development, and lack of the curve of Spee—do however imply that it is not very advanced along this clade. Whether it is within the range of variation of the archaic sapient types of the middle Pleistocene of Europe and Africa will become clearer if and when more specimens are discovered: inferentially, however, it is not.

## 7.4  THE TAXA OF *PARANTHROPUS*

The latest commentator on the 'robust australopithecines' (Grine 1985) divides them into three species: *P. robustus* (Kromdraai), *P. crassidens* (Swartkrans), and *P. boisei* (East African sites). The separation of an East African from a South African species is not unusual and has been fairly recently placed on an unassailable basis by Rak (1983); but the division between the Kromdraai and Swartkrans samples is unusual. Robinson (1954) distinguished the Kromdraai and Swartkrans forms at subspecific level, giving diagnostic features; Howell (1978) also separated them, but did not specify his reasons; and Grine is in fact the first author to substantiate the differences claimed by Robinson, and to add further ones. Briefly, the Kromdraai form is different from the Swartkrans one in the following character states:

a. Characters in which Kromdraai is primitive, Swartkrans and East African sharing a more derived condittion: maximal convexity of $dc_1$ is at mid-crown level, not at apex; apex of $dc_1$ is central, not mesial, in position; tuberculum molare present (if poorly developed) on $dm_1$; mesial cusp slightly higher than distal on $dm_1$; protoconid equal in size to metaconid on $dm_1$, instead of smaller; area of $C_1$ nearly 60 per cent that of $P_4$; proportion of palate that overlaps the articular-eminence-to-zygomatic-tuberosity distance is only 37 per cent instead of above 43 per cent; and buccal groove fairly well-developed on $P^3$.

b. Characters in which Kromdraai is more derived than either Swartkrans or East African: mesioconulid on $dm_1$ absent; tuberculum intermedium present on $M_1$ (though it is also present on three out of 20 from Swartkrans); Carabelli complex absent or very slightly developed on upper molars.

c. Characters in which Kromdraai shares the derived condition with Swartkrans, the East African specimens being primitive: presence of a maxillary fossula (Rak 1983).

d. Characters in which Kromdraai shares the derived condition with the East African specimens, Swartkrans being more primitive: bizygomatic breadth 160 per cent of biorbital breadth (it is 164–175 per cent in East African, but only 148–151 per cent in Swartkrans); $P_4$ breadth as great as that of $M_1$; mesiodistal diameter of $I_1$ only 33–34 per cent that of $M_1$ (it is 27–39 per cent in East African, and 38 per cent in two from Swartkrans); $P^4$ with only slight development of a disto-lingual groove.

In this list, the (a) characters certainly outweigh the competing (c) and (d) characters. Rak (1983) has argued that the East African forms might have originally possessed a maxillary fossula, and lost it because of their further facial modification. Some of the (d) characters are not very cogent. In all, there seems little reason to doubt that the Kromdraai taxon is indeed the most primitive; an apparent anomaly, as the evidence seems to suggest that it is later in time than Swartkrans, and later than most of the East African remains as well [see Chapter 6, subsection 6.2.1(h)].

Although the sample size is small, the considerable number of characters in which Kromdraai is at or beyond the edge of the range of Swartkrans does support the view that they represent different taxa: here tentatively regarded as different species, following Grine (1985).

It should be noted that, while *P. robustus* does have its share of derived character states (some of them apparently parallelling the East African representatives of the genus), only one, and that not consistent ($M^1$ usually has a pit at the end of the buccal groove), can be identified in *P. crassidens*. The East African sample all share a number of strongly derived character states: buccal groove on $dm_1$ extends down to neck of tooth; $M_1$ lacks tuberculum intermedium; $C_1$ area less than half that of $P_3$ (usually), and less than 40 per cent that of $P_4$; bizygomatic breadth great (paralleling Kromdraai); palate very deep; visor-like development of face, and other facial characters elucidated by Rak (1983); mandibular bicanine distance less than half the $M_2$–$M_2$ distance; and characters of $I_1$ reduction, $P_4$ broadening, and $P_4$ distolingual groove reduction that parallel Kromdraai. As far as cranial characters are concerned, these features depend on a single specimen from Olduvai, two from Koobi Fora, and one from Chesowanja; but there is a considerably larger sample of mandibles and teeth.

But, is, the East African sample taxonomically homogeneous (Fig. 7.5)? It has been mentioned already (Chapter 6, section 6.3) that some of the coefficients of variation (CVs) of the so-called *boisei* sample are unusually large. This is especially the case for $P_3$, for which tooth the sample is respectable in size; also for the admittedly smaller sample for $I_2$. In the first case, the

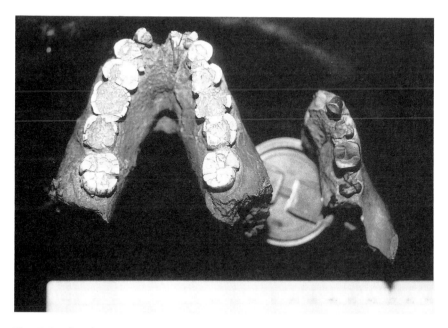

**Fig. 7.5.** Specimens currently referred to *Paranthropus boisei* cannot convincingly be accommodated in a single species. L-7 (left) and L-74 (right), contemporaneous in Omo (Shungura Formation) Member G.

test used by Wood (1985) can be applied. There are just two specimens of $P_3$ from Omo Shungura Member H: L-7-125, a truly enormous specimen, and L-29-43, much smaller. $P_3$ buccolingual breadths are, respectively, 17.5 and 10.2 mm. To have the greatest chance of being sampled from the same population, they would have to come from a population with a mean value midway between, i.e. 13.85. If the CV were 12, which as we have seen is an absolute maximum, the standard deviation (sd) would be 1.66. The two sd limits would be 10.53–17.17, but the values for the two Member H mandibles still lie outside these limits, i.e. the probability of one of them belonging to such a population is 0.05, and of both of them is $0.05^2$, that is, 0.0025—a quarter of one per cent.

A second mandibular specimen from the same level, Omo L-74-21, lacks $P_3$ but possesses $P_4$, whose buccolingual breadth is 13.8 mm and has to be compared to the L-7-125 value of 18.9 (Fig. 7.5). The same test gives not quite as low a probability—about 1 per cent—but the level of separation is still impressive.

If, then, there are two different taxa confounded under the *boisei* label, what are they, and what specimens belong to each? To begin with, they must represent two separate species, as both occur at the same level in Omo. I have attempted to sort out the specimens from Member H times and later in East Africa, with the following result:

a. Large-toothed species: ER403, 404, 818, 1468, 1509, 5877; and Omo L-7-125. The allocation of ER404 and 1509 to this species is provisional only.

b. Smaller toothed species: ER729, 801, 802, 1171, 1477, 3229, 3230; Omo L-74-21, L-29-43; OH30; Peninj (Natron).

Three specimens—two of them only isolated teeth—come from earlier levels at Omo: L-33-9 and 33-508 from Member F, and F-22 from Member G. These do not fit well into either of the samples from the later time span (Table 7.6). There are also the new cranium and mandible from West Turkana, dated at 2.5 MA (Walker *et al.* 1986).

As far as the upper dentition goes, the samples are much smaller and the CV test of Wood (1985) does not reveal significant diversity in any one site. But as far as premolar breadths go, there is a big difference between OH5 and the Chesowanja specimen. ER1804 is almost as big as OH5, and an isolated premolar, ER802, is almost as small as Chesowanja. In as far as molars can be allocated, ER733 would fit better with the large form, and ER1171 with the small form. As OH5 is the type of *Zinjanthropus boisei* L.S. B. Leakey, 1959, this settles this name onto the large East African species.

The positions of the two cranial specimens from Koobi Fora, ER406 and 732, are of extreme interest here (Fig. 7.6). The first of these has no surviving teeth, but approximations can be made to the measurements (with an error of about $\pm$ 1 mm). $P^4$ breadth would be about 17.5 mm, which is within the range of the large species (i.e. *P. boisei* s.s.). A fragment of the crown of $P^4$

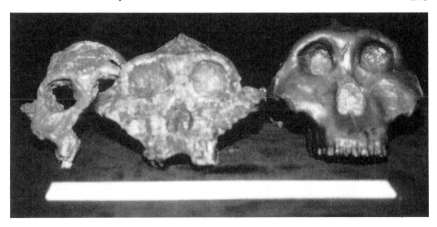

**Fig. 7.6.** [Casts of] ER732, ER406, and OH5. All three are currently referred to *Paranthropus boisei*, ER732 being thought of as a female, the other two as males; it is proposed here to refer the first of these to a separate species.

remains on ER732; the breadth of the complete crown is unlikely to have been more than 14.5 mm, which is smaller than that of Chesowanja, and places this specimen in the smaller toothed species. Though a hypothesis of strongly marked sexual dimorphism can be plausibly entertained on the basis of the comparative cranial morphology of these two specimens (for example, Wolpoff 1980b), these dental comparisons suggest that two different species are involved, and confirm that dental size is reflected in overal skull size.

On this basis, I worked out means and standard deviations for these two species. The results are shown in Table 7.5. Both *P. boisei* and its smaller relative are more megadont than either of the two South African species. The small East African species lacks the extreme molar/premolar broadening of *P. boisei*; its mesiodistal measurements are much the same, or even slightly larger. In Table 7.6, I have also reproduced data from Chamberlain and Wood (1985) on mandibular corpus dimensions; these show that *P. boisei* has the largest mandibles, the small East African species and *P. crassidens* cover an intermediate range, and *P. robustus* is the smallest.

I am well aware that I have in a way 'invented' a new species to explain observed high levels of variability. I am also aware that there are no differences between them apart from (1) degree of premolar/molar broadening and (2) skull and jaw size. The characters of the facial skeleton, for example, are the same (except for the apparently greater zygomatic flare of *P. boisei*): there is no question of the smaller species being one of the South African species in a northerly range extension, nor yet of the contast between the long/narrow face of OH5 and the short/broad face of ER730 being explicable on this basis. Yet the variability, if the samples are combined, really is unacceptably high; and this fact must be faced by any commentator

**Table 7.5**   Maxillary dental measurements (mm) in *Paranthropus*.

| | P.robustus | P.crassidens | P.cf.boisei | P.boisei | WT-17000 |
|---|---|---|---|---|---|
| I¹m–d | | 9.33 | | 10.0 | |
| | | 0.81 | | — | |
| | | 8 | | 1 | |
| b–1 | | 7.53 | | 8.0 | |
| | | 4.46 | | — | |
| | | 6 | | 1 | |
| I²m–d | | 6.91 | | 7.0 | |
| | | 1.07 | | — | |
| | | 9 | | 1 | |
| b–1 | | 6.66 | | 7.6 | |
| | | 0.66 | | — | |
| | | 7 | | 1 | |
| C'm–d | | 8.56 | 8.2 | 8.7 | |
| | | 0.41 | — | — | |
| | | 11 | 1 | 1 | |
| b–1 | | 9.51 | 8.4 | 9.7 | |
| | | 0.89 | — | — | |
| | | 11 | 1 | 1 | |
| P³m–d | 10.0 | 0.75 | 10.0,11.7 | 10.8,10.9 | 11.5 |
| | — | 0.76 | — | — | — |
| | 1 | 15 | 2 | 2 | 1 |
| b–1 | 13.7 | 14.19 | 13.7,14.0 | 16.0,17.0 | 16.2 |
| | — | 0.75 | — | — | — |
| | 1 | 15 | 2 | 2 | 1 |
| P⁴m–d | 10.3 | 10.73 | 11.7 | 11.3,11.8 | |
| | — | 0.81 | — | — | |
| | 1 | 16 | 1 | 2 | |
| b–1 | 15.2 | 15.42 | 15.5 | 17.4,18.0 | |
| | — | 0.82 | — | — | |
| | 1 | 16 | 1 | 2 | |
| M¹m–d | 13.7 | 13.51 | 14.4 | 14.73 | |
| | — | 0.70 | — | 0.81 | |
| | 1 | 15 | — | 3 | |
| b–1 | 14.6 | 14.65 | 14.8 | 16.67 | |
| | — | 0.87 | — | 0.96 | |
| | 1 | 16 | 1 | 3 | |
| M²m–d | 13.8 | 14.64 | 15.5,18.1 | 15.5,17.2 | |
| | — | 0.59 | — | — | |
| | 1 | 12 | 2 | 2 | |
| b–1 | 15.9 | 16.09 | 16.0,17.5 | 18.2,21.0 | |
| | — | 0.94 | — | — | |
| | 1 | 12 | 2 | 2 | |
| M³m–d | 14.4 | 15.23 | 17.2 | 15.7 | |
| | — | 1.08 | — | — | |
| | 1 | 16 | 1 | 1 | |
| b–1 | 16.2 | 16.91 | 16.5 | 21.4 | |
| | — | 0.80 | — | — | |
| | 1 | 16 | 1 | 1 | |

**Table 7.6** Mandibular dental measurements (mm) in Paranthropus.

| | P.robustus | P.crassidens | P.cf.boisei | P.boisei | Omo, > 2 m.a. | WT-16005 |
|---|---|---|---|---|---|---|
| $I_1$m–d | 4.80 | 5.41 | 5.50 | 4.00 | | |
| | — | 0.20 | 0.40 | — | | |
| | 1 | 7 | 3 | 1 | | |
| b–l | | 6.27 | 6.73 | 6.50 | | |
| | | 0.56 | 0.68 | — | | |
| | | 7 | 4 | 1 | | |
| $I_2$m–d | 5.60 | 6.32 | 6.28 | 4.60 | | |
| | — | 0.47 | 0.39 | — | | |
| | 1 | 5 | 4 | 1 | | |
| b–l | | 7.17 | 7.00 | 8.20 | | |
| | | 0.64 | 0.87 | — | | |
| | | 6 | 3 | 1 | | |
| $C_1$m–d | 9.40 | 7.84 | 7.97 | 7.6,7.8 | | |
| | — | 0.63 | 0.58 | — | | |
| | 1 | 12 | 6 | 2 | | |
| b–l | 8.80 | 8.34 | 8.62 | 9.0,9.6 | | |
| | — | 1.03 | 1.16 | — | | |
| | 1 | 12 | 6 | 2 | | |
| $P_3$m–d | 9.8,10.3 | 9.78 | 11.07 | 11.20 | | 10.7 |
| | — | 0.46 | 1.08 | 1.56 | | — |
| | 2 | 15 | 6 | 3 | | 1 |
| b–l | 12.2,12.7 | 11.74 | 12.50 | 17.50 | | 13.8 |
| | — | 1.05 | 1.45 | — | | — |
| | 2 | 14 | 6 | 1 | | 1 |

**Table 7.6** *contd*

| | P.robustus | P.crassidens | P.cf.boisei | P.boisei | Omo, > 2 m.a. | WT-16005 |
|---|---|---|---|---|---|---|
| P$_4$m–d | 10.8,10.9 | 10.96 | 13.98 | 12.65 | 12.9 | (12.0) |
| | — | 0.62 | 0.76 | 1.27 | — | — |
| | 2 | 15 | 8 | 4 | 1 | 1 |
| b–l | 11.4,12.9 | 12.91 | 15.13 | 16.0,18.9 | 12.5 | (15.0) |
| | — | 0.70 | 1.07 | — | — | — |
| | 2 | 15 | 7 | 2 | 1 | 1 |
| M$_1$m–d | 13.78 | 14.45 | 16.43 | 15.78 | | 15.7 |
| | 0.85 | 0.74 | 1.13 | 1.17 | | — |
| | 4 | 23 | 7 | 5 | | 1 |
| b–l | 12.50 | 14.07 | 14.77 | (16.2 + ) | | 14.3 |
| | 0.74 | 0.74 | 1.44 | (1.88) | | — |
| | 4 | 22 | 6 | 4 | | |
| M$_2$m–d | 15.0,15.4 | 16.06 | 18.35 | 17.20 | 19.0 | (17.0) |
| | — | 0.82 | 1.12 | 1.68 | — | 1 |
| | 2 | 19 | 6 | 3 | 1 | |
| b–l | 14.4,14.8 | 14.83 | 17.00 | (17.60 + ) | 17.2 | 16.7 |
| | — | 0.76 | 1.02 | (0.69) | — | — |
| | 2 | 19 | 6 | 3 | 1 | 1 |
| M$_3$m–d | 16.10 | 16.90 | 19.80 | 20.97 | 19.9,20.2 | |
| | 0.26 | 1.15 | 1.68 | 1.95 | — | |
| | 3 | 13 | 5 | 6 | | |
| b–l | 14.27 | 14.68 | 16.80 | 16.86 | 15.1,18.0 | |
| | 0.46 | 1.15 | 1.33 | 2.12 | — | 2 |
| | 3 | 13 | 5 | 5 | 2 | |
| Mandibular corpus area | 660 | 691–1140 | 660–1084 | 1126–1640 | | |
| | 1 | 4 | 17 | 7 | | |

on the specimens. I have proposed a solution, which I believe has to be seriously considered while fully recognizing that others [such as the proposal of 'unexpected directions of sexual dimorphism' of Oxnard (1983)] may be ultimately preferred.

A final observation concerns the earlier specimens from the Omo and West Turkana deposits (Table 7.6*b*). The cheek teeth are narrower than those of any of the four later species, while being at least as large in mesiodistal dimensions.

A full taxonomy of *Paranthropus*, in the context of the Hominini as a whole, needs now to be laid out. The consideration of the pattern of the group's evolution must be left to the next (final) chapter.

## 7.5 A TAXONOMY OF THE HOMININI

In what follows, I have used the standard checklist referencing system, which is for purposes of priority verification, not for bibliographic purposes. The format, and style of abbreviation, follows that of checklists such as Ellerman and Morrison-Scott (1951).

### 7.5.1 Genus unnamed (Fig. 7.7 (1))

This is plesiomorphic sister group to all other Hominini, retaining the following conditions of which the others share the alternative, derived state: upper incisors with large basal tubercles; extreme prognathism, prosthion angle being under 30°, with strong alveolar projection; lateral anterior crest of humerus large; acetabulum small, its diameter less than 60 per cent of pubis length. The braincase appears to be relatively rounded, with no supra-orbital torus, and small: AL162–28 has a cranial capacity of between 350 and 400 cc (Falk 1985). McHenry (1984) finds the shoulder joint of AL288–1 to be very like that of the orang-utan; the elbow, between human and gorilla; the wrist is more human; the pelvis and hip joint are close to *Australopithecus africanus*; the knee joint is like that of a chimpanzee or orang-utan. AL129–1 is very close to AL288–1 in comparable parts, but AL137–48A is somewhat different, though not very far from AL288–1 either.

#### 7.5.1(a) Gen. indet. antiquus Ferguson, 1984

1980 *Australopithecus africanus aethiopicus* Tobias, *Palaeont. afr.*, **23**, 14. Type, by subsequent designation (Olson, 1985*b*: 116), AL 228–1, "Lucy", from the Hadar Formation. Unavailable: proposed conditionally.
1984 *Homo antiquus* Ferguson, *J. Hum. Evol.*, **25**, 527. Type A.L. 288–1, 'Lucy', from the Hadar Formation.
1985 *Homo aethiopicus* Olson, in Delson, ed., *Ancestors: The Hard Evidence*, 116. Type AL 228–1.

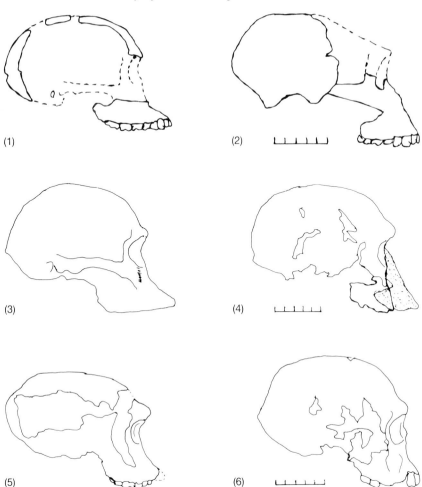

**Fig. 7.7.** Crania of *Australopithecus* and early *Homo* spp., and of gen. indet. *antiquus*. (1) Gen. indet. *antiquus*. After Schmid (1983) and Ferguson (1983). Based on AL288-1, AL162-28, and AL199-1. (2) *Homo* sp., the 'large Hadar species'. After Ferguson (1983), modified after Schmid (1983). (3) *Australopithecus africanus*. Sts 5 ('Mrs Ples'). (4) *Homo* sp. (unnamed). ER1470. The stippled outline represents the probable original orientation of the face; after Wolpoff (1980b). (5) *Homo habilis*. OH24 ('Twiggy'). (6) *Homo ergaster*. ER1813.

1987 *Australopithecus africanus mioddentatus* Ferguson, *Primates*, **28**, 264. Type AL266-1, from Hadar.

This taxon would include at least the following specimens from Hadar: AL288-1, 199-1, 200-1; 162-28 (Schmid, 1983), 128-1 (Stern & Susman, 1983); and 138-48, 322-1, and 129-1 (Senut and Tardieu, 1985). But there is no evidence opposing the provisional assignment to this taxon of all the Hadar fossils with the exception of the AL333 group and, perhaps, AL400–1; including AL266–1, which has recently been made the type of a new taxon by Ferguson (1987). Other specimens that could be analysed in the context of this taxon include Lothagam, Tabarin, Maka and Belohdelie. McHenry (1984) implies that the Kanapoi elbow may not be far removed from this group, but Senut and Tardieu (1985) find that it is much more modern.

### 7.5.2 Genus *Paranthropus* Broom, 1938 (Fig. 7.8 (A), (B))

1938 *Paranthropus* Broom, *Nature*, **142**, 377. Type *Paranthropus robustus* Broom.
1959 *Zinjanthropus* Leakey, *Nature*, **184**, 491. Type *Zinjanthropus boisei* Leakey.

A genus of Hominini with the following derived character states: P4 (in both jaws) expanded, generally broader than either P3 or M1; canines small;

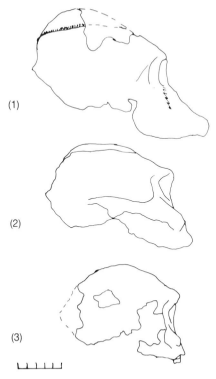

**Fig. 7.8(A).** Crania of species of *Paranthropus*. (1) *P.*sp.indet.: WT17000 ('the black skull'); (2) *P.boisei*: ER406; (3) *P.*cf. *boisei*: ER732.

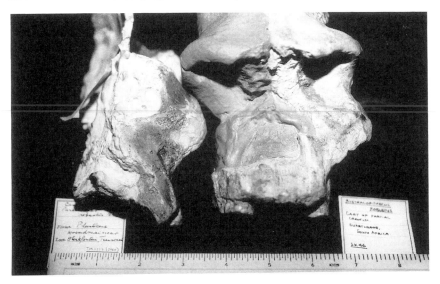

**Fig. 7.8(B).**   Skulls of [cast of] the Holotype of *Paranthropus robustus* (left) and of SK-46, a specimen of *P.crassidens* (right).


$DM_{1,2}$ protoconid not mesial to metaconid; $DM_1$ talonid enlarged; $DM_1$ and $M_1$ buccal groove deep, ending in a pit; $DM_1$ fovea anterior reduced to a fissure; $M^1$ buccal groove very deep: $I_1$ lingual marginal ridges reduced; $I_2$ incisive edge without distal slope; lingual relief of canines lost; canines symmetrical, not projecting beyond occlusal line, their tips rounded not pointed; lower molars oval, their cusps centrally placed; Carabelli complex and protostyle reduced or absent; $P^3$ rounded, its crown not tapering towards occlusal surface; $P_4$ buccal grooves lost; premolar root type B occurs; $M_1$ erupts nearly coevally with $I_1$; pre-maxilla anteriorly straight; curve of Spee enhanced; nasion at glabella; nasals broadened superiorly; vomer inserts on nasal spine; facial buttressing pronounced; petrous axis at 40–55° to bicarotid line; tympanic plate very long; occipito-mastoid crest lost; occipital marginal venous sinus developed; humerus with lateral epicondyle salient. Two of the above characters (relative eruption times of M1 and I1, and petrous axis inclination) converge with *Homo*.

The genus retains the primitive states of the characters whose derived conditions are listed under *Australopithecus* and *Homo*.

Skelton *et al.* (1986) argue that *Paranthropus* and *Homo* form a clade characterized by facial shortening. It has been argued above that the characters used to support this conclusion are too general, and that when broken down further they serve rather to demonstrate that facial shortening has occurred independently in the two genera. Dean (1985) notes that I1/M1 eruption timing and cranial base shortening are similarities between the two, and that these two characters being functionally independent constitute much better evidence for the sister-group status of the two genera. This line of argument is more cogent, but the large number of synapomorphies between *Homo* and *Australopithecus* tip the evidence against it.

The considerable body of supporters of the concept of the members of this genus as 'robust australopithecines', generally incorporating them into the genus *Australopithecus*, must be acknowledged. The arguments of Rak (1983), that facial buttressing distinguishes such a combined genus, have already been mentioned; the style of buttressing is however different in the three species recognized by him, and it would be difficult to point to any detailed derived similarities, beyond a generalized buttressing such as can be found also in primitive members of the genus *Homo*. Corrucini and McHenry (1981) recognize an inclusive *Australopithecus* clade distinguished by dental features; but these features are so susceptible to parallelism, as they readily admit, that it would be hard to sustain the concept against a weight of opposing evidence. Ashton *et al.* (1981) include the 'robusts' in *Australopithecus*, but their analysis is largely concerned with the 'graciles'; Oxnard (1984*b*) again regards the 'robusts' as australopithecines, but stresses that divisions within the australopithecines are deep and ancient.

## 7.5.2(a) *Paranthropus robustus* Broom, 1938

1938 *Paranthropus robustus* Broom, *Nature*, **142**, 377. Type from Kromdraai.

Characterized by relatively larger canines and smaller, especially narrower, M2 and M3 than other species, in which it is relatively plesiomorphic. It is also the most plesiomorphous species of the genus in various non-metrical features (see section 7.4 above). The facial skeleton, as far as can be seen, appears to resemble *P. crassidens* (Rak 1983). No specimens are known of this species except for those from Kromdraai.

## 7.5.2(b) *Paranthropus crassidens* Broom, 1949

1949 *Paranthropus crassidens* Broom, *Nature*, **163**, 57. Type from Swartkrans.
1952 *Paranthropus crassicorporal* Weinert, *Z. Morph. Anthrop.*, **43**, 325. Renaming of *crassidens*.

More megadont than *P. robustus*, and with (in the main) more derived dental states; the facial skeleton has been analysed by Rak (1983), and is characterized by several unusual, probably autapomorphous (contra Rak's assessment) features: form of inferior orbital margin, maxillary fossula, and other aspects of facial moulding.

Only specimens from Swartkrans have been referred to this species. It was recognized as a subspecies of *P. robustus* by Robinson (1954), but generally synonymized with the latter by subsequent commentators; the species was revived by Howell (1978), and has been substantiated by Grine (1985).

### 7.5.2(c) Paranthropus boisei (Leakey, 1959)

1959 *Zinjanthropus boisei* Leakey, *Nature*, **184**, 491. Type, Olduvai hominid 5.

An enormously megadaont species, especially in its broadened cheek teeth; front teeth are extremely reduced. It appears to share some derived states with *P. crassidens* and not with *P. robustus* (see section 7.4). The facial skeleton is inflated, with a suborbital 'visor', and entirely lacks the sculptured condition of *P. robustus*; Rak (1983) interprets this as a derived condition, but as this requires a considerable degree of evolutionary reversal a presently more plausible hypothesis would seem to be that they are both derived, in opposite directions.

I have indicated in section 7.4 what specimens I would include in this taxon. That both OH5 (the type) and ER430 belong to it indicates a range of facial shapes, both long-narrow and short-broad, as in for example the gorilla.

Known sites for the taxon include Olduvai, Koobi Fora, and Omo.

### 7.5.2(d) Paranthropus cf. boisei

As indicted in section 7.4, certain specimens from East Africa are most unlikely to belong in *P. boisei*, having cheek teeth that average longer (except for $M_3$) but distinctly narrower.

Specimens from Peninj, Chesowanja, Koobi Fora, and Omo would fit here. ER732 from Koobi Fora is considerably smaller than ER430, which can be allocated to *P. boisei*, and lacks a sagittal crest.

### 7.5.2(e) Paranthropus sp. indet.

Specimens from levels more than 2 million years old, in the Shungura Formation at Omo and West Turkána, have yet narrower cheek teeth. The cranium, briefly described by Walker *et al.* (1986), is strikingly prognathous (more so than *P. robustus*?), and so more primitive than the later species, and with less robustly developed cheek teeth, shallower mandibular fossa, shallower palate, and weakly flexed cranial base, all primitive traits for the genus. It lacked a maxillary fossula, but is said to have the *crassidens* form of the superior and inferior orbital margins. Milk teeth from the White Sands levels, even older, are compared by Howell (1969) to *P. boisei*.

Walker *et al.* (1986) suggest that this taxon is conspecific with *Paraustralopithecus aethiopicus*, but this is very clearly not the case. An earlier assessment by one of the same authors (Leakey 1974) is far more plausible [see subsection 7.5.4(b) below].

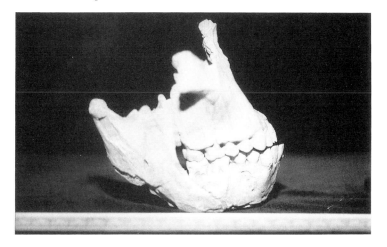

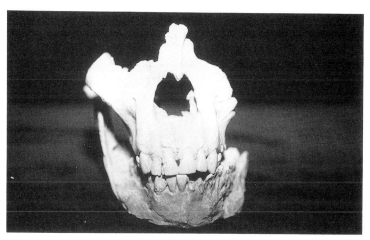

**Fig. 7.9(A).** *Australopithecus africanus*: [casts of] Sts-52a and -b, from (a) side view (above and (b) anterior view (below).

### 7.5.3 Genus *Australopithecus* Dart, 1925 (Fig. 7.7 (3), Fig. 7.9 (A), (B))

1925 *Australopithecus* Dart, *Nature*, **115**, 198. Type *Australopithecus africanus* Dart.

1938 *Plesianthropus* Broom, *Nature*, **142**, 377. Type *Australopithecus transvaalensis* Broom.

?1948 *Praanthropus* Hennig, *Naturw. Rundschau*, **1**, 311. Type, the Garusi hominid. No specific name provided.

?1955 *Praeanthropus* Senyurek, *Turk Tarih Kur. Bell.*, **73**, 55. Type *Meganthropus africanus* Weinert, 1950.

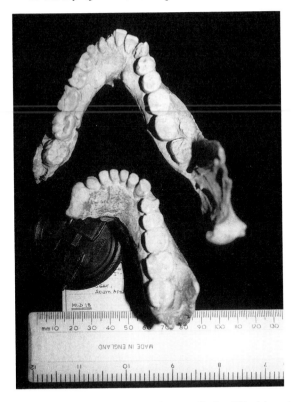

**Fig. 7.9(B).** *Australopithecus africanus*: [casts of] Sts-52b (above) and MLD-1 (below). The former is *V*-shaped due to distortion; the rounded, *Homo*-like shape of the lower mandible is more natural.

A genus of Hominini with the following states synapomorphous with *Homo*: mesial buccal groove more conspicuous than distal on $P_3$, equally developed on $P_4$; hypoconulids reduced on lower molars; nasals raised from plane of face; upper end of nasal aperture level with inferior orbital margins; a sill separating naso-alveolar clivus from nasal floor; transverse buttress reduced; face height less than 75 per cent of upper face breadth; foramen ovale posterior to margin of pterygoid plate; digastric scar in notch on basicranium; biporionic breadth greater than biparietal; frontal angle greater than 35°; femoral distal epiphysis index greater than 75; femoral inter-condylar notch wider than high; femoral condyles symmetrical; tibial spine index greater than 50; humero-femoral index less than 75; intertrochanteric line present; femoral shaft diameter less than 60 per cent of head diameter. And with the following autapomorphic states: $I^1$ crown area more than 169 per cent that of $I^2$; transverse facial buttress absent; zygomatic prominence bulbous (covergent with *Paranthropus*); occipital index greater than 70; femoral platymeria reduced; pelvic torsion apparently increased.

This genus shares so many synapomorphies with *Homo* that it becomes almost a matter of taste whether to keep it separte or to combine the two. The latter course has been followed by Robinson (1972), Groves and Mazak (1975), and Olson (1981*b*, 1985*a,b*) among others. Considering, however, that disputes will go on, and are bound to persist for some time; that the name *Australopithecus* is the one nomen in palaeoanthropology that is widely known to the general public; and, more scientifically, that there are clear synapomorphies defining *Homo* s.s., it seems more reasonable on balance to keep the two apart.

### 7.5.3(a) Australopithecus africanus Dart, 1925 (see Tables 7.7 and 7.8)

1925 *Australopithecus africanus* Dart, *Nature*, **115**, 198. Type, the Taung skull.
1936 *Australopithecus transvaalensis* Broom, *Nature*, **138**, 486. Type from Sterkfontein.
1948 *Australopithecus prometheus* Dart, *S.Afr. J. Sci.*, **1**, 201; *Amer. J. Phys. Anthrop.* **6**, 259. (Priority of these two references is uncertain.) Type from Makapansgat.

Diagnosis as for genus.

Robinson (1954 distinguished two subspecies in this species: *A. a. africanus* from Taung, and *A. a. transvaalensis* (synonym *prometheus*) from Sterkfontein and Makapansgat. The (single available) specimen from Taung does fall outside the range of variation from the other sites in several features: it has a tuberculum sextum and a post-metaconulid on $M_1$, the mesiobuccal groove on $M_1$ ends in a pit, the hypocone on $DM^2$ is reduced, and the anterior fovea on $M^1$ is well-developed. It might therefore be preferable to keep them apart, as Robinson did; but until another geologically late specimen is found the differences are difficult to interpret. Some of the characters of Taung seem to converge on *Paranthropus*, although there is no question but that the Taung skull is fundamentally *Australopithecus*; the large anterior fovea is opposite to the *Paranthropus* condition.

### 7.5.3(b) Australopithecus afarensis Johanson, 1978

1950 *Meganthropus africanus* Weinert, *Z. Morph. Anthrop.*, **42**, 139. Type from Garusi ( = Laetoli).
1978 *Australopithecus afarensis* Johanson, in Hinrichsen, *New Sci.*, **78**, 571. Type, Laetoli hominid 4 (by subsequent designation: Johanson *et al.* 1978). Day *et al.* (1980) show that the correct reference to the name is as above.

A poorly known species, referred provisionally to *Australopithecus*, and distinguished from *A. africanus* by its larger canines on average, $P_4$ broader than $P_3$, the common presence of a diastema, and greater platymeria; also the premolar roots may be of type D, while the palate is narrower and the prognathism is greater (but not as much as in Hadar, AL200–1) (Puech *et al.* 1986).

The Laetoli hominid, despite having been made the type population of *A. afarensis*, is still somewhat mysterious, being known from comparatively few anatomical parts. If it is part of the *Homo* clade, as here proposed, then it is 'below' the AL333 sample from Hadar in its P4 shape (upper and lower), vault bone thickness, and flatness of femoral neck, at least; in all these characters it resembles South African *A. africanus*. Compared to the latter, the Laetoli taxon differs as mentioned above; the differences are in the main primitive (such that it may turn out to be the sister group of *Australopithecus* plus *Homo*, rather than a member of *Australopithecus*), though the premolar relationships are a convergence with *Paranthropus*.

### 7.5.4 Genus *Homo* Linnaeus, 1758 (Tables 7.7–7.10)

1758 *Homo* Linnaeus, *Syst. Nat. ed. 10*, **1**, 20. Type, *Homo sapiens* Linnaeus.

1894 *Pithecanthropus* Dubois, 'Pithecanthropus erectus eines menschenähnliche Uebergangsform aus Java'. Type, *Anthropopithecus erectus* Dubois.

1909 *Palaeanthropus* Bonarelli, *Riv. ital. Paleont.*, **15**, 26. Type *Homo heidelbergensis* Schoetensack.

1909 *Pseudhomo* Amenghino, *An. Mus. Nac. Buenos Aires*, **19**, 109. Type, *Homo heidelbergensis* Schoetensack.

1911 *Notanthropus* Sergi, L'Uomo, pp. ? (not seen). Type, *Notoanthropus eurafricanus* Sergi ( = *Homo sapiens*, fossil).

1915 *Archanthropus* Arldt, *Fortschr. Rassenk.*, **1**, 1. Type, *Homo neanderthalensis* King.

1926 *Anthropus* Boyd-Dawkins, *Rep. Brit. Ass.*, **94**, 386. Type, *Homo neanderthalensis* King.

1927 *Sinanthropus* Black, *Palaeont. Sinica, D,* **7**, 21. Type *Sinanthropus pekinensis* Black, 1927.

1928 *Cyphanthropus* Pycraft, Rhodesian Man and Associated Remains, **1**. Type, *Homo rhodesiensis* Woodward, 1921.

1932 *Praehomo* von Eickstedt, *Z. artzl. Fortbild.*, **29**, 608. No type designated: *Praehomo asiaticus* ( = *Homo erectus*) and *P. europaeus* ( = *H. heidelbergensis*), both of von Eickstedt, listed.

1932 *Javanthropus* Oppenoorth, *Wet. Meded. Dienst Mijnb. Ned.-Ind.*, **20**, 49. Type, *Homo (Javanthropus) soloensis* Oppenoorth.

1933 *Metanthropus* Sollas, *J.R. Anthrop. Inst.*, **63**, 389. No type designated (cf. *Homo neanderthalensis* King).

1935 *Africanthropus* Dreyer, *Proc. Acad. Sci. Amsterdam*, **38**, 119. Type *Homo (Africanthropus) helmei* Dreyer, 1935.

**Table 7.7**  Breadth:length indices of mandibular cheek teeth in *Australopithecus* and *Homo*

| Species | $P_3$ | | | $P_4$ | | | $M_1$ | | | $M_2$ | | | $M_3$ | | |
|---|---|---|---|---|---|---|---|---|---|---|---|---|---|---|---|
| Hadar small sp. | 116.3 | 8.79 | 9 | 115.6 | 8.10 | 9 | 95.9 | 4.92 | 6 | 98.2 | 5.30 | 8 | 89.0 | 3.84 | 4 |
| Laetoli | 100.7 | 13.83 | 9 | 106.1 | 2.35 | 3 | 97.8 | 1.80 | 4 | 90.1 | 91.7 | 2 | 99.9 | 18.15 | 3 |
| Hadar large sp. | 113.4 | 13.50 | 6 | 114.4 | 14.34 | 4 | 97.7 | 3.09 | 4 | 95.0 | 6.67 | 6 | 95.3 | 6.54 | 4 |
| *A.africanus* | | | | | | | | | | | | | | | |
| Makapansgat | 121.4 | 12.03 | 5 | 123.5 | 13.25 | 4 | 100.5 | 8.56 | 5 | 93.9 | 2.55 | 5 | 93.8 | 2.93 | 5 |
| Sterkfontein | 130.3 | 13.07 | 7 | 128.7 | 14.80 | 4 | 95.6 | 13.35 | 7 | 95.7 | 9.23 | 6 | 91.4 | 6.63 | 8 |
| *H.aethiopicus* | 98.8 | 116.3 | 2 | 118.9 | 13.06 | 3 | 100.8 | 7.33 | 4 | 88.6 | 6.93 | 3 | 93.8 | | 1 |
| *H.habilis* | 107.5 | 7.12 | 6 | 102.9 | 6.72 | 5 | 88.7 | 3.08 | 3 | 90.3 | 5.08 | 3 | 83.8 | 8.00 | 4 |
| *H.ergaster* | 116.8 | 121.7 | 2 | 129.1 | 132.1 | 2 | 89.9 | 90.8 | 2 | 92.8 | 94.6 | 2 | 84.7 | 96.1 | 2 |
| *H.pre-erectus/sapiens* | | | | | | | | | | | | | | | |
| Koobi Fora | 112.2 | | 1 | 105.6 | | 1 | 91.0 | 106.2 | 2 | 92.7 | 101.7 | 2 | 89.2 | | 1 |
| Swartkrans | 119.8 | | 1 | — | — | | 97.5 | 100.0 | 2 | 92.5 | 95.7 | 2 | 86.7 | | 1 |
| *H.erectus* | | | | | | | | | | | | | | | |
| Olduvai | 94.0 | | 1 | 107.8 | 118.8 | 2 | 90.8 | 92.0 | 2 | 91.3 | | 1 | — | | — |
| Java | 116.2 | 123.0 | 2 | 121.1 | 123.5 | 2 | 100.0 | 101.6 | 2 | 97.9 | 4.43 | 3 | 93.7 | 7.18 | 3 |
| Lantian | — | | — | 133.3 | | 1 | 91.3 | | 1 | 104.0 | | 1 | — | | — |
| Zhoukoudian | 114.3 | 8.52 | 12 | 113.9 | 16.04 | 9 | 94.2 | 3.59 | 15 | 96.8 | 5.22 | 11 | 94.3 | 8.63 | 9 |
| *H.sapiens* | | | | | | | | | | | | | | | |
| *heidelbergensis* | 106.3 | 13.59 | 8 | 118.8 | 9.25 | 6 | 93.8 | 3.53 | 8 | 93.4 | 7.45 | 8 | 95.4 | 8.12 | 10 |

1943 *Maueranthropus* Montandon, L'homme préhistorique et les prehumains. Type *Homo heidelbergensis* Schoetensack.

1944 *Meganthropus* Weidenreich, *Science*, **99**, 480. Type, *Meganthropus palaeojavanicus* Weidenreich.

1948 *Nipponanthropus* Hasebe, *J. Anthrop. Soc. Nippon*, **60**, 32. Type, *Nipponanthropus akasiensis* Hasebe (probably = *Homo sapiens*).

1949 *Telanthropus* Broom and Robinson, *Nature*, **164**, 322. Type, *Telanthropus capensis* Broom and Robinson.

1950 *Europanthropus* Wust, *Z. Morph. Anthrop.*, **42**, 1. Type, *Homo heidelbergensis* Schoetensack.

1950 *Europanthropus* Weinert, *Z. Morph. Anthrop.*, **42**, 8. Type, *Homo heidelbergensis* Schoetensack.

1955 *Atlanthropus* Arambourg, *C.R. Acad. Sci. Paris*, **239**, 893. Type, *Atlanthropus mauritanicus* Arambourg.

1965 *Tchadanthropus* Coppens, *C.R. Acad. Sci. Paris*, **260**, 2869. Type, *Tchandanthropus uxoris* Coppens.

?1968 *Paraustralopithecus* Arambourg and Coppens, *S. Afr. J. Sci.*, **1968**, 59. Type, *Paraustralopithecus aethiopicus* Arambourg and Coppens. The introduction of this name a year earlier (Arambourg and Coppens 1967) is an unavailable reference as the name is given provisionally.

A genus of Hominini sharing with *Australopithecus* numerous synapomorphous states, as listed above, under that genus; and with the following autapomorphic states: $P_4$ shapes index below 96; $P^4$ shape index above 75; vomer inserts well-behind inferior nasal spine; mandibular fossa deepened, with well-developed articular eminence; cranial vault bones thickened; foramen magnum less back-sloping (convergent with *Paranthropus*); femoral neck index above 75. In addition, one species (the unnamed Hadar species) is relatively poorly known; the following features are synapomorphic for the remainder of the genus, and may or may not turn out to be applicable to the Hadar species, and so synapomorphic for the genus as a whole: $M_1$ erupts at about the same time as $I_1$, alveolar projection reduced, petrous axis 40–55° (these states converge upon *Paranthropus*); maxillo-alveolar index above 100, palate broader than long; frontal process of maxilla faces laterally; post-orbital constriction reduced; bizygomatic breadth less than 130 per cent of biorbital; face height less than biorbital breadth.

### 7.5.4(a) Homo sp. (unnamed): Hadar

As far as can be presently deduced, this species includes only the AL333 series (including AL333w) from Hadar; possibly also AL400. Included in *Australopithecus afarensis* by Johanson *et al.* (1978), but shown by Coppens (1983*b*), Senut and Tardieu (1985), and others to have *Homo*-like features.

If this assessment is correct, this species is the sister group of others in the genus, lacking the following character states which are seen in all the other species except *H. aethiopicus*: loss of protoconid cingulum on lower molars, posterior cranial elongation (both features convergent on *Paranthropus*); loss of premolar root type A; M1 > M3; deep, narrow mandibular fossa; nasals raised from facial plane, broad, but sending a wedge into frontals, and narrowing inferiorly—both traits convergent with *Paranthropus*; forward-sloping foramen magnum; enlargement of cranial capacity; two menisci on proximal tibia.

The specimens assigned to this species are large in size and robust in build, and converge in a number of features on *Paranthropus*, whence Olson (1981*b*), 1985*b*) has even proposed to include them in that genus: buccal groove on $DM_1$ and $M_1$ ends in a pit; $I_2$ edge does not slope distally; no nasal sill; frontal squama flat; venous drainage commonly (not invariably) by occipital marginal sinus. But the AL333 series are not *Paranthropus*; not only do they share the *Homo/Australopithecus* (and the peculiarly *Homo*) synapomorphies listed above, but the incisors ($I^1$ buccolingual, $I^2$ mesiodistal, $I_2$ buccolingual) and canines ($C^1$ and $C_1$ buccolingual) are the largest in the Hadar/Laetoli sample, a very un-*Paranthropus*-like trait; the idea that the premolars are especially broadened has been criticized by Kimbel *et al.* (1985). The post-canine dental metrics do not in general stand out among the Hadar/Laetoli sample, except that the single $M^3$ in the 333 series is by far the largest from the two sites.

The cranial capacity of one specimen, AL333–45, is 500 cc (Holloway 1983), which would be large for *Australopithecus* (though not outside the known range), but small for more advanced species of *Homo*; Falk (1985), however, recommends caution as this determination was based on an endocast whose frontal portion was wholly missing, and notes that AL333–105 has a capacity estimated at 320 cc which, even taking its infant status into account, is even smaller than that of *A. africanus*.

A note on the Hadar and Laetoli sample: the thesis that all the specimens from these two sites are assignable to a single species, *Australopithecus afarensis*, has been ably defended by White and *et al.* (1981); Kimbel *et al.* (1985). All specimens from the two sites have large front teeth with robust, curved roots; the naso-alveolar clivus is curved, extending well into the nasal cavity; $C^1$–$P^3$ juga are very prominent; canine fossae are present, deep (as in *Homo*); the palate is shallow and narrow; diastemata often occur; the tooth-rows converge posteriorly; the posterior temporalis fibres are enlarged, the superior temporal lines not diverging until lambda is reached; articular eminences are virtually absent; mental foramina are low-placed, below mid-corpus height; the superior transverse torus is usually weakly expressed; the inferior transverse torus is at the base of the symphysis; $DM_1$ is narrowed, tapering, asymmetrical; $P_3$ is a long, narrow oval shape and often unicuspid.

Tobias (1980) disputed that the Hadar and Laetoli fossils are anything but subspecies of *Australopithecus africanus*. Restricting himself to the characters cited in the original diagnosis (Johnson *et al.* 1978), he noted the following: alveolar prognathism is as great in some Sterkfontein specimens as in Hadar ones (though, in his Fig. 3 (p. 6), the convexity over the incisor roots is still much greater in AL200–1a than in Stw 73, and the incisor roots are much more advanced over the canine roots); dental arcade shapes like those from Hadar and Laetoli can be seen in Sterkfontein and Makapansgat too; MLD–1 has inferior temporal and superior nuchal lines only 1 mm apart, not unlike some Hadar examples; mandibular fossae are shallow in all australopithecines; the disposition of the symphyseal transverse tori is very like that in *A africanus* in general; and some *A. africanus*, such as MLD–2, have exactly the same $P_3$ form as cited for Hadar and Laetoli. Tobias's solution was to include both samples in *A. africanus*, but, given that the dental metrics of the two samples seemed to differ, he referred them to different subspecies, restricting the name *A. africanus afarensis* to the topotypical sample from Laetoli and conining the name *A. a. aethiopicus* for Hadar. [The latter taxon was erected (a) without nominating a type specimen, which was unfortunate but does not affect its availability; but (b) only on a provisional basis, which does affect its availability.]

Olson (1981*b*) studied the basicrania of fossil hominins in the light of those of apes and modern humans, distinguishing clearly between a *Homo* type (including *A. africanus*), in which the digastric originates in a notch medial to the mastoid, and a *Paranthropus* type in which its origin occupies the medial slope of the mastoid itself. His study of Hadar AL333–45 convinced him that this specimen showed the *Paranthropus* morphology. Turning to the dentitions, he divided the Hadar sample into two groups, assigning the Laetoli sample to the Hadar group which also included the AL333 series. Thus Laetoli specimens and the names awarded to them (*Praeanthropus africanus* and *Australopithecus afarensis*) were allocated like AL333 to *Paranthropus*, albeit to an early, primitive species, for which the prior name would be *P. africanus* (Weinert); the characters of the remaining Hadar specimens would be those of *Homo*. In his 1981*b* paper Olson called this species simply *Homo* sp., but in a 1985*b* sequel he referred to it as *Homo aethiopicus* Tobias, 1980 (nominating AL288–1, 'Lucy', as the type). Because Tobias's name is unavailable (see above), Olson's usage is the prior available one.

Kimbel *et al.* (1985) took great exception to Olson's analysis, showing that the AL333–45 basicranium is distorted and, if anything, is *Homo*-like. (In retrospect, it seems likely that Olson (1981*b*, 1985*b*)—and Ferguson (1981, 1984), see below—was misled by its heavy pneumatization.) They showed, in addition, that the dental characters do not divide in the manner proposed by Olson.

For Ferguson (1983, 1984) the Hadar sample contains three different taxa: *Sivapithecus* sp. *Australopithecus africanus*, and *Homo antiquus* (new species). That each of the three taxa includes specimens from the 333 series tends to bring this scheme into question, but it does draw attention to the variability of the sample. In particular, de Bonis *et al.* (1981) had also stressed the primitive nature of the sample in their comparisons with *Ouranopithecus macedoniensis*, but the two samples are similar, not identical, and even AL200–1, Ferguson's prime candidate, could not really be mistaken for a *Sivapithecus* (it has, for example, distinctly smaller canines, more marginal placed molar cusps, and a more lingually situated $M^3$); moreover, AL333–45, which is said by Ferguson to have an ape-like nuchal area, is the one debated by Olson (1981*b*, 1985*b*) and Kimbel *et al.* (1985). Similarly, the difference between Ferguson's *A. africanus* and *H. antiquus* series is simply one of size, and of size-correlated characters (degree of tooth-row convergence, bucco-lingual narrowing, $P_3$ unicuspidy).

An interesting feature of Ferguson's 1984 paper is his critique of the reconstruction of the *A. afarensis* skull (based, in fact, entirely on 333 specimens) as first published in Johanson and Edey (1981). He points out that in the reconstruction the occlusal plane anteriorly converges with the Frankfurt plane, whereas in all other hominoids, extant and fossil, it is anteriorly divergent or, at most, parallel. The reconstruction is therefore incorrect, a point made independently by Schmid (1983); correcting this error makes the resulting reconstruction a great deal more like *A. africanus*.

Coppens (1983*a,b*), basing himself in part on the work of Senut and Tardieu (1985), distinguishes two quite different taxa within the Hadar/Laetoli sample: one very archaic, the other 'astonishingly modern', the archaic form being far the more abundant. Neither conforms to Tobias's (1967*a*) diagnosis of the genus *Australopithecus*.

Certainly the post-cranial data are absolutely clear, and split the Hadar sample into two divisions, which differ in the form of the elbow joint (Senut 1981), the knee (Tardieu 1981), and the ankle (Jungers and Stern 1983), at the very least. The differences in the distal humerus are not of the sort which one can reasonably ascribe to sexual dimorphism (Senut 1986).

I have indicated above that, when interpreted cladistically, the 'small' Hadar type, as typified by 'Lucy', comes out as the plesiomorphic sister group of all other Hominini, wheras the 'large' Hadar type, comprising the 333 series, is part of *Homo*, albeit the sister group of other species. [The *Paranthropus*-like features, such as the form of the nasals (Olson 1985*b*), must in this light be regarded as convergent—perhaps connected with the robusticity of the 333 individuals.] The position of the Laetoli sample is less-satisfactory; while it must be borne in mind that this, too, may turn out to be heterogeneous, at present there is little point in trying to distinguish different morphological forms among the relatively poor material. Conveniently it may be assigned to the genus *Australopithecus*, which it does seem to resemble most, as the one and only *A. afarensis*. The composite *A. afarensis*, as defined by Johanson *et al.* (1978) and discussed, for example, by Rak (1983), is distinguished entirely by possession of symplesiomorphic features, which of course is not evidence for the monophyly of any putative taxon; when apomorphic character states are sought, they are different in different individuals, and this is the key to their relationships.

*7.5.4(b) Homo aethiopicus (Arambourg and Coppens 1968) (Figs. 7.10, 7.11; Table 7.8)*

1968 *Paraustralopithecus aethiopicus* Arambourg and Coppens, *S. Afr. J. Sci.*, Omo, Shungura Member C.

Diagnosis: a poorly known species (Figs. 7.10 & 7.11), probably belonging to *Homo*, possessing the derived traits of posterior cranial elongation (if the Omo cranial specimen is correctly attributed), loss of premolar root type A, and loss of protoconid cingulum, but lacking the derived conditions of $M_3$ reduction and symphyseal reduction.

It was remarked above (section 7.3). that there are two somewhat different mandible types in the lower levels at Koobi Fora, typified by the well-preserved specimens ER1482 (V-shaped) and ER1802 (U-shaped). Apart from this basic difference, and an associated difference in symphyseal buttressing, there seem to be few differences between the two types, either metric or non-metric. The type of *Paraustralopithecus aethiopicus* resembles ER1482, and if this association is accepted the name *Homo aethiopicus* will be available for the species.

This species is known from Koobi Fora, below the KBS Tuff (i.e. earlier than 1.89 million years ago), and from contemporary or even older levels at Omo. Apart from the Omo mandibles, the Member E juvenile braincase, L-338y-6, must be considered in this context; its small cranial capacity (though just how 'small' this really is will depend on an assessment of its personal age) and platycephaly recall *H. ergaster* (below).

Walker *et al.* (1986) consider Arambourg and Coppens's (1968) name to apply to a primitive *Paranthropus*. The dental measurements alone would preclude this identification.

*7.5.4(c) Homo rudolfensis (Alexeev 1986): Koobi Fora and Omo (Lower)* (Fig. 7.12)

1986 *Pithecanthropus rudolfensis* Alexeev, *The Origin of the Human Race*, 93 Koobi Fora, below KBS tuff. Diagnosis: a species of the genus *Homo* lacking ridged or arched nasals; midface broader than upper face; endinion at same level as ectinion; Carabelli complex on upper molars well-developed. (These are all primitive traits.) With *Homo habilis* it shares the derived conditions of narrow $P_3$ (mesiodistal length over 90 per cent, usually over 100 per cent, of buccolingual), and reduction of fovea anterior on $P_3$. It is uniquely derived in that the superior surface of the posterior zygomatic root is flat; the mandibular symphysis is rounded and lacks internal buttressing; and the crown area of the lower canine is more than 90 per cent (usually more than 100 per cent) of that of $P_4$. Cranial capacity of ER1470 is 770 cc; that of ER1590 is even larger.

ER1470 and 1590 have been, up to now, almost universally referred to *Homo habilis*, but it is worth emphasizing that there has never been any attempt to justify this assignment. There is no doubt that, on a purely

**Table 7.8** Mandibular dental measurements in *Homo* (early phase samples).

| | Australopithecus africanus | | Homo | Homo habilis | | Homo |
| | Makapansgat | Sterkfontein | aethiopicus | Olduvai | Sterkfontein | ergaster |
|---|---|---|---|---|---|---|
| $I_1$m–d | 5.0 | 5.9–6.3 | — | 6.5–7.2 | | — |
| | 1 | 2 | — | 2 | | — |
| b–1 | 6.0 | 6.1–8.1 | — | — | | — |
| | 1 | 2 | — | — | | — |
| $I_2$m–d | 5.5 | 6.63 | 5.8 | 6.60 | | 7.1 |
| | — | 0.99 | — | 1.40 | | — |
| | 1 | 3 | 1 | 3 | | 1 |
| b–1 | 8.0 | 7.80 | 8.4 | 7.33 | | 7.0 |
| | — | 0.89 | — | 0.31 | | — |
| | 1 | 3 | 1 | 3 | | 1 |
| $C_1$m–d | 7.7–9.0 | 9.35 | 7.7–8.5 | 8.60 | 8.4 | 9.0 |
| | — | 0.66 | — | 1.02 | — | — |
| | 2 | 6 | 2 | 4 | 1 | 1 |
| b–1 | 8.9–10.7 | 10.27 | 11.4 | 9.45 | 9.0 | 9.2 |
| | — | 1.05 | — | 1.04 | — | — |
| | 2 | 7 | 1 | 4 | 1 | 1 |
| $P_3$m–d | 10.13 | 9.50 | 8.70 | 9.15 | | 9.3 |
| | 1.31 | 0.72 | 1.23 | 1.17 | | — |
| | 3 | 6 | 5 | 6 | | 1 |
| b–1 | 12.33 | 12.18 | 11.03 | 9.93 | | 11.5 |
| | 0.83 | 1.24 | 1.72 | 0.87 | | — |
| | 3 | 6 | 4 | 6 | | 1 |
| $P_4$m–d | 9.70 | 10.00 | 9.37 | 9.19 | | 8.6 |
| | 1.09 | 0.65 | 1.16 | 1.17 | | — |
| | 4 | 5 | 6 | 9 | | 1 |
| b–1 | 11.88 | 12.28 | 11.07 | 10.61 | | 11.4 |
| | 0.32 | 0.79 | 1.73 | 0.83 | | — |
| | 4 | 4 | 6 | 9 | | 1 |
| $M_1$m–d | 12.98 | 14.00 | 13.17 | 13.30 | | 12.3 |
| | 1.44 | 1.49 | 1.26 | 0.72 | | — |
| | 4 | 8 | 6 | 7 | | 1 |
| b–1 | 13.15 | 13.36 | 12.70 | 11.87 | | 10.9 |
| | 0.68 | 1.25 | 1.29 | 0.86 | | — |
| | 4 | 8 | 5 | 7 | | — |
| $M_2$m–d | 15.40 | 15.10 | 14.46 | 13.93 | | 12.8–13.8 |
| | 0.85 | 0.91 | 1.76 | 1.46 | | — |
| | 5 | 6 | 7 | 6 | | 2 |
| b–1 | 14.56 | 14.57 | 13.80 | 12.92 | | 12.3–12.8 |
| | 0.59 | 1.38 | 1.44 | 1.19 | | — |
| | 5 | 6 | 5 | 6 | | 2 |
| $M_3$m–d | 15.56 | 15.55 | 14.5–14.5 | 15.17 | | 13.1–14.4 |
| | 1.33 | 0.91 | — | 0.60 | | — |
| | 5 | 8 | 2 | 3 | | 2 |
| b–1 | 14.44 | 14.44 | 13.75 | 13.30 | | 12.2–12.5 |
| | 5 | 8 | 4 | 3 | | 2 |

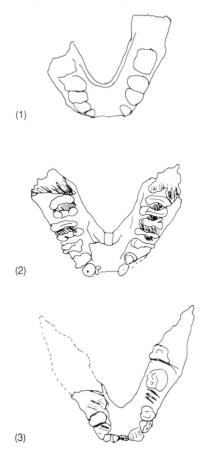

**Fig. 7.10.**   What is *Paraustralopithecus aethiopicus?* (1) WT16005, assigned to this taxon by Walker *et al.* (1986). (2) Holotype of *P.aethiopicus*, from Omo-Shungura Mbr.C. (3) ER1482, considered to belong with the *P.aethiopicus* mandible by Leakey (1974) and by Wood (1978). This alternative is here considered more likely: note for example the shape of the lingual contour and the (inferred) canine size.

phenetic level, they are rather similar, and if the characters of $P_3$ which they share are genuinely synapomorphic then there would indeed be no reason not to associate them in the one species (for which the name *habilis* would have priority). But *H. habilis* (as here restricted) shares a number of derived traits (of teeth, face shape, nasal shape, and occipital morphology) with the later species of *Homo*, traits which 1470 and 1590 lack; so that the $P_3$ features are more plausibly interpreted as either parallelisms, or symplesiomorphic traits (in which case a reversal would be involved in the evolution of the later species of *Homo*), or a legacy of a polymorphic common ancestor.

*7.5.4(d)  Homo habilis Leakey et al. 1964 (Fig. 7.13; Table 7.8)*

1964 *Homo habilis* Leakey *et al.*, *Nature*, **202**, 7. Olduvai Bed I.

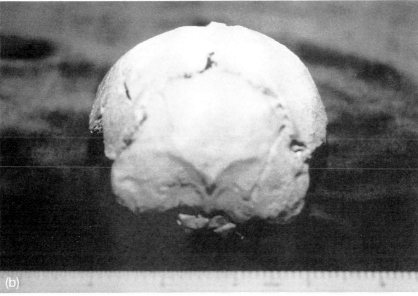

**Fig. 7.11.** [cast of] Omo (Shungura Formation) L-338y-6, from Member E: (a) side view, (b) posterior view. It is here interpreted as *Homo aethiopicus*, at least provisionally; but there is no reason why it could not in fact represent *Australopithecus* sp. (as indeed might the whole species *aethiopicus*).

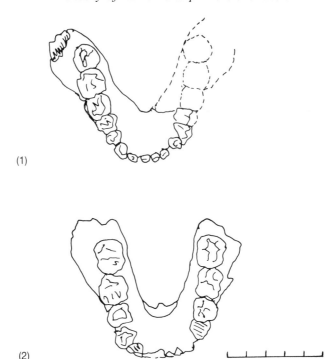

**Fig. 7.12(A).** Mandibles of *Australopithecus* and primitive *Homo*. (1) *Australopithecus africanus*. MLD-18 (completed by symmetrical imaging). (2) *Homo rudolfensis* ER1802.

Diagnosis: a species of the genus *Homo* sharing, with all other species except *H.* sp. unnamed (Hadar) and *H. aethiopicus*, the following derived character states: nasal bones ridged or arched, upper face broader than midface, endinion lower than ectinion, Carabelli complex (of upper molars) reduced; sharing with *H. aethiopicus* a narrow $P_3$ with reduced anterior fovea; and uniquely derived in the following character states: subsidiary cusps of $P_3$ virtually absent; supramastoid crest absent; basion well in front of bitympanic line; $M^1$ mesiobuccal groove ends in a pit; lingual relief of lower canine reduced; upper canine crown area less than 70 per cent that of $P^3$; $P_4$ mesiodistal length more than 90 per cent (usually more than 95 per cent) its buccolingual breadth.

The cranial capacity of the type, OH7, was calculated at 657 cc by Tobias (1964, 1971), giving an extrapolated adult value of 687; Holloway (1980) approached the calculation afresh, and arrived at a probable range of 700 to 750 cc. For OH13 and OH16, Tobias gives calculated values of 650 in each case. Leakey *et al.* (1971) give a figure of 590 cc for OH24. All these figures are admittedly somewhat dubious, and have to be taken as giving an approximate level of encephalization only.

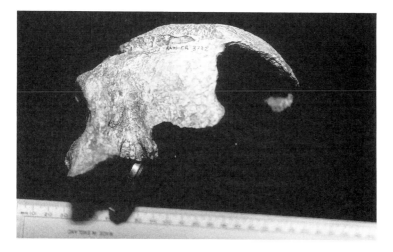

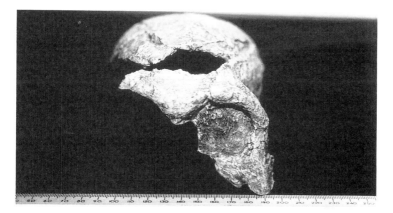

**Fig. 7.12(B).** *Homo rudolfensis* from the Lower Member, Koobi Fora: [cast of] ER3732, (a) side view (above), (b) from above (centre), (c) front view (below).

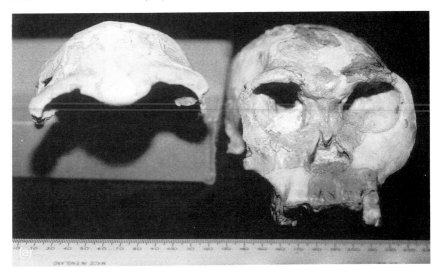

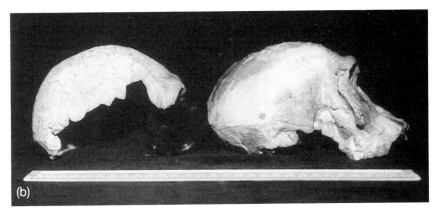

**Fig. 7.13.** *Homo habilis*: [casts of] OH16 (left), OH24 (right), from (a) side view, (b) front view.

Stringer (1986) has lately queried whether the Olduvai sample does form a single species. In the present analysis, evidence of heterogeneity was sought but not found: all the specimens here assigned show, where comparable parts are available, the same derived traits.

The cheek teeth average narrower than those of *H. aethiopicus*, and in the main are smaller in size, though $M_3$ is larger; the front teeth are of equivalent size. The teeth are, however, somewhat larger than those of the *Homo* species from Hadar.

Clarke (1985a) has argued that specimens from Sterkfontein Mbr 5 are assignable to *H. habilis*, a conclusion which is accepted here [see subsection 7.3(a) above].

*7.5.4(e) Homo ergaster Groves and Mazak, 1975 (Figs. 7.14–7.16; Table 7.8)*

1975 *Homo ergaster* Groves and Mazak, *Casopis Min. Geol.*, **20**, 243. Type KNM-ER992, from Upper Member, Koobi Fora Formation, Kenya.

Diagnosis: a species of the genus *Homo*, sharing with all others except *H. rudolfensis* (*H. aethiopicus* being largely unknown) the derived features of arched or ridged nasals, broad upper face, endinion lower than ectinion, and reduction of Carabelli complex; and with all except this (*H. aethiopicus* again largely unknown) and *H. habilis* the derived features of protruded midface such that malar region is posterior to pyriform aperture, absence of anterior pillars, presence of occipital torus, and potential occurrence of super-numerary bones at lambda. Presumed uniquely derived traits are the flattening of the cranial vault and the premolar broadening, such that P³

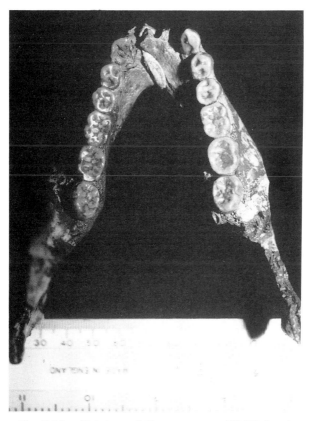

**Fig. 7.14.**   Holotype of *Homo ergaster*, ER992 [cast].

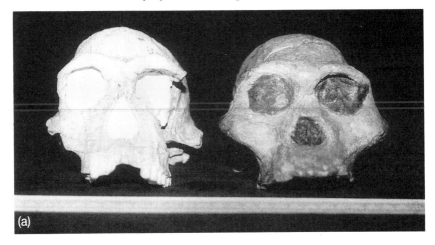

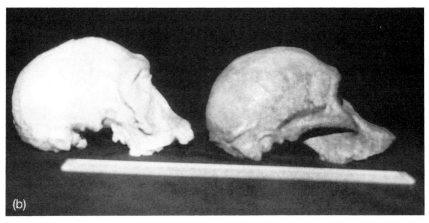

**Fig. 7.15.**   [cast of] ER1813 (left) and Sts-5 (right), from (a) front view, (b) side view. ER1813 is certainly not *A.africanus*.

mesiodistal diameter is less than 72 per cent of its buccolingual diameter. The cranial capacity of ER1813 is 510 cc; of ER1805, 582 cc.

There seems to be some uncertainty about the stratigraphy of 1813, but it originates from the Upper Member or from the level of the KBS Tuff itself, and is thus a maximum of 1.89 million years old. [This now appears to be incorrect: it is from below the KBS Tuff (I. McDougall, pers. comm.)] 1805 is younger, originating (Dean and Wood 1982) from just below the Okote Tuff, i.e. about 1.6 million years old.

The cranial base of 1813 resembles that of specimens ascribed to *H. erectus* (Dean and Wood 1982), while that of 1805 is longer. Neither, however, resembles *A. africanus*, to which 1813 had previously been attributed (Walker and Leakey 1978): the basicranium is shortened, the sphenoid

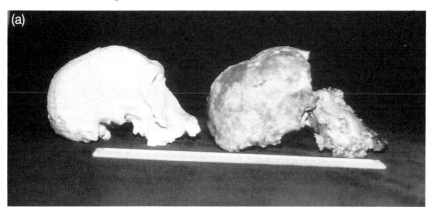

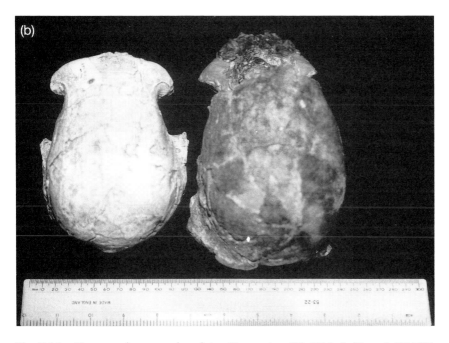

**Fig. 7.16.** Two specimens assigned to *H.ergaster*: ER-1813 (left) and ER1803 (right). (a) side view, (b) from above.

widened, the tympanic and petrous parts angled. The 1805 facial skeleton is considerably distorted; its reconstruction by Walker (1981*b*) gives it a much closer resemblance to 1813, albeit it is noticeably larger.

Walker (1981) also noted a resemblance of 1813 to an Olduvai specimen, OH13, and Stringer (1986) combined them in his small-brained 'cf. *habilis*' category. I believe the similarity to be gradistic only. Among the features whereby they differ are: the much larger cranial capacity of OH13 (650 cf. 510 cc.), and its higher, more rounded vault; the persistence of a supramastoid crest (admittedly small) in 1813, its loss in OH13; and the presence in the former of lambdoid ossicles, its much shallower mandible fossa (depth: breadth index only 18, cf. more than 30), its higher occipital index (78 vs. 71), and its broader (maxillary) premolars and much larger canine.

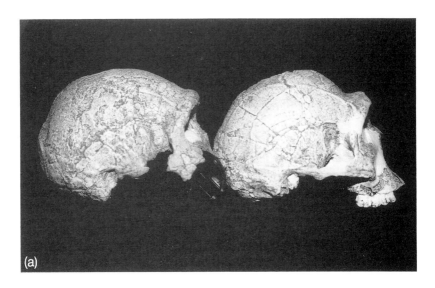

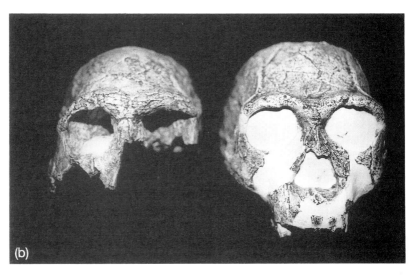

**Fig. 7.17.** *Homo* sp. (Species 4f of text): [casts of] (left) ER3733, (right), ER3883, (a) side view, (b) front view.

Reasons for ascribing the mandible ER992 to this species rather than the following were given above (section 7.3).

### 7.5.4(f)  Homo sp. (unnamed): Koobi Fora (Upper) (Figs. 7.17 and 7.18)

1949 *Telanthropus capensis* Broom and Robinson, *Nature*, **164**, 322. SK45 mandible from Swartkrans, Transvaal, South Africa. If transferred to the genus *Homo*, this name becomes a secondary homonym of *Homo capensis* Broom 1917 ( = the Boskop calvaria, *Homo sapiens).*
1976 *Homo erectus* Leakey and Walker, *Nature*, **261**, 572. KNM-ER3733 from Koobi Fora, Kenya.

Diagnosis: a species of the genus *Homo* sharing the derived traits of

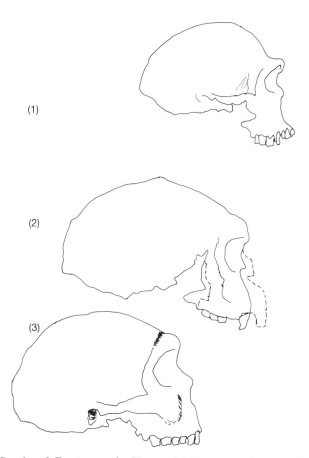

**Fig. 7.18.**   Crania of Erectus-grade *Homo*. (1) *Homo* sp. (unnamed). ER3733. (2) *Homo erectus erectus*. Sangiran 17 ( = Pithecanthropus VIII). The dotted line indicates the probable original facial orientation; after Thorne and Wolpoff (1981). (3) *Homo sapiens heidelbergensis*. The Petralona skull.

*H. ergaster* as given above, and in addition the following derived traits which it shares with *H. erectus* and *H. sapiens*: alveolar projection index below 108; supraorbital bar thickened; occipital scale shorter than nuchal; cranial vault bone thickened; mastoid process more than 12 mm long; tympanic length less than 24 per cent of bitympanic breadth; palate length less than 62 per cent of biorbital breadth; planum alveolare reduced; $I^1$ crown area less than 140 per cent that of $I^2$; symphyseal transverse tori reduced. There are no detectable uniquely derived traits: the species does not correspond precisely to the *erectus-sapiens* morphotype, in that—although it appears to possess to autapomorphic traits—*H. erectus* and *H. sapiens* share further derived traits not found in the present species. If this judgement is correct, then the following conditions can be predicted for future discoveries: $M^3$ length less than 90 per cent that of $M^1$; $M_1$ tuberculum sextum absent or reduced.

At the time of the preliminary description of the first discovered specimen, ER3733 (Leakey and Walker 1976), phrases like 'the unequivocal occurrence of *H. erectus* from the Koobi Fora Formation' were used (loc. cit., p. 572). 3733 was described as 'strikingly like' Zhoukoudian *H. erectus*; indeed, 'Such orthodox anthropometric comparisons that can be made at present fall well within the range of the Peking specimens' (loc. cit., p. 573). This is not in fact the case: glabella to opisthocranion length is given as 183 mm, whereas the smallest Zhoukoudian cranium measures 188 mm (Weidenreich 1943, p. 106); biauricular breadth of 3733 is given as 132 mm, but the smallest value for Zhoukoudian is 141 mm; the least distance between the temporal lines is 60 mm, compared to a minimum of 86 among the Zhoukoudian specimens. This latter comparison is so much below the Zhoukoudian range that every indication was given that the cranial capacity would turn out to be small: it is in fact 848 cc, the Zhoukoudian range being 915 to 1225. Thus the comparison with Asian *H. erectus* would seem to have been greatly overdone. A careful comparison of ER3733 has now shown that it does not possess the autapomorphic traits of *H. erectus* and so cannot be placed in that species at all (Wood 1984).

Apart from 3733, the only other published cranial capacity for a specimen of this species is for ER3883, which is 804 cc.

Walker (1981*b*) and Clarke (1985*a*) draw attention to the close similarity of the Swartkrans and Koobi Fora representatives of this species. The former (Tables 7.9 and 7.10) would seem, on the evidence of admittedly small samples, to have broader premolars and all-round larger molars (except for $M_1$) than the latter.

### 7.5.4(g) *Homo erectus* Dubois 1892

For synonymy, see under subspecies.

Diagnosis: a species of the genus *Homo* sharing the derived traits of the previous species, with in addition the following derived traits shared with *H. sapiens*: nasals broad, more than 65 per cent as broad as high; nasals

**Table 7.9** Maxillary dental measurements (mm) in later *Homo*.

| | *Homo* sp. Koobi Fora | Swartkrans | *Homo erectus* erectus | officinalis | pekinesis | *Homo sapiens* heidelbergensis |
|---|---|---|---|---|---|---|
| I¹m–d | — | — | 10.0 | 11.5 | 9.63 | 8.0 |
| | — | — | — | — | 1.48 | — |
| | — | — | — | 1 | 6 | 1 |
| b–1 | — | — | — | 8.1 | 7.83 | 9.0 |
| | — | — | — | — | 0.33 | — |
| | — | — | — | 1 | 6 | 1 |
| I²m–d | 7.9 | 7.8 | 8.0 | — | 7.77 | 8.10 |
| | — | — | — | — | 0.84 | 1.15 |
| | 1 | 1 | 1 | — | 3 | 3 |
| b–1 | 8.5 | 8.0 | 10.4 | — | 8.10 | 8.23 |
| | — | — | — | — | 1.00 | 0.25 |
| | 1 | 1 | 1 | — | 3 | 3 |
| C'm–d | 8.5 | 10.7 | 8.9–9.5 | — | 9.40 | 9.68 |
| | — | — | — | — | 0.73 | 0.88 |
| | 1 | 1 | 2 | — | 5 | 5 |
| b–1 | — | 10.3 | 8.4–11.8 | — | 10.06 | 9.98 |
| | — | — | — | — | 0.31 | 0.04 |
| | — | 1 | 2 | — | 5 | 5 |
| P³m–d | 8.50 | 8.5–9.6 | 7.90 | — | 8.42 | 7.94 |
| | 0.20 | — | 0.89 | — | 0.63 | 0.56 |
| | 3 | 2 | 4 | — | 6 | 5 |
| b–1 | 12.33 | 13.4–13.9 | 10.80 | — | 11.81 | 10.72 |
| | 1.04 | — | 1.10 | — | 0.82 | 0.98 |
| | 3 | 2 | 4 | — | 7 | 5 |
| P⁴m–d | 8.1–8.2 | — | 8.27 | — | 7.76 | 7.60 |
| | — | — | 0.12 | — | 0.59 | 0.71 |
| | 2 | — | 3 | — | 11 | 4 |
| b–1 | 11.5–12.1 | — | 10.97 | — | 11.35 | 10.78 |
| | — | — | 1.10 | — | 0.61 | 0.67 |
| | 2 | — | 3 | — | 11 | 4 |
| M¹m–d | 12.2–12.5 | 14.0 | 10.8–12.2 | 12.8 | 11.18 | 11.72 |
| | — | — | — | — | 1.00 | 0.66 |
| | 2 | 1 | 2 | 1 | 8 | 6 |
| b–1 | 11.8 | 13.3 | 12.6–13.6 | 13.7 | 12.44 | 12.98 |
| | — | — | — | — | 0.80 | 1.02 |
| | 1 | 1 | 2 | 1 | 7 | 6 |
| M²m–d | 12.40 | 14.2 | 10.7–13.6 | 11.0 | 10.76 | 11.60 |
| | 0.90 | — | — | — | 0.76 | 0.38 |
| | 3 | 1 | 2 | 1 | 8 | 5 |
| b–1 | 13.17 | 14.1 | 12.5–15.2 | 13.1 | 12.59 | 13.20 |
| | 0.67 | — | — | — | 0.51 | 0.32 |
| | 3 | 1 | 2 | 1 | 8 | 5 |
| M³m–d | — | 12.7 | 10.17 | 9.5 | 9.63 | 10.80 |
| | — | — | 0.71 | — | 0.51 | 1.60 |
| | — | 1 | 3 | 1 | 10 | 4 |
| b–1 | — | 15.4 | 13.47 | 13.0 | 11.55 | 10.93 |
| | — | — | 1.19 | — | 0.76 | 1.35 |
| | — | 1 | 3 | 1 | 10 | 4 |

# 278    A theory of human and primate evolution

**Table 7.10**  Mandibular dental measurements (mm) in later *Homo*.

| | *Homo* sp. Koobi Fora | Swartkrans | *Homo sapiens* erectus | officinalis | pekinesis | *Homo sapiens* heidelbergensis |
|---|---|---|---|---|---|---|
| I¹m-d | — | — | 7.4 | — | 6.36 | 5.65 |
| | — | — | — | — | 0.32 | 0.24 |
| | — | — | 1 | — | 5 | 4 |
| b-1 | — | — | 6.5 | — | 6.26 | 7.0–7.1 |
| | — | — | — | — | 0.37 | — |
| | — | — | 1 | — | 5 | 2 |
| I²m-d | — | — | — | — | 6.87 | 6.0–7.0 |
| | — | — | — | — | 0.37 | — |
| | — | — | — | — | 7 | 2 |
| b-1 | — | — | — | — | 7.01 | 7.5–7.8 |
| | — | — | — | — | 0.29 | — |
| | — | — | — | — | 8 | 2 |
| C₁m-d | 8.8 | — | — | — | 8.56 | 7.60 |
| | — | — | — | — | 0.37 | 0.58 |
| | 1 | — | — | — | 8 | 6 |
| b-1 | 9.0 | — | 8.9 | 9.2 | 9.14 | 9.58 |
| | — | — | — | — | 0.83 | 0.80 |
| | 1 | — | 1 | 1 | 8 | 6 |
| P³m-d | 9.0–9.0 | 8.6 | 8.7–9.9 | 10.0 | 8.67 | 8.69 |
| | — | — | — | — | 0.55 | 0.46 |
| | 2 | 1 | 2 | 1 | 14 | 7 |
| b-1 | 9.5–10.1 | 10.3 | 10.7–11.5 | 10.6 | 9.77 | 10.19 |
| | — | — | — | — | 0.56 | 0.94 |
| | 2 | 1 | 2 | 1 | 13 | 7 |
| P⁴m-d | 9.0 | — | 9.03 | 7.2 | 9.04 | 8.60 |
| | — | — | 0.56 | — | 0.50 | 0.63 |
| | 1 | — | 4 | 1 | 9 | 9 |
| b-1 | 9.5 | — | 11.23 | 9.6 | 10.04 | 10.09 |
| | — | — | 0.47 | — | 1.23 | 0.74 |
| | 1 | — | 4 | 1 | 10 | 9 |
| M¹m-d | 11.3–12.2 | 11.3–12.0 | 13.50 | 12.6–14.5 | 12.59 | 12.78 |
| | — | — | 1.23 | — | 1.09 | 0.96 |
| | 2 | 2 | 5 | 2 | 15 | 12 |
| b-1 | 11.1–11.8 | 11.5–12.0 | 13.15 | 11.5–12.2 | 11.78 | 11.78 |
| | — | — | 0.30 | — | 0.84 | 1.09 |
| | 2 | 2 | 4 | 2 | 15 | 12 |
| M²m-d | 11.6–12.4 | 12.6–13.0 | 14.06 | 14.43 | 12.48 | 13.04 |
| | — | — | 0.61 | 1.26 | 0.55 | 1.15 |
| | 2 | 2 | 5 | 4 | 11 | 12 |
| b-1 | 11.5–11.7 | 12.3–13.2 | 13.40 | 13.68 | 12.07 | 12.20 |
| | — | — | 0.70 | 0.50 | 0.65 | 1.08 |
| | 2 | 2 | 5 | 4 | 11 | 12 |
| M³m-d | 13.0–13.6 | 14.3 | 14.35 | — | 11.84 | 12.15 |
| | — | — | 1.08 | — | 1.44 | 0.80 |
| | 2 | 1 | 4 | — | 10 | 12 |
| b-1 | 11.6–12.0 | 12.4 | 12.70 | — | 11.13 | 11.58 |
| | — | — | 0.26 | — | 0.77 | 0.93 |
| | 2 | 1 | 3 | — | 10 | 12 |

arched; biparietal breadth greater than biporionic; angular torus present; tympanic plate vertical; petrotympanic angle less than 35°; anterior foveae of upper molars reduced; palate length less than 50 per cent of biorbital breadth; palate evenly deep from front to back; buccal grooves on $P^3$ reduced; mandibular robusticity less than 60; $P_4$ talonid reduced; $P^4$ talon reduced; and with the following uniquely derived traits: supraorbital bar developed into a straight torus, interrupted only slightly at glabella, not following rounded contours of upper margins of orbits, and strongly flared into lateral 'wings'; a strongly guttered superior surface to temporal root of zygomatic arch; large occipito-mastoid crest; frontal angle below 45°; occiput strongly angulated, with a strong occipital torus; angular torus and supramastoid crest strong; cranial vault exceedingly flattened; a mastoid/petrosal fissure; tympanic plate thickened.

Certain characters cannot be assessed in the first listed (most primitive) subspecies, which is known by only a single specimen lacking a facial skeleton: curve of Spee absent; face height less than 70 per cent of bifrontal breadth. These two characters may therefore characterize either the species as a whole, or only its Asian section.

### 7.5.4(g)(i) *Homo erectus olduvaiensis Tobias, 1968*

1963 *Homo leakeyi* or *Homo erectus leakeyi* Heberer, *Z. Morph. Anthrop.* **53**, 176. Olduvai hominid 9. 'Provisional name', hence invalid.
1968 *Homo erectus olduvaiensis* Tobias, in Kurth, ed., *Evolution und Hominisation*, 2nd edn, p. 188. Same type specimen.

Diagnosis: a subspecies of *H. erectus* lacking the derived traits which unite all other subspecies. Interorbital breadth is less than 25 per cent of bifrontal breadth; there is no parietal keeling or coronal ridge; a vaginal crest is present; the vault is not excessively flattened. There are no certainly autapomorphic traits in this subspecies, although the only known specimen is extremely robust, with heavy, prominent supraorbital torus and enormous glenoid fossae, and these may turn out to be uniquely derived traits if future discoveries shown them to be characteristic of a whole population. On the other hand this robusticity might also be a primitive trait of the species, which would certainly explain the 'deflated' appearance of the braincase of the other lower Pleistocene representatives of the species.

This form is the only non-Asian representative of *H. erectus*. It may or may not be slightly earlier than the earliest Javanese representatives.

### 7.5.4(g)(ii) *Homo erectus erectus Dubois, 1892 (Fig. 7.19)*

1892 *Anthropopithecus erectus* Dubois, *Verslag. Mijnb. Batavia*, 3rd quarter, 10. Trinil 2.
1896 *Homo javanensis primigenius* Houze, *Rev. Univ. Bruxelles*, 1896 issue, p. 1.
1896 *Homo pithecanthropus* Manouvrier, *Bull. Soc. anthrop. Paris*, **7**, 396.

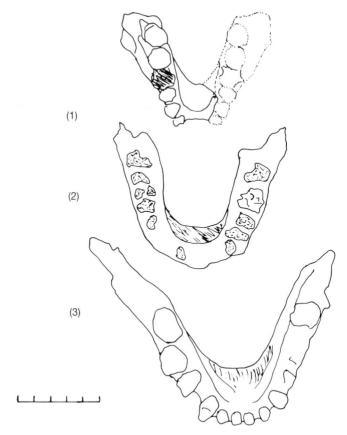

**Fig. 7.19.** Mandibles of *Homo erectus.* (1) *H.e.erectus.* Sangiran 9 ( = Pithecanthropus C), completed by symmetrical imaging. Much incisive alveolar bone is missing, giving a misleading impression of the symphyseal contour; but still the jaw is unexpectedly robust and narrow. (2) *H.e.pekinensis.* Zhoukoudian, 1959 jaw. (3) *H.e.officinalis.* The Lantian jaw.

1899 *Hylobates giganteus* Bumuller, *Das menschliche Femur nebst Beitragen zur Kenntnis der Affenfemora*, **12**, 124.

1901 *Pithecanthropus duboisii* Morselli, *Il Precursore dell'Uomo*, p. 1.

1909 *Hylobates gigas* Krause, in Bardeleben, *Handb. Anat. Menschen*, Vol. 1, issue 3.

1922 *Homo trinilis* Alsberg, *Das Menschheitsratsel*, p. 398.

1932 *Homo (Javanthropus) soloensis* Oppenoorth, *Wet. Meded. Dienst Mijnb. Ned.-Ind.*, issue 20, 49. Ngandong 1.

1932 *Homo primigenius asiaticus* Weidenreich, *Z. Anat. Entwicklungsges*, **99**, 212. Ngandong.

1936 *Homo modjokertensis* von Koenigswald, *Proc. k. ned. Akad. Wet. Amsterdam*, **39**, 1007. Perning 1.
1944 *Meganthropus palaeojavanicus* Weidenreich, *Science*, **99**, 480. Sangiran 6.
1944 *Pithecanthropus robustus* Weidenreich, *Science*, **99**, 480. Sangiran 4.
1949 *Pithecanthropus dubius* von Koenigswald, in Van Bemmelen, *Geology of Indonesia*, **1**, 110. Sangiran 5.
1975 *Pithecanthropus ngandongensis* Sartono, in Tuttle, *Palaeoanthropology*, p. 346. Substitute for *soloensis*.
1975 *Pongo brevirostris* Krantz, in Tuttle, *Palaeoanthropology*, **369**. Sangiran 4 (in part) and 5.
1976 *Homo erectus trinilensis* Sartono, *Abstr. UISPP, Nice*. Trinil 2 and some Sangiran specimens.
1976 *Homo palaeojavanicus sangiranensis* Sartono, *Abstr. UISPP, Nice.*, Sangiran 6.

Diagnosis: a subspecies of *H. erectus* sharing with all others except *olduvaiensis* a broad interorbital breadth ( > 25 per cent of bifrontal breadth), parietal keeling, coronal ridge present, vaginal crest absent, extreme platycephaly. Uniquely derived are the possession of a supratoral plane and angular torus. As a primitive state the molars, especially the posterior ones, are very large, and there is a clear M3 >2 >1 diminution sequence in the lower jaw.

All the Java *H. erectus* fossils have been included in this subspecies. That is not to say there are not differences among them. The division between 'Jetis' and 'Trinil' representatives has now to be abandoned (Watanabe and Kadar 1985). The division by Jacob (1976) into 'robust' and 'gracile' species, *Pithecanthropus erectus* and *P. soloensis*, seems likely to be one of sexual dimorphism, as does the division by Sartono (1976a) into *Homo erectus* and *H. palaeojavanicus*. The Ngandong sample, possibly upper Pleistocene [see Chapter 6, subsection 6.2.2(a)], does represent a slightly different population, differing mainly by the enlarged cranial capacity [1055 to 1300 cc (n = 6)], as against 815 to 1059 cc for the middle Pleistocene specimens (n = 9), which as Fig. 8.1(b) shows (Chapter 8, p. 300) is largely a matter of skull size increase. This could be taken to indicate the need for a separate subspecies, but a simple anagenetic advance would seem indicated, and a smooth gradual increase is expected if further, chronologically intermediate, specimens are discovered. Former placement (e.g. Weidenreich 1943) of the Ngandong specimens in *H. sapiens* was based purely on this cranial capacity increase; Jacob (1976), Sartono (1976) and Santa Lucca (1980) have in different respects all demonstrated that the Ngandong specimens are fundamentally *Homo erectus erectus*; indeed, they bear a very close similarity to such 'robust' middle Pleistocene specimens as Sangiran 4 (Jacob, Santa Lucca) and Sangiran 17 (Sartono, Jacob).

*7.5.4(g)(iii)  Homo erectus officinalis* von Koenigswald, 1952 (Fig. 7.19)
1952 *Sinanthropus officinalis* von Koenigswald, *Anthrop. Papers Amer. Mus. N.H.*, **43**, 308. Hong Kong pharmacist teeth, thought to be associated with *Stegodon-Ailuropoda* fauna.
1964 *Sinanthropus lantianensis* Wu, *Scientia Sinica*, **13**, 809. Chenjiawo mandible. Provisional name, hence unavailable.
1965 *Sinanthropus lantianensis* Wu, *Scientia Sinica*, **14**, 1035. Gongwangling cranium. (Available nomen.)
1973 *Homo (Sinanthropus) erectus yuanmouensis* Hu, *Arta Geol. Sin.*, **1**, 67. Shangnabang, Yuanmou, Makai Valley, Yunnan: two incisor teeth.

Diagnosis: a subspecies of *H. erectus* with apparently the same form of supratoral plane as in nominotypical subspecies, but differing in smaller cranial capacity (780 cc) in the single measurable specimen, smaller $P_4$, larger $M^1$, and smaller $M^3$ (Table 7.7). Mandible V-shaped. This subspecies must be regarded as provisional only. It is here made to incorporate all the early Chinese specimens [Lantian cranium and mandible, Hongkong teeth, Yuanmou teeth, and the so-called 'australopithecine' teeth of Gao Jian (1975)], the justification being only that they do not belong to *H. e. pekinensis*. They may, in fact, prove on the basis of further sudy to be attributable to nominotypical *H. e. erectus*: present metrical data place them only just outside the Javanese range.

The Lantian specimens span the Brunhes–Matuyama boundary; the Hongkong teeth are associated with supposedly lower Pleistocene Yellow Earth; the 'australopithecine' teeth are said to be lower Pleistocene. The age of the Yuanmou specimens is controversial; though generally supposed to be of early middle Pleistocene date, it has recently been argued that they are, after all, 1.7 million years old, just as the original palaeomagnetic survey suggested (Qing Fang 1985).

*7.5.4(g)(iv)  Homo erectus pekinensis Black, 1927 (Fig. 7.19, 7.20)*
1927 *Sinanthropus pekinensis* Black, *Palaeont. Sin.*, **7**, 21. Zhoukoudian (Choukoutien), Lower Cave, locality 1. Molar tooth.
1929 *Sinanthropus sinensis* Black, *Science*, **69**, 674. (Lapsus for pekinensis.)
1982 *Homo erectus hexianensis* Huang et al., *Vert. Palas.*, **20**, 251. Hexian, Anhui. Calvaria.

Diagnosis: a subspecies of *H. erectus* lacking the angular torus and supratoral plane traits of the nominotypical subspecies, but possessing the derived states of a prominent frontal convexity, emphasizing the orphryonic groove; a bony convexity spearating digastric groove from stylomastoid foramen; separation of occipital torus into upper and lower divisions well medial to asterion (see Santa Lucca 1980, pp. 129–30); small incisors and molars (especially third molars, which are the smallest of the series in both jaws). This is a well-distinguished subspecies, with its own autapomorphic traits and lacking those of *H. e. erectus*, but sharing with the latter some synapomorphies distinguishing from *H. e. olduvaiensis*. The time span is middle Pleistocene, hence contemporaneous with, or somewhat earlier than, the later (Ngandong) examples of *H. e. erectus*.

The Hexian calvaria shares the derived character states of the Zhoukoudian specimens, but the ophryonic groove is less-pronounced. This, as Huang *et al.* (1982) indeed remark, approaches it to the Java subspecies, implying that the relations between the two may in fact be clinal rather than strictly subspecific.

The cranial capacities of the seven measurable specimens vary from 915 to 1225 cc, with a mean of 1051. Chiu *et al.* (1973) make the case that some evolution can be seen through the Zhoukoudian sequence, mainly involving an increase in cranial capacity.

### 7.5.4(h) *Homo sapiens* Linnaeus, 1758

Diagnosis: a species of *Homo* sharing with *H. erectus* the derived states listed under that species, but lacking the latter's uniquely derived states, and possessing the following characters: high occipital index (convergent with *H. rudolfensis* and *H. ergaster*); inion well-inferior to opisthocranion; vault high, rounded, the height more than 60 per cent of its length; post-orbital constriction reduced, minimum frontal breadth being more than 90 per cent of biorbital breadth; occipital angulation reduced; greatest vault breadth on parietal or squamosal region; symphyseal tori reduced; digastric scars on inferior margin of symphysis tending to be deep and broad; mandibular marginal tori well-developed, everted, so that bimarginal breadth is greater than breadth between middle of molar surfaces (see Stewart 1972).

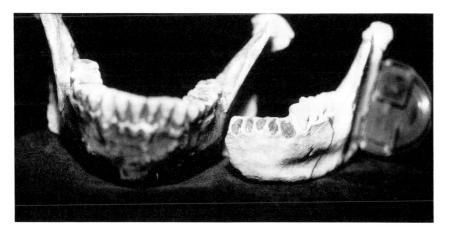

**Fig. 7.20.** Mandibles of *Homo erectus* and *H.sapiens*: Choukoutien (left) and Ternifine 2 (right). Showing the prominent marginal ridge of the latter.

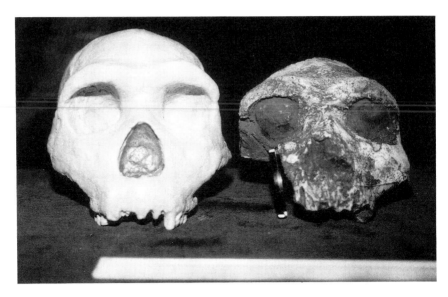

**Fig. 7.21.** *H.s.heidelbergensis*: [casts of] Petralona and Bodo-1.

*7.5.4(h)(i)  Homo sapiens heidelbergensis Schoetensack, 1908 (Figs. 7.20, 7.21 and 7.22)*

1908 *Homo heidelbergensis* Schoetensack, *Beitr. Pal. Menschen*, **1**. Mauer, near Heidelberg, West Germany. Mandible.

1921 *Homo rhodesiensis* Woodward, *Nature*, **108**, 371. Kabwe (Broken Hill), Zambia. Cranium.

1928 *Homo primigenius africanus* Weidenreich, *Natur u. Mus.*, **58**, 1. Kabwe.

1934 *Praehomo europaeus* von Eickstedt, *Rassenkunde und Rassenges- chichte der Menschheit*. Mauer.

1935 *Homo (Africanthropus) helmei* Dreyer, *Proc. Acad. Sci. Amst.*, **38**, 119. Florisbad, South Africa.

1935 *Homo florisbadensis* Drennan, *S. Afr. J. Sci*, **32**, 601. Florisbad.

1935 *Homo kanamensis* Leakey, *Stone Age Races of Kenya*, **9**. Kanam.

1936 *Palaeoanthropus njarasensis* Reck and Kohl-Larsen, *Geol. Rudsch.*, **27**, 401. Lake Eyasi, Tanzania.

1954 *Atlanthropus mauritanicus* Arambourg, *C.R. Acad. Sci. Paris*, **239**, 895. Tighenif (Ternifine), Algeria.

1955 *Homo saldanensis* Drennan, *Amer. J. Phys. Anthrop.*, **13**, 634. Saldanha, South Africa.

1956 *Homo (pithecanthropus) atlanticus* and *ternifinus* Dolinar-Osole, *Archaeol. Vestn.* **7**, 178. Tighenif.

1965 *Tchadanthropus uxoris* Coppens, *C.R. Acad. Sci. Paris*, **260**, 2869. Yayo, Koro Toro, Chad. 'Provisional name': unavailable.

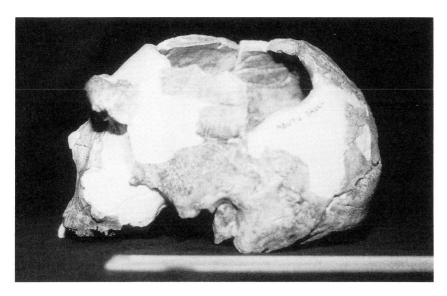

**Fig. 7.22.** *Homo sapiens heidelbergensis*: the Ndutu cranium.

1966 *Homo erectus* (seu *sapiens*) *palaeohungaricus* Thoma, *L'Anthropologie (Paris)*, **530**. Vertesszollos, Hungary.
1975 *Homo petralonensis* Murrill, *Petralona Skull*.
1978 *Homo erectus bilzingslebenensis* Vlcek, *J. Hum. Evol.*, **7**, 249. Steinrinno, Bilzingsleben, East Germany.

Diagnosis: the primitive subspecies of *Homo sapiens*, plesiomorphic sister group of the others. In general, may be characterized by relatively great occipital angulation, well-developed horizontal supramastoid crest, low vault, low frontal angle (below 60°), marked superior temporal line, presence of a sagittal ridge, large supraorbital torus (but following contour of upper orbital margin, and dorso-ventrally thickened above middle of each orbit, unlike *H. erectus*), external auditory meatus above level of glenoid cavity and clasped between the two branches of the zygomatic root (see de Lumley and Sonakia 1985), petrotympanic axis above 100°.

It has proved impossible to differentiate geographic samples of this subspecies from one another. Among the crania Kabwe and Saldanha, from southern Africa; Eyasi and Bodo, from eastern Africa; Tighenif, Sale, and Thomas, from North Africa; and Petralona, Vertesszollos, Bilzingsleben, and Arago, from Europe, all fit into a common pattern. Similarly with the mandibles: the two mandibles from Arago, in particular, are very different from one another, as are the three from Tighenif, and either of these samples encompass a range which would incorporate Mauer, Montmaurin, the

Moroccan mandibles, the two Kapthurin mandibles, Cave of Hearths (Makapansgat), and the fragmentary Kabwe and Saldanha specimens.

The only differentiation that can be made within the group is on the basis of age. The four specimens that seem to be early in age, i.e. terminal lower Pleistocene or beginning of middle Pleistocene (Ndutu and Olduvai hominid 12 from East Africa, Sale and Tighenif from North Africa), have cranial capacities averaging 971.8 (range 727 to 1180 cc, if the Tighenif estimate is accurate), while the six from later in the middle Pleistocene (Arago, Vertesszollos, and Petralona from Europe; Kabwe, Saldanha, and Ndutu from Africa) average 1183 cc (range 1100–1280). No morphological progression seems to accompany this endocranial increase. At the end of the middle Pleistocene, certain specimens (Ngaloba, Omo Kibish) appear which seem to represent a transition towards *Homo sapiens* cf. *sapiens*; these will be treated under that heading.

This taxon has customarily been considered a western variant of *H. erectus*. Stringer, in particular, has long opposed this view, most recently (1984) pointing out that the latter has autapomorphic features which its western contemporaries do not share; Wood (1984) and Andrews (1984) offer additional support for Stringer's hypothesis, with the strong implication that not only are the Eurafrican fossils an archaic form of *H. sapiens* but that the origin of anatomically modern people is no longer to be sought among *H. erectus* as narrowly redefined. In some respects Bodo did appear to form a link between the two, with its greater prognathism, thicker vault bones, more keeled vault, broader face, and less-developed subspinale than, for example, Petralona, Kabwe, or Arago (Stringer 1984); but these are not, on the whole, characters which I have found in this study to be cladistic differentiators of *H. erectus* and *H. sapiens*, while the emerging picture of an Asian race of archaic *sapiens* puts the relationship between the two species in a new light.

*7.5.4(h)(ii)  Homo sapiens mapaensis* Kurth, 1965 (Fig. 7.23)
1965 *Homo erectus mapaensis* Kurth, in Heberer, *Menschliche Abstammungslehre*, p. 383. Maba (Ma-pa), calvaria.
1981 *Homo sapiens daliensis* Wu, *Scientia Sinica*, **24**, 530. Dali, China. Cranium.
1985 *Homo erectus narmadensis* Sonakia, in Delson, *Ancestors: the hard evidence*, p. 337. Hathnora, Narmada Valley, India. Calvaria. (A reference to *Rec. Geol. Surv. India*, **113**, 159–172 is given as the source of the name, but appears not to have been published as yet.)

Diagnosis: a subspecies of archaic *H. sapiens* differing from *H. s. heidelbergensis* in the following derived traits: frontal angle above 60°, vault high (bregmatic height index above 40), supramastoid crest running olbiquely supero-posterior instead of horizontal.

The great similarities between Dali and Hathnora are well brought out by de Lumley and Sonakia (1985), who class them both as 'evolved *Homo*

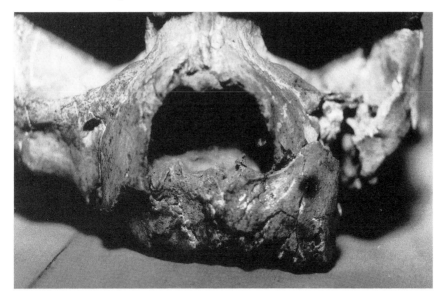

**Fig. 7.23.** *Homo sapiens mapaensis*: the Dali cranium. Close-up of facial skeleton, to show the way the upper jaw has overridden the cheek region on the left and under-ridden it on the right, so giving the impression of an unnaturally short face. (Photo by the author of original in the Institute of Vertebrate Pataeontology and Palaeo-anthropology, Academia Sinica, Beijung. Published by courtesy of Dr Wu Xinzhi and the IVPP.)

*erectus'*. In their paper, Maba is not considered, but reference to Woo and Pang (1959), as well as my own examination of the specimen, courtesy of Professor Wu Rukang, show that it is close to the other two Asian specimens. The high frontal angle (66° in Dali, 70° in both Maba and Hathnora) and bregmatic height index (42.4 in Hathnora, 43.2 in Dali, 45.0 in Maba) would both appear to approach this form to modern *H. sapiens*, but the emerging fossil evidence for the origin of modern humans suggests that Africa, not Asia, was the scene of this event [see Subsection 7.5.4(h)(iv) below]. The form of the supramastoid crest is quite unique.

Dali and Maba are certainly later in time than sympatric *Homo erectus*. Woo and Pang (1959) and Wu (1981) make the case that Maba and Dali, respectively, are descended from *H. e. pekinensis*, citing for example the transverse naso-frontal suture which does not bulge into the frontals but continues the line of the fronto-maxillary suture. This view will be discussed again below, but in the meantime it will be noted that this form of the suture is commonly seen in many *Homo*, living and fossil, and may indeed be the primitive form in the genus.

*7.5.4(h)(iii) Homo sapiens neanderthalensis King, 1864 (Fig. 7.24)*

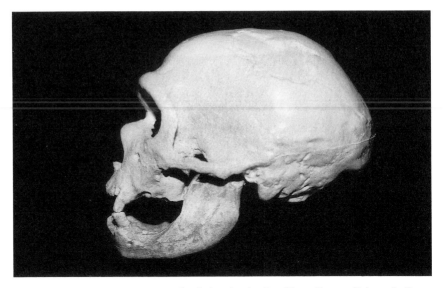

**Fig. 7.24.** *Homo sapiens neanderthalensis*: the La Chapelle-aux-Saints skull, an individual of extremely advanced age.

1864 *Homo neanderthalensis* King, *Quart. J. Sci.*, **1**, 88. Neandertal, Dusseldorf, West Germany.

1895 *Protanthropus atavus* Haeckel, Syst. Phyl. Wirbeltiere, **3**. Neandertal.

1898 *Homo primigenius* Wilser, *Naturw. Wochenschr.*, **13** 1. Neandertal.

1902 *Homo neanderthalensis var. krapinensis* Gorjanovic-Kramberger, *Mitt. Anthrop. Ges. Wien*, **32**, 189. Krapina, Yugoslavia.

1908 *Homo transprimigenius mousteriensis* Forrer, *Urgesch. Europaers*. Le Moustier, France.

1909 *Homo mousteriensis hauseri* Klaatsch and Hauser, *Arch. f. Anthrop.*, **35** (NF), 287. Le Moustier, France.

1909 *Homo spyensis* or *Homo priscus* Krause, *Anat. Menschen*, **1**, 3. Spy, Belgium.

1910 *Palaeanthropus europaeus* Sergi, *Scientia*, **8**, 475. Neandertal.

1911 *Homo chapellensis* Buttel-Reepen, *Naturw. Wochenschr.*, **26**, 177. La Chapelle-aux-Saints, France.

1911 *Homo breladensis* Marett, *Archaeologia*, **62**, 449. La Cotte de Saint Brelade, Jersey, UK.

1911 *Homo calpicus* Keith, *Nature*, **87**, 313. Forbes Quarry, Gibraltar. For author and date of this name, see Campbell 1965, p. 18.

1916 *Homo naulettensis* Baudouin, *C.R. Acad. Sci., Paris*, **162**, 519. La Naulette, Belgium.

1928 *Homo heringsdorfensis* Moller, in Werth, *Fossile Mensch.*, 201. Ehringsdorf, West Germany. Nomen nudum.

1931 *Homo galilensis* Joleaud, *Rev. Scientif.*, **69**, 466. Zuttiyeh, Galilee, Israel. Nomen nudum.

1932 *Homo galilaeensis* Hennig, *Petermann's Mitt.*, **78**, 134. Zuttiyeh. Nomen nudum.

1935 *Homo neanderthalensis* var *aniensis* Sergi, *Anthrop. Anz.*, **12**, 281. Saccopastore, Italy.

1936 *Homo steinheimensis* Berckhemer, *Forsch. Fortschr.*, **12**, 349. Steinheim, West Germany.

1936 *Homo murrensis* Weinert, *Z. Morph. Anthrop.*, **35**, 518. Steinheim.

1940 *Homo marstoni* Paterson, *Nature*, **146**, 15. Swanscombe, Kent, UK. Nomen nudum.

1940 *Homo ehringsdorfensis* Paterson, *Nature*, **146**, Ehringsdorf. Nomen nudum.

1941 *Homo kiik-kobensis* Bonch-Osmolovskii, *Paleolit. Kryma*, **1**. Kiik-Koba, Crimea, USSR.

1942 *Homo swanscombensis* Kennard, *Proc. Geol. Assoc.*, **53**, 105. Swanscombe. Nomen nudum.

1943 *Homo sapiens proto-sapiens* Montandon, L'Homme prehist. prehum. Swanschombe (selected by Campbell 1965, p. 24).

1944 *Protanthropus tabunensis* Bonarelli, *Ultima Miscallanea, 1*, **4**, 1. Tabun, Mt. Carmel, Israel. *Nomen nudum.*

1957 *Homo sapiens shanidarensis* Senyurek, *Anatolia*, **2**, 49. Shanidar (infant), Iraq.

Diagnosis (after Stringer and Trinkaus 1981; see also Rak 1986): a subspecies of *Homo sapiens* in which the frontal angle is high (57–72°) as in *H. s. mapaensis*, higher than in *H. s. heidelbergensis*; vault relatively low, the bregmatic height index being below 43 as in *H. s. heidelbergensis*; a strong pre-lambdatic depression is present; supramastoid crest horizontal; frontal sinus enormous, extending laterally in supraorbital ridges (more so, on average, than in other taxa); occipito-mastoid crest large; anterior mastoid tubercle present; external auditory meatus behind zygomatic root; articular tubercle generally well-developed; lateral margins of nasal aperture out-turned; nasal bones prominent; maxillary sinus large; greatest breadth of vault in squamosal suture region, lateral walls of vault convex; mental foramen far back on mandible, behind level of $P_4$; large retromolar space. In post-cranial skeleton, clavicles are long, distal extremities shortened, metacarpals long, hand phalanges very short and robust, deeply guttered for muscular insertion (Musgrave 1973). Except for the first two characters, this subspecies is highly derived in all these respects. Cranial capacity large.

The famous Neandertal race occupies a well-circumscribed space-time zone: Europe and western Asia, east as far as Uzbekistan (site of Teshik-Tash), south to Israel (Tabun, Amud, Zuttiyeh); from the late middle Pleistocene (Ehringsdorf, Swanscombe, Steinheim) until nearly as late as 30 000 BP (Saint-Cesaire, France). It is, very broadly, true to say that a late,

European ('classic') type can be distinguished from an early European and a late Wester Asian type, the subspecific peculiarities being accentuated in the classics. Specimens intermediate in time (Saccopastore) and space (Krapina) are intermediate in type; and the earlier specimens from Shanidar, too, are less-extreme in their characters than the later ones (Trinkaus 1983).

The early members of this subspecies, especially Steinheim and Swanscombe, have not usually been included here [Wolpoff (1980) regards them as females of the Petralona type, here called *H. s. heidelbergensis*], but Stringer (1985) calls them the 'early Neandertal' group, showing their derived traits to be those of the later Neandertalers, and finds no difficulty in deriving them from *heidelbergensis*. Trinkaus (1983) and Stringer and Trinkaus (1981), focusing on Shanidar, draw attention to the (relatively few) features in which the West Asian sample differs, on average, from the Europeans: less-elongated clavicle (claviculo-humeral index 46.7 to 49.0, as against 51.9 to 53.0 in the Europeans), less-reduced distal limb bones (Brachial index 74.2 to 78.0, cf. 69.9 to 76.1), taller skeletons, and perhaps larger cranial capacity in males.

Sexual dimorphism is marked, the male's skeleton being in every respect larger than the female's, with greater calculated statures [males ($n$ = 10), 163.8 to 173.5; females ($n$ = 5), 155.2 to 163.4]. Cranial capacity is 1550–1681 in four males from Europe, 1305 to 1350 in two females; two West Asian males, Shanidar 1 and Amud, are 1600–1740, and a West Asian female, Tabun 1, is only 1271 (Trinkaus 1983). Taking all measurable or calculable specimens into account, whatever their sex, provenance, or chronological age, we get a grand mean of 1462.0 ± 173.35 (sd) ($n$ = 16); but the considerations mentioned above, especially sex, indicate that such a procedure is misleading.

The Neandertalers are the earliest *Homo* for which there is good evidence of what might be broadly called their 'essential humanity': burials, including grave goods, and nurturing of disabled individuals [such as the appalingly handicapped Shanidar 1, immortalized as 'Creb' by Auel (1980)]. Vallois (1960) found only one out of 39 specimens examined to be even approaching 60 years of age at death, but I suspect his estimates, based as they are solely on the notoriously unrealiable method of aging by suture closure (Powers 1962; Bocquet-Appel and Masset 1982), to be in the main far too low. The 'old man' of La Chapelle-aux-Saints, for example, shows all the indications of advanced senility, and there is no theoretical reason (Cutler 1975) why one should be surprised to encounter the remains of Neandertalers who had indeed achieved great ages.

### 7.5.4(h)(iv) Homo sapiens cf. sapiens Linnaeus, 1758
The nomenclature of this putative subspecies is replete with names given to modern 'races'. Only those awarded to fossil examples will be given here; in the main (as for *H. s. neandertalensis*, above), these are taken from Campbell (1965).

1899 *Homo spelaeus* Lapouge, L'Aryen. Cro-Magnon, France.

1899 *Homo priscus* Lapouge, *loc. cit.* Chancelade, France.

1906 *Homo grimaldii* Lapouge, *Polit.-Anthrop. Rev.*, **4** 16. Grotte des Enfants, Grimaldi, France.

1910 *Homo aurignacensis hauseri* Klaatsch and Hauser, *Prahist. Z.*, **1**, 273. Combe Capelle, France.

1911 *Notanthropus eurafricanus recens* Sergi, L'uomo. Secondo le orgine, l'antichite levoriazioni e la distribuzione geographic. Combe Capelle selected by Campbell, 1965, p. 16.

1911 *Notanthropus eurafricanus archaius* Sergi, 1911, Predmost, Czechoslovakia.

1915 *Homo mediterraneus fossilis* Behm, *Prometheus*, **26**, 161. Oberkassel, West Germany.

1915 *Homo fossil proto-aethiopicus* Giuffrida-Ruggeri, *Arch. Antrop. Etnol.*, **45**, 292. Combe Capelle.

1917 *Homo capensis* Broom, *Amer. Mus. J.*, **17**, 141. Boskop, South Africa.

1920 *Homo predmostensis* Absolon, in Klaatsch and Heilborn, Werdegang d. Menschheit und die Enstehung der Kultur . . . Predmost.

1921 *Homo sapiens cro-magnonensis* Gregory, *Origin and evolution of human dentition*. Cro-Magnon.

1921 *Homo wadjakensis* Dubois, *Proc. k. Ned. Akad. Wet. Amsterdam*, **23**, 1013. Wajak, Java, Indonesia.

1925 *Homo larteti* Pycraft, *Man*, **25**, 162. Cro-Magnon.

1929 *Homo australoideus africanus* Drennan, *J. Roy. Anthrop. Inst.*, **59**, 147. Cape Flats, South Africa.

1931 *Homo drennani* Kleinschmidt, Der Urmensch. Cape Flats.

1932 *Palaeanthropus palestinus* McCown and Keith, *Proc. 1st. Int. Congr. Prehist. Protohist. Soc., London*. Skhul, Israel (Skhul 1 selected by Campbell 1965, p. 12).

1940 *Homo leakeyi* Paterson, *Nature*, **146**, 12. Kanjera, Kenya.

1948 *Nipponanthropus akasiensis* Hasebe, *J. Anthrop. Soc. Nippon*, **60**, 32. Nishiyagi, Japan.

Diagnosis: a subspecies, or group of subspecies, of *H. sapiens* lacking the autapomorphic traits of *H. s. neandertalensis*, but possessing the following autapomorphic traits [based in part on Day and Stringer's (1982) 'working definition']: cranium short, high, height-to-length index above 70; parietal arch long, high, inferiorly narrow, superiorly broad, parietal angle above 138°, bregma-asterion chord to biasterionic breadth > 1.19; frontal angle (Schwalbe's) above 72°; supraorbital torus not continuous, divided into lateral and medial portions; occiput well curved, occipital angle above 114°; a pronounced mental eminence; limb bones gracile, thin-walled, with relatively small articular surfaces. Supramastoid crest reduced, often virtually absent. Lateral walls of cranium more or less vertical, not markedly convex.

Femoral cortex thin (anteroposterior cortical index at mid-shaft below 60); femur relatively stout (minimum breadth index above 50) (Kennedy 1984); limb bones, including distal extremities, elongated. (The indices given by Kennedy distinguish the femur from that of *H. s. heidelbergensis*; the elongation distinguishes them from *H. s. neanderthalensis*.)

Modern *Homo sapiens* is a controversial concept, both taxonomically and phylogenetically. The problem of whether there is a single modern subspecies or several is intractable. On the one hand there are those who insist that there are distinguishable, definable human races, equivalent to zoological sub-species: but how many there are is argued, and with rather few exceptions (notably Coon 1963) the proponents of this view tend to assume that even if races are poorly definable today, they were at one time much more distinct, but have since interbred to such a degree that they had merged into a continuum: a view that goes back at least to Huxley (1870), and was the unspoken assumption of almost all anthropologists up until the 1930s. Some of the supporters of the races-as-subspecies viewpoint have gone as far as to apply Linnaean nomenclature (Baker 1974), a practice fraught with hazards, not least the necessity to fix a nominotypical subspecies and to interpret Linnaeus's terms *americanus, europaeus, asiaticus,* and *afer*—are they binomial or not? [I myself think not; not only are they written in a way unparalleled in the binomial list, though suggesting that they are subordinate to the specific name *sapiens*, but there are two others in the list: *ferus* ('wolf-boys' and the like) and *monstrosus* (teratological and deformed indivi-duals)—hardly meant to be taxa!]

It is much commoner today in anthropological texts to present races in a kind of half-hearted, apologetic way, or to deny their real existence altogether, lumping all modern peoples as *Homo sapiens sapiens*. Like the old 'pure races' model, this is really more of an assumption than a worked-out scheme, and has become widespread apparently from a feeling that the general public cannot be trusted with a race concept. One of the few profes-sionals to examine seriously the question of whether there actually are definable races (Schwidetzky 1974) concludes that there probably are (different lines of evidence tend towards similar conclusions; differentiation between population pairs very commonly amply satisfies the 75 per cent rule for subspecies), but that the problem is so complex and multi-stranded that no more than pointers emerge from studies up to the present. And that is where the question remains, even today.

This is a separate problem to that of the origin of 'racial' features. Old polygenic vs. monogenic arguments were refined by Weidenreich (1939) into the much more reasonable one of the polycentric vs. monocentric origin of modern *Homo sapiens*. Polycentrism can be looked at in many different forms: are the races/subspecies of *Homo sapiens* separately descended from those of *Homo erectus* (Weidenreich 1939; Coon 1963); do the geo-graphically corresponding races of the two species show genetic continuity

(Wolpoff *et al.* 1984); or did *Homo erectus* evolve into *H. sapiens* on a broad front, such that regional characteristics were likely to have survived (the 'spectrum' theory of Weiner 1958)? Monocentrism, on the other hand, requires that *Homo spaiens* (of modern type) had a unique origin, spread out, and differentiated afresh into geographic variants, with or without inter-breeding with its precursors (Bräuer 1984).

So, is there a evidence for regional continuity? Weidenreich (1939) and Coon (1963) thought so, and the case has been argued in detail most recently by Wolpoff and colleagues (1984). Weidenreich noted traits in common between *Homo erectus pekinensis* and modern Mongoloid peoples: mainly high frequencies of particular polymorphisms (shovel shaped incisors and so on), but also a few continuous variables. He also argued a connection between Javanese *Homo erectus* and modern Australoids, but this was more on the grounds of an apparent transition series (Sangiran–Ngandong–Wajak–modern Australians) than on the common possession of traits. Coon revived Weidenreich's idea, but it is remarkable how it is only in the *pekinensis*-Mongoloid case that he cites actual physical features in support: the rest of the book consists in descriptions of fossils and their arrangement into regional continua, without actually explaining what it is that, say, the Kabwe skull is supposed to have in common with modern 'Congoids' [contenting himself with statements such as 'on the whole, its face is mostly Negro' (1963, p. 626), which cuts (1974) little ice as a logical argument.] Larnach and Macintosh (1974) performed for the Australoid lineage what Weidenreich had performed for the Mongoloid: namely, they took a putative precursor form (the Ngandong sample) and compared it point by point with Australoids (Australia and New Guinea). Wolpoff (1980a) compared, if much more briefly, Petralona with modern Europeans. Finally Wolpoff *et al.* (1984) have revived, and vastly refined, the arguments of Weidenreich and of Larnach and Macintosh, and presented an evolutionary model to account for regional continuity in general. It is worth considering their arguments here, at least briefly.

1. Shovel shaped incisors. Following Weidenreich (1939), Wolpoff *et al.* (1984) note that incisor shovelling characterizes East Asian hominins as far back as the middle Pleistocene (Yuanmou, in fact). There would thus be con-tinuity in this characteristic, which would have been preserved as a regional feature across the transition from *Homo erectus* to *H. sapiens*. Certainly, it is true that the frequencies of marked (full) shovelling are highest in modern 'Mongoloid' peoples, although it should be noted that there are differences within this general 'racial' complex, from 98 per cent frequency in Pima Amerindians down to 63 per cent in Aleuts, and to only 19 per cent in unspeci-fied 'Polynesians' (if it is accepted that these are of broadly East Asian derivation) (Cadien 1972). Moreover a mild degree of shovelling, corre-ponding to the 'trace' or 'semi'-grades, is characteristic of fossil hominins in general, from australopithecines to Neandertalers (Robinson 1956,

pp. 23–34): it is the total absence of shovelling, the condition common only in 'Caucasoids' and Africans, that is unusual in the Hominini as a whole. We could thus make a case that modern East Asians have simply preserved, and in some cases exaggerated, a symplesiomorph hominin trait.

2. Early reduction of posterior dentition (including third molar agenesis). This is not in fact an East Asian characteristic. Tables 7.7 and 7.8 shows that the first molar, in both jaws, is as reduced in *Homo sapiens heidelbergensis*, as well as in *Homo sp.* (unnamed), Koobi Fora (Upper), as it is in *Homo erectus pekinensis*, while the second molar is almost as small. As far as modern peoples are concerned, East Asians (but not Southeast Asians) share with 'Caucasoids' a reduction of the cheek teeth, while tropical peoples tend to have larger cheek teeth; the only middle or lower Pleistocene taxon with large cheek teeth is *Homo erectus erectus*. It is true that M3 agenesis has a high frequency both in *H. e. pekinensis* (and its predecessor, *H. e. officinalis*, in the Lantian jaw) and in modern Mongoloids, but its 24 per cent frequency in modern Europeans is not far below the 40 per cent in Eskimos; and its earliest occurrence is, of course, in an Omo mandible!

3. Facial flatness. From Weidenreich's Fig. 165 (1943 p. 388), designed to show how facial flatness links Zhoukoudian and Mongoloids, it is visible that apes share the same morphology! (Sangiran 17 seems to be similar as well). Part of the facial flatness complex is that the junction of the lower maxillary and zygomatic (malar) margins are angular when viewed from below; but this too is clearly characteristic of Sangiran 17 and SK847. I propose that facial flatness is another symplesiomorph hominin feature.

4. Sagittal keeling, diminishing through time. Sagittal keeling is, of course, characteristic of *Homo erectus* in general, and is found in 'Australoid' crania too. It is, on the other hand, certainly not a primitive character. Here, then, might be a character that really does support regional continuity; though over the whole East Asian region, not just China.

5. 'Inca bones' (i.e. lambda and lambdoid ossicles), again becoming less frequent through time. Subsequent discoveries have tended to rob this feature, one of Weidenreich's (1939) trump cards, of much of its significance. From ER1813 on, Inca bones recur sporadically throughout Pleistocene and recent *Homo*. Well outside the East Asian theatre, they are multiply represented in such fossils as Petralona and Vertesszollos. Among modern populations, they are (according to my figures) even more frequent in Australia (40 to 65 per cent) and Melanesia (58 per cent) than in East Asia and the Americas (30–54).

6. Small frontal sinus, becoming even smaller with time. Weidenreich's (1939) own figures were somewhat equivocal on this one: according to him, the frontal sinus was totally absent in 30 per cent of Australians and 38 per cent of Melanesians; though admittedly its absence is most frequent (19 out of 22, i.e. 86 per cent) in a Buriat (Mongol) sample. The vicissitudes of frontal

sinus development in fossil *Homo* are poorly documented, though it is known that some European specimens (Petralona; Neandertalers) had unusually large ones, and this seems to me likely to be the highly derived condition.

7. Fronto-nasal and fronto-maxillary sutures in a straight line. This is certainly another primitive feature; Olson (1978, 1985*a*) has argued that nasal bones which do not wedge into the frontals is a diagnostic feature of the *Homo* (here, the *Homo/Australopithecus*) lineage. Indeed, it characterizes almost all members of the genus *Homo* until the evolution of some exceptionally beaky nosed populations (Neandertalers, modern 'Caucasoids'): whether it is somehow a by-product of nasal prominence, and whether indeed it supports Coon's (1963) alliance of Neandertalers with Caucasoids, can be argued.

8. Maxillary, mandibular, and auditory exostoses. The significance of these, even whether they are genetic at all, is unknown; Pardoe (in preparation) considers auditory exostoses to be a purely environmentally induced phenomenon. For what it is worth, SK847 also appears to have maxillary exostoses.

Now, Wolpoff *et al.* (1984) do not dispute that these features occur in other Pleistocene populations; but state that they occur at much lower frequencies, and are distributed discontinuously. This may be true, but it is also true that these other Pleistocene populations are in no case samples as the Zhoukoudian one is—or, less emphatically, the Sangiran and Ngandong ones are—and, given the often somewhat scrappy condition of the specimens, there is often not much chance for the characters to be distributed any other way than discontinuously! What is at issue is, in any case, not the presence or absence of particular genes (or gene complexes), but rather whether they can be maintained continuously at high frequencies over long periods of time. It has been seen (above) that, in any case, many of the *erectus/sapiens* continuities that really do exist through time in East Asia are demonstrably primitive retentions, so cannot be used for phylogenetic reconstruction.

Wolpoff *et al.* (1984) go on to make a case for *H. e. erectus*/Australoid continuity. I will not analyse this case here; suffice it to say that it, too, appears to rely on the persistence of primitive traits. The same might be said of at least some of the features cited by Wolpoff (1980) as in common between Petralona and modern Europeans.

What appears to be going on is that each of the 'polar types', in the spectrum of cross-cutting clines that is modern *Homo sapiens*, has retained some primitive characteristics and developed some uniquely derived ones. Examples of retained primitive features would include Mongoloid facial flatness and incisor shovelling; Australoid brow-ridge development; and Caucasoid hairiness.

Part of the problem may be the tacit assumption that *Homo erectus* is the ancestor of *Homo sapiens*. Under a unilineal model it is of course axiomatic that this is so: the very nomen *Homo erectus* would refer simply to a segment along a continuum, and relatively few authors (Wolpoff 1980*b*; Rightmire 1984) even seemed prepared to acknowledge that there might be geographic variation prior to modern times. Most recently, such a view has been strongly challenged: *Homo erectus* has autapomorphic character states, and is contemporary with other fossils, some of them at one time or another mistakenly referred to it, which do not share these traits (Wood 1984; Stringer, 1984; Andrews 1984). This viewpoint would push *H. erectus* to one side, and derive modern humans from its neglected contemporaries, which thereby acquire the status of *Homo sapiens*. I have argued above that this view is correct: that the line leading to modern people goes through a western (Eurafrican) stock, at least until the latest middle Pleistocene.

But does such a view really exclude *Homo erectus* genes from contributing to modern *H. sapiens* ancestry? I believe not: a number of untested assumptions have got by unnoticed along with the spread of cladistic methodology. Some of these have been mentioned above (Chapter 1, section 1.5); but another, relevant in the present context, is the downward extension, to species level and even below, of the same criterion as is used for the higher categories. There has therefore been a logical jump, from the demonstration that a species is cladistically valid to the assumption that it is also thereby reproductively isolated. The demonstration, therefore, that *Homo erectus* is cladistically distinct from *Homo sapiens* does not amount to a denial that the two formed a genetic continuum across the inhabited world. A cladospecies, then, is not automatically a biospecies.

The taxonomy proposed above involves replacement at certain points: of *Homo erectus* in China by *Homo sapiens mapaensis*; of *Homo erectus* in Java by (as far as we know) *Homo sapiens* cf. *sapiens*; and of *Homo sapiens neanderthalensis* in Europe and the Middle East by, again, *Homo sapiens* cf. *sapiens*. (Perhaps other cases too.) There is no inherent reason to suspect that the first two of these cases of replacement had to have been qualitatively different from the third: it happens that, for cladistic reasons, they are, formally, replacement of one species by another while the third case is replacement of one subspecies by another in the same species, but I cannot think that this taxonomic formality meant anything to the people involved. In each of the three cases some factor—cultural superiority or whatever it was—enabled one group of people to spread at the expense of another, until the latter ceased to exist as a homogeneous entity; but the door is still open to envisage interbreeding as a channel for allowing some of the genes of the old group to filter into the new one, if it is thought that the evidence demands it. I have indicated above that there is some, arguable, suggestion that modern East Asians might share a few derived conditions with their precursors; if this evidence holds up, then interbreeding, rather than the regional continuity model, would seem to provide the most plausible explanation.

Where, then, did *Homo sapiens* of modern type evolve? I noted above that *Homo sapiens mapaensis* has a couple of features that seem more like *H. s.* cf. *sapiens* than *H. s. heidelbergensis*; on the other hand, it has a peculiar and idiosyncratic form of the supramastoid crest, and there is also the evidence, well brought out by Bräuer (1984), that there is a transition phase between the latter and modern *sapiens* in Africa. The transitional forms are Omo (Kibish) I and II, Laetoli 18 (Ngaloba), and probably Florisbad (see also Clarke 1985*b*). These, as Bräuer demonstrates, are a variable group intermediate between the middle Pleistocene forms (here referred to *heidelbergensis*) and the early upper Pleistocene fossils of modern type such as Klasies River Mouth and Border Cave. Moreover, the two East African sites are both dated at 120 000 and 130 000 BP. Discounting Florisbad for the moment, because it is undated, the evidence here seems to place the origin of modern types in East Africa shortly before the middle/upper Pleistocene transition. Bräuer would have *H. s. neanderthalensis* emerging in Europe at the same time.

# 8
# The progress of human evolution

It is finally time to review the evidence. What sort of a story does it tell? How did human evolution progress, how well does it fit the models proposed earlier, and what special features does it show?

## 8.1 THE EVOLUTION OF THE HUMAN BRAIN

The literature on human evolution is replete with ideas on why we have such a big brain. It is only recently that some (Hemmer 1967; Godfrey and Jacobs 1981; Passingham 1982; Hofman 1983; Blumenberg 1984) have started to enquire what this increase actually amounts to, what it depends on, and what evolutionary mode it represents.

For Hemmer and Hofman, at least, the picture is one of punctuations: transition from one static allometric slope to another, the slopes themselves being parallel. Hofman (1983) specifies that there are three major transitions (australopithecine to *H. habilis* to *H. erectus* to *H. sapiens*), with no intermediate stages between. The problems in all these transitions are energetic, and brain weight is firmly scaled to body weight (note, however, the degree to which this involves extrapolations: see section 8.3 below). According to Armstrong (1985), brain size scales isometrically to its energy reserves, so that, proportionally, more of the body's energy reserves must be mobilized to serve the brain than in any other taxonomic group. Blumenberg (1984) surveys the regressions that have been calculated for brain weight against body weight, and supports the associations previously argued by Cutler (1975) between brain size and longevity. The brain size increase is entirely post-natal (see also Berge *et al.* 1984), but it does not follow that the gestation of *Homo sapiens* has been shortened: indeed, Blumenberg calculates from the regression equation of Leutenegger (1977) that it should be 223 days; in other words, it has actually been lengthened during evolution.

As I will discuss below, the reconstruction of body weight, to serve as an independent variable, is fraught with such difficulty as to render any results based on it rather questionable. The logical procedure, one would have thought, is to take some directly measurable variable and regress cranial capacity against it. Only Hemmer (1967, 1972, etc.) has done this; he argues that, despite (or, perhaps, because of) the fact that it is part of the braincase itself, the Glabella–Opisthocranion measurement is eminently suitable to act as the independent variable. There will be a slight, but not strongly signi-

ficant, effect of the development of superstructures such as glabellar prominence; but, by and large, an intrapopulation regression line will represent
simply the relationship between brain size and absolute size (head length, the
closest approach to body size standardization one can reasonably hope for),
while a shift of the line will indicate a different relationship (lateral or vertical
enlargement of the brain, or some shape differentiation, as opposed to the
brain being larger simply because the body is).

I have brought Hemmer's (1972) diagrams up to date (Fig. 8.1); in
agreement with Smith (1984) I have kept the figures normal (non-
logarithmic), the regressions falling along straight lines without any need to
turn them into logarithms. Initially, a series of regression lines were
calculated: for different samples of *H. erectus* and *H. sapiens*, and for
different combinations of other taxa for which only small samples are
available.

From data given by Hrdlicka (1928) and Slome (1929), different regressions have been calculated for modern *Homo sapiens* samples (Table 8.1).
The slopes (on, admittedly, rather small samples in the main) vary from 0.031
to 0.041, with one outlier, a sample of Solomon Islanders at 0.062; the
intercepts are between 3.47 and 5.39 (except, of course, for the Solomons
sample). A sample of 23 upper Palaeolithic Europeans and North Africans
gives a somewhat lower slope, 0.024; and five upper Pleistocene(?) Chinese
give 0.022.

Hrdlicka's (1928) samples are all of Africans and Melanesians, so the next
step is to combine the small samples into these two geographic groups [Fig.
8.1(a)]. The Africans ($n$ = 75, including Slome's sample of 17 Khoisan) now
give a slope of 0.039, and the Melanesians ($n$ = 45) give 0.040. The intercepts
differ only slightly: the slightly lower intercept for Melanesians suggesting
perhaps the influence of the prominent glabella. Combining the two samples
with the upper Palaeolithic gives a slope of 0.034, and adding Skhul and
Qafzeh does not change this. The Neandertalers have a slope of 0.035
($n$ = 21), almost identical to the combined *H. s.* cf. *sapiens* sample, and
the intercept is hardly different (glabellar prominence again, perhaps,
accounting for the slightly lower intercept). We can therefore combine all
these upper Pleistocene/Holocene samples into one, giving a slope of 0.032,
intercept 5.054 ($r$ = 0.814, $n$ = 77).

In *H. erectus* [Fig. 8.1(b)] the slope for *H. e. pekinensis* (Zhoukoudian plus
Hexian) is 0.028; for Ngandong plus Sambungmacan, 0.017; for Sangiran,
0.020. As Hemmer (1972) found, these lines are nearly identical, and can be
combined. The slope for all of Asia is 0.030; including OH9 gives 0.029 with
an intercept of 4.35 ($r$ = 0.756, $n$ = 22). All these slopes are slightly lower
than most of those for *H. sapiens*; even so the intercept is somewhat less.

The archaic *sapiens* subspecies (*heidelbergensis* and *mapaensis*) can now be
fitted in: they span the border area between the *erectus* and *sapiens* polygons,
but are mostly within the former. Omo II and Ngaloba fit within the same

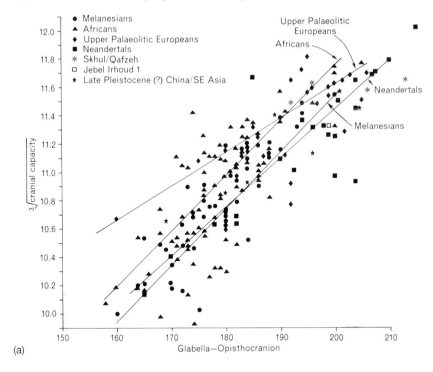

(a)

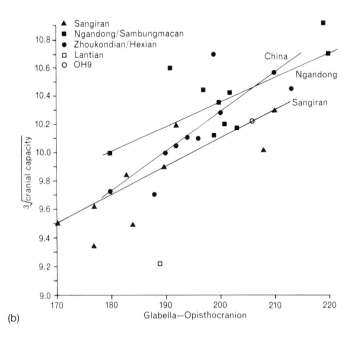

(b)

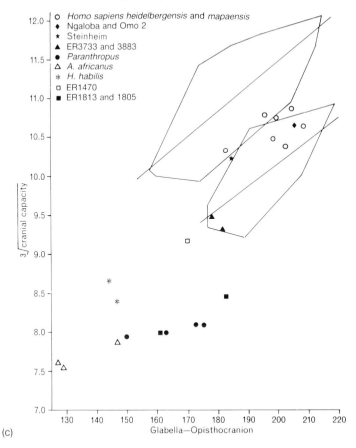

**Fig. 8.1.** Relative scaling of brain size, as represented by cranial capacity, on absolute size, as represented by Glabella-to-Opisthocranion length. For explanation of rationale, see text. In each diagram, abcissa = Glabella–Opisthocranion; ordinate = cube root of cranial capacity.

(a) Late Pleistocene and modern *Homo sapiens*. The data points are all intermingled, and the slopes and intercepts are not significantly different.

(b) *Homo erectus*. The data points are again intermingled, the slopes and intercepts not significantly different. Only Lantian falls outside the range of variation for other samples.

(c) All Hominini. The samples plotted in (a) and (b) are here represented by polygons and regression lines only; the upper polygon and line represents late Pleistocene/modern *Homo sapiens*, the lower represents *Homo erectus*. Archaic taxa of *H. sapiens* scatter across the boundary between the *erectus* and later *sapiens* polygons; some other (mainly early Pleistocene) fossils fall in the *erectus* polygon, or near a downward extension of the latter's regression line, but the australopithecines and *H.ergaster* are well outside even a hypothetical downward extension of the *erectus* range.

**Table 8.1** Cranial capacities of living and fossil hominins.

| | Capacity | | | | Regression on Glabella–Opisthocranion | | | |
|---|---|---|---|---|---|---|---|---|
| | Mean | sd | CV | n | Slope | Intercept | r | n |
| *Homo sapiens cf. sapiens* | | | | | | | | |
| African | | | | | | | | |
| Khoisan | 1159.0 | 118.93 | 10.26 | 15 | 0.034 | 4.584 | 0.555 | 17 |
| Kenya: male | 1401.0 | 80.36 | 5.74 | 14 | | | | |
| female | 1242.0 | 105.61 | 8.50 | 25 | | | | |
| all | 1299.1 | 123.41 | 9.50 | 39 | 0.034 | 4.781 | 0.549 | 39 |
| W. Africa: male | 1360.0 | 64.49 | 4.74 | 7 | | | | |
| female | 1181.7 | 114.96 | 9.73 | 6 | | | | |
| all | 1277.7 | 127.08 | 9.95 | 13 | 0.041 | 3.787 | 0.884 | 13 |
| S. Africa: male | 1402.5 | 102.70 | 7.32 | 6 | 0.031 | 5.385 | 0.829 | 6 |
| Melanesian | | | | | | | | |
| New Guinea: male | 1418.3 | 27.54 | 1.94 | 3 | | | | |
| female | 1190.8 | 86.51 | 7.26 | 6 | | | | |
| all | 1266.7 | 133.44 | 10.53 | 9 | 0.036 | 4.435 | 0.784 | 9 |
| Torres Straits: | | | | | | | | |
| female | 1145.0 | 68.68 | 6.00 | 4 | | | | |
| all | 1195.0 | 126.64 | 10.60 | 5 | 0.036 | 4.421 | 0.724 | 5 |
| New Britain: male | 1311.5 | 131.23 | 10.01 | 13 | | | | |
| female | 1152.3 | 94.27 | 8.18 | 11 | | | | |
| all | 1238.5 | 139.36 | 11.25 | 25 | 0.041 | 3.468 | 0.905 | 24 |
| Solomons: male | 1403.0 | 42.07 | 3.00 | 5 | | | | |
| all | 1370.8 | 87.32 | 6.37 | 6 | 0.062 | − 0.509 | 0.927 | 6 |
| Other | | | | | | | | |
| Upper Palaeolithic, | | | | | | | | |
| Europe/N. Africa | 1482.0 | 112.58 | 7.60 | 28 | 0.024 | 6.830 | 0.641 | 23 |
| ditto, China | 1374.0 | 121.98 | 8.88 | 5 | 0.022 | 6.948 | 0.871 | 5 |
| *Homo sapiens neanderthalensis* | | | | | | | | |
| Europe | 1432.9 | 163.07 | 11.38 | 12 | | | | |
| Middle East | 1549.3 | 198.04 | 12.78 | 4 | | | | |
| Together | 1462.0 | 173.35 | 11.86 | 16 | 0.032 | 5.010 | 0.825 | 16 |
| Pre-neandertal | 1283.0 | 178.75 | 13.93 | 6 | 0.046 | 2.112 | 0.725 | 5 |
| All Neandertal | | | | | 0.035 | 4.450 | 0.800 | 21 |
| All Africans | | | | | 0.039 | 3.950 | 0.682 | 75 |
| All Melanesians | | | | | 0.040 | 3.542 | 0.868 | 45 |
| All moderns | | | | | 0.039 | 3.830 | 0.749 | 120 |
| Moderns + upper Palaeolithic | | | | | 0.034 | 4.571 | 0.798 | 152 |

| | Capacity | | | | Regression on Glabella–Opisthocranion | | | |
|---|---|---|---|---|---|---|---|---|
| | Mean | sd | CV | n | Slope | Intercept | r | n |
| ditto + Skhul | | | | | 0.034 | 4.960 | 0.806 | 156 |
| ditto + Neandertal: i.e. all upper Pleistocene | | | | | 0.032 | 5.054 | 0.814 | 177 |
| *Homo sapiens, archaic* | | | | | | | | |
| Maba, Dali | 1185.0 | | | 2 | | | | |
| Europe | 1164.3 | 36.53 | 3.14 | 3 | | | | |
| Africa early | 953.3 | | | 2 | | | | |
| Africa middle | 990.0 | | | 2 | | | | |
| Africa late | 1235.0 | 40.93 | 3.31 | 3 | | | | |
| *heidelbergensis* | 1108.5 | 172.02 | 15.5 | 10 | 0.016 | 7.331 | 0.785 | 5 |
| *Homo erectus* | | | | | | | | |
| *pekinensis* | 1050.7 | 101.22 | 9.63 | 7 | 0.028 | 4.687 | 0.723 | 7 |
| Ngandong/S'macan | 1130.7 | 93.25 | 8.25 | 7 | 0.017 | 6.962 | 0.510 | 7 |
| Sngiran/Trinil | 923.9 | 79.21 | 8.57 | 8 | 0.020 | 6.103 | 0.740 | 6 |
| ALl Asian | | | | | 0.030 | 4.191 | 0.741 | 21 |
| Asian + OH9 | | | | | 0.029 | 4.347 | 0.756 | 22 |
| *'Habilines'* | | | | | | | | |
| *H.habilis* | 653.8 | 55.28 | 8.46 | 4 | | | | |
| + 1470 | | | | | 0.025 | 4.916 | 0.896 | 3 |
| + *H.ergaster* | | | | | – 3.613 | 8.946 | – 0.229 | 4 |
| All together | | | | | 3.571 | 7.957 | 0.134 | 5 |
| *'Australopithecines'* | | | | | | | | |
| *Paranthropus* spp. | 515.2 | 13.97 | 2.71 | 5 | 0.007 | 6.971 | 0.165 | 3 |
| *A.africanus* | 450.3 | 29.45 | 6.54 | 7 | 0.015 | 5.644 | 0.957 | 3 |
| Together | | | | | 0.012 | 6.123 | 0.965 | 6 |
| Australopithecines | | | | | | | | |
| + 1470 | | | | | 0.021 | 4.801 | 0.732 | 7 |
| ditto + *H.habilis* | | | | | 0.018 | 5.367 | 0.578 | 9 |
| ditto + ditto + | | | | | | | | |
| *H.ergaster* | | | | | 0.015 | 5.852 | 0.555 | 11 |
| *A.africanus* + 1470 | | | | | 0.037 | 2.812 | 0.958 | 4 |
| ditto + *H.habilis* | | | | | 0.037 | 2.875 | 0.892 | 6 |
| *A. africanus* + *H.ergaster* | | | | | 0.015 | 5.608 | 0.987 | 5 |

range, though the former is at the very upper limit of it. Evidently there was a sudden, rapid transposition, at the time of the emergence of the modern and Neandertal grade. ER3733 and 3883 fall in the *erectus* polygon [Fig. 8.1(c)].

The australopithecines, if combined into a single group (*A. africanus* with *Paranthropus*) give an extremely low slope, 0.012. This is far below those for the two *Homo* species, and seems to me to indicate that the two taxa have different degrees of encephalization. No lines are therefore drawn in Fig. 8.1 for the australopithecine level. *H. ergaster* has as low a level as the australopithecines. *H. habilis*, on the other hand, is higher, and would fall on a downward extension of the *H. erectus* slope, which would also bring in ER1470 on the way. In the absence of any comparative ape sample, it is difficult to know what sort of slope to look for, and so which combinations would give a reasonable biological result; as the slope of the *erectus* grade is lower than that of the *sapiens* grade, I hypothesize that among the various slopes for 'australopithecines plus' listed in Table 8.1 only those lower than 0.029 (the combined *erectus* slope) merit further consideration. We may presumably take it that *H. ergaster* has at least an australopithecine grade of encephalization; consequently, no slope higher than 0.015—that connecting *A. africanus* and *H. ergaster*—would seem to be acceptable. This would seem to require that (1) *Paranthropus* has a degree of encephalization little or no less than *Australopithecus* (and *H. ergaster*), as the slope remains the same whether or not *Paranthropus* is included; and (2) *H. habilis* and the 1470 taxon have a higher grade of encephalization; the slope of 0.025, while based on only three specimens (ER1470, OH16, and OH24) would be of the right general magnitude.

Before leaving the question of the measurement of encephalization, a word is necessary about the influential Pilbeam/Gould allometric model (Pilbeam and Gould 1974). It seems doubtful, reading their subsequent publications, whether either of the two authors would now hold to the model proposed, at least in the form in which it was set out; but just because it has been so influential—differences between australopithecines explicable purely by scaling, and so on—it is necessary to return to it, even though it was effectively demolished the following year by Wolpoff and Brace (1975). The two decisive criticisms are the authors' choice of specimens to include in their chosen species, and the uncertainty surrounding body weight estimates. The first point is important: to exclude OH24 from *Homo habilis* while including ER1470 in it (neither decision being argued) is to risk altering the position of the species on the chart. The second point, however, is really crucial: the authors cite a range of body weight estimates for each species before choosing one to use as a basis for curve-fitting, noting as they do so how very difficult it is to reconstruct body weights. Wolpoff and Brace (1975) cynically remark that guesses should not form the basis for an evolutionary scheme dependant on an allometric pattern, and Kay (1975) strenuously disputes at least one of their estimates.

What difference would it make if the body weight estimates were wrong? A great deal. Considering first their cheek-teeth-scaling diagram (Pilbeam and Gould 1974; Fig. 3, p. 896), use of their 'extreme' body weight estimates, or even the less extreme ones favoured by Kay (1975), would push both *A. africanus* and what they call *A. boisei* right off the 'australopithecine' line; while use of McHenry's (1975) weight estimate for the latter would put it on the pongid line! In their Fig. 2 (the cranial capacity diagram; 1984, p. 895), use of McHenry's body weight estimate for *boisei* would again place it on the 'pongid' line. If *Homo habilis* were based on just Olduvai specimens (mean = 650), the *Homo* slope would become noticeably steeper; if it were based just on ER1470 and 1590, it would become shallower.

The proof of the pudding is in the eating? Perhaps, and one cannot but agree that the allometric approach was a remarkably innovative idea in those days; but proof-of-pudding arguments have a hidden circularity in them. This circularity is best brought out by a comment by Pilbeam and Gould (1984) on their choice of weight estimate for *A. africanus*: 'Our results suggest that some of the estimates may not be far wrong'! (p. 900).

This lengthy discussion is in part a justification of the approach used above: we do not know how much these beasts weighed; the only way even to estimate (and that poorly) their weights is on the basis of preserved remains; so, why not reduce the uncertainty by using the remains themselves? And so we return to the problem: what does it all mean?

The most radical approach to modelling the increase in size of the human brain is that of Blumenberg (1983). In essence, this 'stochastic, hierarchical' model reverses the old argumentation of what brain enlargement is for, and asks: What did it make possible? Tool use will not work as a prime mover: the earliest known stone tools precede known brain enlargement. Similarly, no profound ecological change appears to coincide with brain enlargement. On the other hand an enlarged brain confers on its bearer, individual and/or population, a 'profound adaptive potential'; and if meat-eating was becoming a large factor in the hominin diet—and Blumenberg quotes archaeological evidence that it was—then a powerful feedback relationship (via protein and, especially, calcium availability) with brain development would be set up.

What is new and challenging about Blumenberg's (1983) model is its rejection of adaptive models, the 'just-so stories' of Gould and Lewontin (1979). There is absolutely no doubt that a large brain, once developed, has a vast exaptive potential (hence, not strictly 'adaptive', as Blumenberg claims) and, given the enormous metabolic constraints on it as described above, a switch of dietary base can hardly fail to have affected its future development. It is hierarchical in that the cause-and-effect chain operates at several levels: genome, epistatic, phenotypic, group/social.

Blumenberg (1983) makes it clear that the 'cause' of brain expansion is not to be sought in its adaptive significance as such. It seems to me that he comes within an ace of saying that brain enlargement is secondary—to what? I will extend the argument, and in section 8.3 below I will defend the following proposition: brain enlargement is an epiphenomenon of neoteny.

## 8.2 THE EVOLUTION OF HUMAN LOCOMOTION

Anyone who expected—and probably we all did, whether tacitly or explicitly—that our ancestors through time evolved more and more efficient bipedal locomotion, through a series of intermediate stages from then to now, must have had a rude shock over the past ten years or so. Oxnard (1975, 1984*b*, and references therein) has realized it for longer than that; McHenry has been slowly coming to the same conclusion and recently (1986) made it explicit: proto-hominid locomotion was not merely something intermediate between ape and human modes.

Ashton *et al.* (1981) summarized locomotor specialization of *Australopithecus africanus*, as represented by the pelvis, as being quite like the human form in weight transmission, but like apes in the attachment and disposition of muscle blocks. For McHenry (1984) the shoulder closely resembled that of *Pan paniscus*, but the wrist was more human; the pelvis, unlike anything else (small acetabulum, reduced width of bone behind acetabulum, broad iliac blade, great distance between posterior superior and posterior inferior iliac spines, reduced width of sacral surface, long superior pubic ramus); the distal femur was also unlike anything else, though marginally closer to the human form than to any ape. Oxnard (1975) specified that the hip extensors attached relatively more caudally than in humans, the iliac blade being oriented in an intermediate way, but (Zihlman and Hunter 1972) very broad, beaky, and flaring, lacking the anterior S-twist that in *Homo sapiens* facilitates internal rotation.

Lovejoy (1974) saw the *A. africanus* pelvis as a perfectly bipedally functioning one, differing from the modern human pelvis only in the lack of enlargement of the obstetrical 'true pelvis', which became necessary later on to accommodate an enlarged neonatal head size. This proposition has recently been examined by Berge *et al.* (1984), who find that obstetrical problems would in fact have been similar to those of *Homo sapiens*, with rotation and flexion of the neonate in the pelvic cavity: there is the same closing of the distance between sacroiliac and hip joints, dorsal inclination of the sacrum, and open subpubic angle, though the biacetabular diameter is greater, the ischiopubic index is greater (140, compared to under 120 in humans), the ilio-pubic angle is 140°, and the ischio-pubic angle is only 65–70°, the smallest in primates. The differences must therefore be locomotor, not obstetrical, contra Lovejoy. Schmid (1983) draws attention to the attachment of trunk muscles on the iliac crest, suggesting that the disadvantageous internal rotation action of *M. glutaeus medius* pointed out by

Zihlman and Hunter (1972) is balanced by a considerable advantage in the function of *M. latissimus dorsi.*

All this means that the locomotor system of *Australopithecus africanus* was unique—not simply an intermediate stage between us and apes. The immunological approach to primate evolution was cold-shouldered for years in favour of supposedly opposed palaeontological evidence, and has now been vindicated. Could there be a parallel here with the morphometric approach of Oxnard (1975, 1984*b*)?

What, then, of other taxa of australopithecine grade? The Hadar fossils are currently the subject of dispute, although it seems to me that the data of Senut and Tardieu, in particular, are hard to counter. The small Hadar species, as represented by AL288 (Lucy), AL137–48a, and AL322–1, has an elbow joint that is different from humans and resembles the African apes (Senut 1981). The knee joint of Al288 and AL129–1 is ape-like (Tardieu 1983). According to the multivariate analysis of McHenry (1984), in the small Hadar species the shoulder region is closest to *Pongo*, quite far from that both of the African apes and of *A. africanus*; the distal humerus is unlike anything else (though marginally closest to the gorilla); the capitate is rather human, and very close to *A. africanus*; the pelvis is virtually a duplicate of that of *A. africanus*; the proximal end of the femur, with its small head, medio-laterally thick shaft, long neck, and proximally placed lesser trochanter, is unlike anything else; and the distal femur, in agreement with Tardieu, is just like that of chimpanzees and orang-utans, and quite diffeent (but see McHenry 1986) from that of *A. africanus*. Jungers and Stern (1983) find that the lower limb is relatively short, and the pedal phalanges are long and ape-like. Overall, the skeleton seems to be close to that of *A. africanus*: not identical (especially the shoulder and the knee), but close enough to encourage McHenry (1986) to speak of a 'relatively stable adaptation' that persisted over a considerable period of time and space.

The large Hadar species, meaning in the main the AL333 sample, appears to differ from the small species in ways which make it much more human. The elbow has many modern features according to Senut (1981), in which, curiously, it resembles the much older Kanapoi elbow—but Feldesman (1982) strongly disagrees about the latter. The knee joint, too, is human-like (Tardieu 1983), although the lateral meniscus is, inferentially, ring-shaped as in apes not crescentic as in humans (Senut and Tardieu 1985). The legs may have been relatively longer than in Lucy (Jungers and Stern 1983). The calcaneus is however more ape-like (Deloison 1985); the pisiform is rod-like and the phalanges of both hand and foot are curved (Stern and Susman 1983). McHenry (1984) finds the capitate to be most like the human one, and rather different from both AL288 and *A. africanus*; the proximal femur being more like AL288 and the distal unlike anything else (apparently in broad agreement with Tardieu). Stern and Susman (1983) think that there must have been differences in locomotion between large and small Hadar types, a conclusion which McHenry (1986) considers premature.

Before turning to acknowledged members of the genus *Homo*, two other australopithecine-grade fossil groups must be considered. The question of locomotor adaptations in the Laetoli hominid depends, so far, on the famous footprints. Unfortunately, opinions differ on whether these could (Day and Wickens 1980) or could not (Schmid 1983) have been made by a biped in the human fashion. For Deloison (1985) they could have been made by a creature with a far from human calcanear morphology.

The other 'australopithecine' to be considered is of course *Paranthropus*. The knee joint of ER1500 is ape-like, as in the small Hadar species (Tardieu 1983). The distal humeri of ER739, 740, 1504, and 3735 also resemble the small Hadar form in some features, but are more human in many respects (Senut 1981). The Kromdraai talus is most similar to the orang-utan; the Swartkrans metacarpal is even further from humans and apes than these are from each other (Oxnard 1975). The arm-leg ratio of ER1500, according to McHenry (1978), seems intermediate between ape and human, or more like the gorilla; that of Kromdraai (with a humero-femoral index of 84) would be similar; as would be that of ER1503–4 (estimated humero-femoral index, 76–90). The pelvis is more specialized than in *A. africanus* or Lucy, and further from *Homo* (Berge 1984). In general, the evidence does tend to support Robinson's (1972) much-maligned contention that *Paranthropus* (a) differs from *Australopithecus* and (b) is less human-like. Many basic features of McHenry's 'relatively stable (australopithecine) adaptation' are none the less there, as they are in at least the small Hadar form; in the light of the results of the cladistic analysis of the previous chapter, this does support McHenry's (1986) contention that there was indeed such a configuration, which however, could, vary somewhat from one taxon to another.

There has been a considerable rethink of *Homo habilis* since the classic paper of Leakey *et al.* (1964) with its brief analysis of the post-cranial skeleton; beginning with a reconsideration of the hand by Susman and Creel (1979), followed by new analyses of the foot by Lewis (1981) and by Susman and Stern (1982). OH8 would seem to be unlike any other taxon: differing from *H. sapiens* in its short anterior calcaneus, the narrow talar facet on the navicular, the high navicular facet on the lateral cuneiform, and the ventro-medial 'beak' of the cuboid, lodged into a deep excavation of the calcaneus restricting the amount of supination. The subtalar axis is flat and oblique as in *Pan*, not reoriented toward the long axis of the foot, and more vertical, as in *Homo sapiens*; the hallux would still have been somewhat divergent, even though considerably approximated to the other toes (Lewis 1981). These quite un-human features are presumably reflected in the multivariate findings of Oxnard (1975), that the OH8 talus resembles Kromdraai, as does the OH10 terminal phalanx. It is a pity that so few of the skeletal parts of *H. habilis* correspond to those of other taxa. The overall impression is, however, of a post-cranial skeleton definitely more human than in any 'australopitherine', though far from perfectly modern; and, from the evidence we have, perhaps like the large Hadar species.

Nor has *Homo erectus* been permitted to remain, unexamined, as post-cranially *sapiens*-like. Apparently independently, both Day (1982) and Sigmon (1982) came to the conclusion that there had for too long been an untested assumption that the post-crania of *erectus* and *sapiens* are identical. The pelvis has an exaggerated acetabulo-cristal buttress, comparatively large acetabulum, small auricular surface, forwardly raked anterior superior iliac spine and, where preserved, a medially rotated ischial tuberosity (Day 1982). The iliac blade is medially deflected anterior to the acetabulo-cristal buttress, and the medial fossa is not strongly excavated (Sigmon 1982). The femur has strongly developed platymeria, at least in the proximal part of the shaft; a very low position of the minimum shaft breadth; a large head; and a very thick cortex (Kennedy 1983).

The interesting thing about this morphology is that it is not restricted to *Homo erectus* as cladistically defined. Specimens possessing the above morphology are the Zhoukoudian specimens, which are certainly *H. erectus*, and OH28, which may be; also Arago XLIV and, as demonstrated to me by C.B. Stringer, Kabwe, both of them here classed as *Homo sapiens heidelbergensis*; ER1808, a pathological skeleton from above the KBS Tuff at Koobi Fora, referable to the 3733 species; and, quite unexpectedly, ER3228 (pelvis) and 1481A (femur) from below the KBS Tuff. Kennedy (1983) regards the femoral morphology as diagnostic of *H. erectus*; hence all these specimens belong to that species. Trinkaus (1984) disagrees, noting that the features regarded as those of *H. erectus* by Kennedy are more those of archaic *Homo*, and in modified form persist in Neandertals, constituting an excellent example of evolutionary stasis [he regards ER1481A as *H. habilis*, and, consistently, Rose (1984) regards ER3228 as belonging to this species as well].

Trinkaus's (1984) point is certainly well-taken, and consistent with McHenry's (1986) model of australopithecine stasis cited above; although, as other fossils from below the KBS Tuff at Koobi Fora are not here classed as *H. habilis*, nor can ER1481A be. The near identity (not perfect identity, as Trinkaus shows) of the sub-KBS femur with those of *H. erectus* and archaic *H. sapiens* raises an interesting point: Could the Olduvai distal limb specimens have fitted, functionally speaking, onto a limb which began with an *erectus*-like pelvis and femur? Since the '*erectus*' pattern consists of pelves and femora, the *habilis* pattern of a foot with (apparently) innocuous tibia and fibula, are the two patterns really the same, a pattern lasting from the inception of the genus *Homo* until the emergence of Neandertal on the one hand and anatomically modern people on the other?

The above summary certainly cannot do justice to the complexity of the issues and the diversity of the views expressed by different workers. I have not dealt at all with the distinctive Neandertal morphology, as described, for example, by Heim (1976) and Trinkaus (1983). What seems in general to be the story is a succession of morphologies: an australopithecine one, in which bipedal locomotion was only part of a pattern which also incorporated a sophisticated climbing ability; a habiline one, in which some climbing ability was present but a more terrestrial mode, not however a developed striding bipedalism, had taken over; an erectus-grade one, which may be the same as the habiline one but is represented by different body parts and shows a robust pelvic and femoral build, perhaps implying a heavy, barrel-shaped trunk; and a sapient one, in which body build has become more linear and a full-striding gait with internal rotation and tripod foot has developed. Each 'phase' was long-lasting; the australopithecine phase showed much internal variation, some of it in the direction of later phases; the erectus-grade one apparently showed little variation.

This is certainly a very different story from the traditional one in which modern locomotion has been achieved by the *habilis* level, and the australopithecines were not-yet-perfected intermediate stages, well out of the trees and waddling across the savannah with their eyes fixed on that distant day when their inefficient present would become a fully adapted future and it would all have been worth it.

## 8.3 HETEROCHRONY IN HUMAN EVOLUTION

In Chapter 3, section 3.1, I briefly described the idea of heterochrony, elaborated in its modern form by Gould (1977) and Alberch *et al.* (1979), and noted that it could explain much about the kinds of changes that occurred in evolution. In particular it is genetically economical, requiring not a series of point mutations but perhaps one or a few very simple alterations in rate. I argued in section 3.1 that there are really three distinguishable heterochronic modes: progenesis, in which ontogeny is cut short at a given point; hyper-morphosis, in which it is prolonged; and neoteny, in which dissociation occurs. In a way, therefore, neoteny is in part the co-occurrence in the same organism of the two other processes.

If the important point about neoteny is dissociation of different developmental processes, then it becomes valid to look at an organism—or rather, a taxon—character by character, in the manner of cladistic analysis, to evaluate the ontogenic status of each. This is not to say that some general morphogenetic state might not characterize the whole organism, with further degrees of heterochronic fine-tuning. The most advanced example of neoteny among mammals appears to me to be the order Cetacea. (Doubtless all marine mammals are neotenous to some degree.) Here, the general body form is that of an artiodactyl embryo at about the time the limb buds are forming: the fusiform body shape, with no strong demarcation between body and tail;

the absence of external appendages, such as ears; the glabrous skin; and the relatively undifferentiated condition of the skeleton, with hardly any hind-limb skeleton at all, manus not yet pentadactyl, vertebral types not diffe-rentiated, ribs poorly formed and no sternum, cranial elements retaining much kinesis, tooth buds simple in form. The general picture is of develop-mental arrest at a particular stage; as a consequence of this the growth rate characteristic of that stage is maintained, and cetaceans are all of large size. Yet in other respects—and it cannot be otherwise if a functioning organism is to result!—development continues: the skeleton ossifies, the soft parts develop. A hypothesis like this enables one to ask questions: are larger species more neotenous than small ones? Are Mysticeti as a whole, in which the teeth fail to erupt, more neotenous than Odontoceti, in which they do at least erupt however simple in form they remain? (I am aware that, looked at simplis-tically, these two propositions are mutually incompatible, as the smallest mysticetes are much smaller than the largest odontocetes; but neoteny, as emphasized above, is not a simple nor a unitary process.) On the other hand, neotenous processes are bound to have their exaptive consequences; in the cetacean case, the prolongation of foetal growth rates cannot but involve the organ that is growing fastest at this time: the brain. I am greatly enamoured of the idea that the humpback whale is an overgrown pig embryo with knobs on; and I find it revealing that in these neotenous mammals the large brain is most plausibly viewed as not selected for as such, but is large by being, so to speak, swept along with the predominating trend. Because this is exactly what Blumenberg (1984) maintained about the human brain (see section 8.1 above); and it has been proposed, most lately by Gould (1977), that neoteny is the guiding principle in human evolution too.

I now propose to examine the characters differentiating the taxa and lineages of Hominini, to see which fit the neoteneous pattern. Up to now there has been much general talk of modern humans being neotenous with respect to other hominoids, but no examination of just where the neotenous events occurred. In the origin of anatomically modern *Homo sapiens* alone? Of the genus *Homo*? Or of the tribe Hominini? Or stepwise, in many places?

We know too little of the small Hadar species, gen.indet.*antiquus*, to determine whether neoteny had occurred. The short canines are the only indication that it might have done so.

There are a number of neotenous traits among those shared by *Homo, Australopithecus*, and *Paranthropus*. The reduction of prognathism, facial and alveolar, is the most important of these. Reduction of the basal tubercles of the maxillary incisors is another. A case could be made that the reduction of the lateral anterior crest on the humerus is neotenous.

The genus *Paranthropus* is characterized by several neotenous conditions among its derived traits: small size of canines, and their symmetrical shape; reduction of lingual relief on lower incisors and canines; oval shape and central placement of cusps on lower molars; enhancement of curve of Spee;

further reduction of prognathism; loss of occipito-mastoid crest; and basicranial shortening.

A few neotenous features occur in the line leading to *Australopithecus* and *Homo*: the shortened face, both relative to its breadth and in bringing the upper end of the nasal aperture up between the orbits; and the high frontal angle. These few features persist in *Australopithecus*, which shows no increase in neoteny in those (relatively few) traits in which it may be autapomorphic. Nor is there much evidence for neoteny in the *Homo* line prior to the separation from it of the large Hadar species (if correctly placed here), although the deepened mandibular fossa could be such a character.

In the line leading to other species of *Homo*—the firmly placed species, above the separation from the Hadar form—there is a great deal of neotenous change. The reduction of third molars; broad palate and maxillo-alveolar index above 100; reduced alveolar prognathism; deep, narrow mandibular fossa; lateral facing frontal process of maxilla; reduced post-orbital constriction; reduced angulation of petrous axis; forward sloping foramen magnum; and further reduced facial height; all these features represent an ontogenetic regression and are hence neotenous.

If my preferred cladogram—*rudolfensis* splitting off first—is correct, then I cannot identify neotenous changes in the lineages within *Homo* until we reach *H. habilis*; but in the case of this latter species, almost all its uniquely derived character states are neotenous. These are the simplified morphology of $P_3$ and the canines; reduction of upper canine size; the gracile cranial form, without a defined supramastoid crest; and shortened basicranium, with basion well in front of bitympanic line. Only the groove pattern on $M^1$ and the elongation of the lower premolars cannot be definitely ascribed to neoteny.

It is of considerable interest that *Homo habilis* is more strongly neotenous than its sister group, the lineage of other *Homo* species. The features defining the other lineage, mainly relating to midfacial prominence and occipital remodelling, are possibly somewhat hypermorphic; so is the vault flattening characteristic of *H. ergaster*. But after the separation of this latter species, neoteny begins to characterize the lineage again, and with increasing intensity, stage by stage. The *Homo erectus/sapiens*/3733 lineage has reduced alveolar prognathism and reduction of alveolar planum, reduced symphyseal tori, shortened basicranium, facial broadening, third molar reduction, equalization of maxillary incisor sizes, and simplification of lower molar crowns by reduction of tuberculum sextum. The Koobi Fora species (3733 type) continues the neotenous process, with nasals broadened relative to their height, broadened parietals, further basicranial shortening (vertical tympanic plate; much reduced petrotympanic angle), shortened palate of even depth, reduced mandibular robusticity.

Then the line splits, and once again we have a contrast between a hypermorphic species (*Homo erectus*, with its enlarged superstructures) and a neotenous one (*H. sapiens*, with its high, rounded, gracilized vault, reduced post-orbital constriction, and further reduced symphyseal tori). Finally, within *H. sapiens* there is another split, again between a hypermorphic form (*H. s. neanderthalensis*) and a further neotenous one (*H. s.* cf. *sapiens*).

There is, thus, a trend towards ever greater neoteny in the modern human condition; were this not carefully analysed, it could easily be mistaken for a case of gradualism, but close examination shows that it was in fact a jerky process. Nor was it the line leading to modern humans that was always the one characterized by neotenous changes. *Homo* is more neotenous than *Australopithecus*, the *erectus/sapiens* lineage more neotenous than *H. ergaster*, *H. sapiens* more neotenous than *H. erectus*; but *Paranthropus* is more neotenous than the *Australopithecus/Homo* lineage, and *H. habilis* more neotenous than the *ergaster/erectus/sapiens* lineage. It is, however, very noticeable how neotenous conditions arise with increasing frequency at each quasi-speciation event along the lineage from *H. ergaster* to modern humans.

How does brain size fit into this? Is it, as I have postulated for whales, dragged along in the wake of neoteny? We have seen that at the 'lower end' of human evolution the evidence is very poor. There are three endocranial volume determinations for Hadar specimens, two of them—the two regarded with some confidence by Falk (1985)—being extremely small (one for the small form, the other for the large form); unfortunately there are no cranial lengths to go with them, so their relative sizes cannot be assessed. We cannot, therefore, tell whether the not-too-different(?) encephalization levels of *Australopithecus* and *Paranthropus* are unchanged from their common ancestor, or acquired in parallel from a lower level. As aspects of the neotenous changes in both lineages—basicranial shortening in *Paranthropus*, forehead slope in *Australopithecus/Homo*—concern braincase features, it is quite possible that there has been parallel development.

Somewhat of a puzzle is that, although neotenous neurocranial changes occur in the main-line early *Homo* lineage (after separation from the Hadar 333 form), *Homo ergaster* has a degree of encephalization which is certainly not above that of *Australopithecus africanus*. Perhaps there has been a reduction in encephalization of *H. ergaster*?—I suppose this to be unlikely, or perhaps I, like everyone else, am too bemused by the magnificence of the human brain to contemplate even the possibility that any evolutionary line could ever regress once it had begun on the path to its acquisition. The preferred model must therefore be that the basicranial shortening of main-line *Homo*, which is what the neurocranial neoteny at that point mainly amounts to, did not bring endocranial enlargement in its wake.

Both *Homo rudolfensis* and *Homo habilis* have a degree of encephalization above that of *Australopithecus africanus*. If the cladogram is correct, then this enlargement has been independent of both each other and of later enlargements along the *erectus/sapiens* lineage. The alternative is the unthinkable, mentioned above: that *H. ergaster* has actually reduced its level of encephalization. *H. rudolfensisis* is not otherwise neotenous, or perhaps one should specify that it is not known to be so, though the species is rather incompletely known; but *Homo habilis* is, very strongly (both neurocranially and facially). The *erectus/sapiens* lineage is strongly neotenous, and has a degree of encephalization well above that of its sister group, *H. ergaster*; *H. sapiens* is strongly neotenous, and is still further encephalized compared to *H. erectus*.

The proposed neoteny/brain enlargement programme seems to work sometimes, but not at other times. Certainly it works at the later levels, where our information is best; and where the neotenous changes cluster around the neurocranium it certainly does seem to work. This does not of course resolve the question, why? There is a correlation, pointed out by Cutler (1975) and abundantly noticed by other authors since then, between brain size and longevity; brain size is related to neoteny; neoteny is correlated by Gould (1977) with K-selection, and this in turn is part of a package which involves longevity and prolonged growth stages. What, then, is causing what? I think we have a glimmer of some answers, but we are far from knowing the questions to which they may respond! I wonder whether, in our present state of understanding at least, the questions and the answers may not be in some sense mutually exclusive (Adams 1982, 158).

## 8.4 ECOLOGY AND HUMAN EVOLUTION

The ecological setting of human evolution is of great interest, even if it is difficult to derive the right questions from it. The 'killer ape' hypothesis of Dart (1953) and Ardrey (1961), given a more acceptable face by Washburn and Lancaster (1968), ascribed all the innovations of human evolution to the adoption of a hunting life style: upright posture, tool use (read: weapon use), canine reduction, brain enlargement, language. The model was criticized by Jolly (1970), and caustically—and aptly—dissected by Morgan (1972); Jolly saw proto-human traits as due to feeding on small, hard objects (grass seeds), while Tanner and Zihlman (1976) argued that the major features appealed to as hunting adaptations could at least as well, and in some cases (especially material culture) better, be explained as adaptations to food gathering by females, a sex whose existence had been forgotten by killer ape proponents. Isaac (1976) tried to balance things out by offering a division-of-labour model, in which return to a home base was suggested as a crucial human innovation; a point expanded on, with comments on social organization (and critique of the view of human social organization as 'pair-bonding'), by Groves and Sabater Pi (1985).

Most authors have seen 'down from the trees' and its extension, out into the savannah, as important ecological (and literal) steps in human evolution, though Morgan (1982) prefers to see many human features—bodily hairlessness but scalp hypertrichosis, subcutaneous fat, manual dexterity, even upright posture—as relicts of an aquatic past. African apes are partly (chimpanzees) or almost wholly (gorillas) terrestrial, and australopithecines are turning out (see Section 8.2 above) to have had an arboreal function in their post-cranial skeleton that was more than just residual; so 'down from the trees' just will not do, at least in the initial stages of human evolution; as we do not know what the habiline post-cranial morphology meant, nor even whether it was the same as the erectine one or not, we cannot appeal to a shift along the arboreal–terrestrial spectrum at any stage. What we do know, especially from the work of Behrensmeyer (1985), is that at Koobi Fora *Paranthropus* remains are concentrated in lake–shore environments, whereas those of *Homo* are more widely but more thinly spread. What we do not know is what these creatures were eating: big game, small game, or plant foods only (see especially Binford 1981); this being so, we can hardly insist on any one particular model of proto-hominid diet as a guide to the meaning of morphological or behavioural changes, although it must be acknowledged that vegetable food is far more likely to have figured large in the equation than is meat.

Blumenberg (1983) draws attention to the differential abundance of *Homo* and *Paranthropus* at different sites. Why do the respective percentages (*Homo/Paranthropus*) vary from 80/20 at Olduvai to 40/60 at Koobi Fora and to 10/90 at Swartkrans? The question may or may not be meaningful, but it would seem to be one that could help to shed light on the problem of extensive hominin sympatry.

The rather unusual model of Geist (1978) ought to be considered here. Based on the same author's 'dispersal theory', whereby different selective regimes can be expected in different environments, partly as a result of degrees of intraspecific competition, Geist sees human evolution as a succession of alternating competition-free and competition-bound phases. He is also, unlike many other writers, alive to the importance of seasonality. The gracile australopithecines (read: genera *Australopithecus* and *Homo* early taxa) would, on present-day mammalian analogies, have subsisted on small packages of food of high-nutrient content in dry savannah, while the robusts (*Paranthropus*) would subsist on meristematic plant parts (not seeds, contra Jolly 1970), and would be on wet savannah and near flooding rivers, and never far from trees. Geist does not cite Behrensmeyer's (1975) work here, but the model is remarkably consistent; and the low biomass thereby predicted for the 'graciles' compared to the 'robusts' squares well with their relative densities at Swartkrans and Koobi Fora (though not at Olduvai!). In the dispersal theory the evolution of new adaptive syndromes is largely a consequence of dispersal into new, unoccupied areas; so the *erectus/sapiens* lineage must have evolved outside Africa and replaced its precursors by carnivory at a later stage. Neoteny is due to the rapid selection for prolonged growth and larger body size in the empty, but rapidly filling, environment following dispersal; it comes to an end as the new environment is filled up, but continues at the edge of the range where the population is still dispersing into new environments. The model is complex and, I think, does not entirely fit with the facts (if that is what they are!) that have been deduced, both by archaeology and by analyses such as the present one, but is very stimulating and closely argued—on analogies with other mammals, not on suppositions of what people like us, only perhaps less-intelligent, might do.

I have given much less space to ecological modelling than is usual in a work of this nature. This is partly because I do not wish to stray too far into archaeology and theoretical ecology, and partly because the dispassionate observer cannot but get disheartened in surveying what has been done up to now. Jolly, Tanner and Zihlman, Isaac, Morgan, Geist: all have brought up valid points and proposed theses that are at least partly testable. But all such theses quickly threaten to get out of hand, and the cagey suggestions, 'proto-hominids might have done . . .', quickly nosedive into mere speculation, especially in the hands of writers with less personal experience with the actual evidence. In the end the most that can be said for a given model will be that it is consistent with the archaeological evidence; the most productive models generate predictions that can then be tested archaeologically, and it may well be that ultimately it will be possible to obtain some consensus; but so far I get the distinct impression that much more has been said than is warranted by the evidence.

## 8.5 A NOMOGENETIC THEORY OF HUMAN EVOLUTION

In Chapter 3, section 3.4 I drew together the threads of the evolutionary argumentation of the past 15 years or so, and proposed that Berg's (1926) concept of nomogenesis be revived to cover the non-Darwinian modes that do, after all, seem to predominate in the major features of evolution. A list of eight points was made; the eight points are not meant to cohere to the extent that they stand or fall as a unit, but all relate to recent advances in evolutionary theory which do seem at least to be plausible. It is now time to test these on the story of human evolution.

1. If evolution proceeds largely by speciation, then human evolution—which has been very rapid, and led to a very highly differentiated modern species—should be a bush rather than a ladder. Moreover, as a highly autapomorphic form, modern *Homo sapiens* itself should be (a) a product of many sequential quasi-speciation events and (b) geologically young.

These predictions are all part of a single package; the whole thrust of section 7.5 in Chapter 7 was that it is absolutely correct. Indeed, I surprised myself in the degree to which the Hominini appeared to be speciose; not least in the number of species in *Paranthropus*—which should have been expected; after all, the best known of them (*P. crassidens* and *P. boisei*) are both highly derived in their morphologies.

As for modern *Homo sapiens* itself, it is the product of at least ten quasi-speciation events; and all indications are that it is geologically young, about 120–130 000 years old.

2. Fast-evolving (i.e. highly speciose) lineages are stenotopic and demic.

This prediction is difficult—at this stage, virtually impossible—to test. In that there is widespread sympatry of hominin species (up to four co-occurring at any one site), stenotopy is at any rate implied. Many of the species were evidently vicarious, occurring at only one or a few sites, suggesting a demic pattern.

3. New species appear suddenly, and begin by being very variable.

Certainly new taxa do appear suddenly in the fossil record; where gradual changes can be traced (as in the Java sequence), even after a million years or so they have not added up to the equivalent of a different species. The only fossil occurrence that is plausibly interpreted as marking the actual origin of a new taxon is that of *Homo sapiens* cf. *sapiens*. Here, apparently without any prior notice being given, two contemporary fossils from the Kibish Formation at Omo suddenly appear, documenting as far as I can see the appearance of the taxon to which we belong. The two are quite unlike each other: one almost (but not quite!) a typical *H. s. heidelbergensis*, the other almost (but again, not quite!) a typical anatomically modern *sapiens*. As if to confirm what is going on, there is the Ngaloba skull from Tanzania, approximately contemporary with the Kibish skulls and falling morphologically between them. We do not have sufficient dated specimens from that general period to indicate just how sudden the appearance of the modern human taxon was, but every indication is that it was sudden; and, if the differences among the three of them are real, then this initial phase was very variable—far more so than any taxon during the periods of stasis.

4. New species develop in the centre of the ranges of the parent species (centrifugal speciation); there will be many sympatric or parapatric quasi-species; a new species may arise from just one subspecies of its parent species (heteromorph speciation).

There is clearly, in present circumstances where the boundaries of proto-hominid distribution are not known, difficulty in deciding just where the centre of a given species' range may be; although some inferences can be made. Until the early Pliocene, the incomplete state of rifting in the Red Sea region meant that the Arabian peninsula was part of the African land-mass, so that the Afar Triangle was not 'peripheral'; after this period, connections betwen Arabia and the Horn of Africa were broken, and the Afar did now lie at the edge of the zone of continuous proto-hominid distribution (Whybrow

1984). The position of this crucial region thus changed during the time span under consideration. On the other hand, South Africa has never been anything but peripheral; when hominins spread throughout the Old World the position changed yet again, the Horn of Africa was now again 'central' in position, while Europe and the Far East took on peripheral roles.

The earliest levels at Hadar have two taxa: the primitive Lucy type (gen. indet.*antiquus*), and Coppens's (1983*b*) 'astonishingly modern' 333 type (here called *Homo* sp., albeit the sister group of all other *Homo*). The presence of the latter, the highly derived species, is entirely consistent with the centrifugal model: a new species has evidently been generated in a central region of the proto-hominid range. The latest Hadar level has only Lucy herself: a primitive species, hanging on at the edge of the proto-hominid range. At the other end of Africa a species only somewhat less-primitive, *Australopithecus africanus*, is present at the same time or somewhat later and, if the dates for Taung are to be believed, persists into the Pleistocene. A prediction from this is that *A. africanus* remains will be found in East and/or North-east Africa in the early Pliocene: perhaps, indeed, they are already there mixed in with the *antiquus* remains at Hadar while, as we have seen, the Laetoli remains are also very *A. africanus*-like.

On the Plio-Pleistocene boundary, *H. habilis* and *H. rudolfensis* replace each other as if they were subspecies; it is a violation of cladistic principles to acknowledge this while arranging one species (*H. habilis*) higher on the cladogram than the other, but on the contrary it follows from the hetero-morph speciation principle. Both Olduvai and Koobi Fora being presumably 'central' in position, no prediction follows from geographic data as to where the speciation event occurred that generated the *ergaster/erectus/sapiens* lineage; but from the cladogram it would be expected adjacent to the range of *H. habilis* and in the latest Pliocene—since *H. habilis* is strongly autapo-morphic and is not itself that ancestor. Meantime *H. habilis*, or a form very close to it, survived (as the centrifugal model would predict) longer into the lower Pleistocene at Sterkfontein in South Africa.

Wherever the new species arose, it is present as *H. ergaster* in the Upper Member of Koobi Fora, where it has replaced the *H. rudolfensis* and the primitive, but poorly known, *H. aethiopicus*. Suddenly it is joined by a new species, the 3733 species, for which there is as yet no evidence from Olduvai though it must have spread that far and beyond, as it is present—a surviving peripheral species—at Swartkrans, in South Africa, at the lower/middle Pleistocene boundary. The origin of the 3733 species need not be sought beyond the Koobi Fora region itself; obviously the presence of this ancestor-descendant pair at that single site is a sampling accident and their ranges were, at some stage, wider; but the sudden appearance of the derived taxon in a continuous depositional sequence is highly consistent with, even if it does not prove, the sympatric (stasipatric?) version of the centrifugal model.

This model predicts, too, that the 3733 species will be found at Olduvai (it had to occur there at some point, as it recurs later in South Africa), and suggests that it will occur there at a later date than at Koobi Fora. This prediction is necessary because it is at Olduvai, not at Koobi Fora, that the next speciation occurred: the origin of *H. erectus*. It is at Olduvai, too (in Bed IV) that the earliest confirmed occurrence of *H. sapiens* occurs; *H. erectus* had meanwhile extended its range to eastern Asia, and become extinct in Africa, so that the two species, *erectus* and *sapiens*, were now vicarious.

It would appear from the fossil evidence to date that *Homo erectus* was replaced in eastern Asia by two separate entries of *H. sapiens* into its range. In the first case, *H. e. pekinensis* is replaced by *H. s. mapaensis*, apparently between 150 000 and 200 000 BP; in the second, *H. e. erectus* is replaced by *H. s.* cf. *sapiens* in the upper Pleistocene. The nature of these replacements, and whether they were accompanied by gene-flow or not, was discussed earlier [Chapter 7, Subsection 7.5.4(h)].

Even the origin of anatomically modern *H. sapiens* was centrifugal. Given the distribution of the species in the late middle Pleistocene—Africa, Europe, and most of habitable Asia (i.e. except the South-east)—no region was more central than North-east Africa. And it is there—Omo and Laetoli—that the origin of the modern taxon can be demonstrated (see point 3. above).

The history of *Paranthropus* is more difficult to interpret, but the same principles can be observed: in particular, the most primitive species, *P. robustus*, continuing in South Africa while much more derived forms were coexisting, sympatrically, in East Africa, much as *H. habilis* and *Homo* sp. (3733 type) continue in South Africa apparently after they have been replaced in East Africa.

In all this there is no case of new, strikingly advanced taxa arising in a distinctly allopatric zone: it all happened in the centre, as the centrifugal model predicts; and there is no evidence against the notion that some of it, at least, happened sympatrically.

5. Extensive parallelism is to be expected; if gene transfer is real, it will be expected that parallelism will be greater between sympatric pairs.

The cladistic analysis presented in chapter 7, exactly like those of Skelton *et al.* (1986) and Wood and Chamberlain (1986), was replete with parallelism, to the degree that some cladistic placements—especially those of the Hadar fossils—were made without too much confidence. Parallel enlargement of cranial capacity has already been mentioned: unless *H. ergaster* reduced its capacity, then the *H. habilis/rudolfensis* pair must be allowed to have enlarged their capacities in parallel with the 3733/*erectus/sapiens* lineage, and perhaps with each other too (unless they were, truly, conspecific).

There is, however, no real evidence for gene transfer, at least such as to affect morphological characters. It is true that a megadont *Paranthropus* and *Homo* pair coexisted in the Lower Member at Koobi Fora, while a more microdont pair coexisted (later in time) at Swartkrans; but at Koobi Fora the megadont *Paranthropus* became more megadont, while the megadont *Homo* spp. dropped out and were replaced by microdont forms, and the microdont *Homo* at Swartkrans is an immigrant, not indigenous. It is impossible to think of any convincing example where a sympatric pair appear to have differentiated from their vicars in parallel directions.

6. Most marked evolutionary jumps will be heterochronic, especially neotenous; with increasing frequency as generation length increases if the Krassilov effect is true.

It has already been seen (section 8.3 above) that neotenous events occurred at many points in human evolution; it is also clear that they really begin to dominate the picture at the top of the cladogram—at the origins of the *3733/erectus/sapiens* lineage, of *Homo sapiens* itself, and of *H. s.* cf. *sapiens*. As has been discussed above (under point 4.), the first of these events is, if arguably, sudden; it is, at any rate, the most plausibly documented of any actual speciation event in the Hominini. The origin of the species *Homo sapiens* is not fully documented. The origin of the modern form of *H. sapiens* is, as I have strongly argued, sudden as well. This would seem consistent with the punctuation/heterochrony association. Both Javanese and, probably, Chinese forms of *H. erectus* can be shown to have had a gradual increase in size (first claimed as an increase in cranial capacity: but see the discussion in section 8.1, above), and this can be taken as a documentation of a non-heterochronic change which was not a punctuation. The correlation is therefore plausible, if not proved.

As for the Krassilov effect, if the longevity/encephalization association is valid, then it would seem very nicely illustrated in the overal picture just elucidated: neoteny demonstrably does enter more and more into the evolutionary process as encephalization (hence longevity?) increases in the human lineage.

7. New traits are not necessarily adaptive for a new environmental condition, but may, as exaptations, open up new niche possibilities.

Here I will refer once again to Blumenberg's (1984) model of brain evolution. I would like, too, to draw attention to the African origins of both *H. erectus* and *H. sapiens*: their new traits arose in the same old environment and apparently enabled them to enter new environments and to extend their ranges—the distinguishing characteristics did not arise, as traditional adaptationist/allopatric models would hold, in segments of ancestral species that had become isolated in a new habitat. Again, the origin of the modern type of *H. sapiens* was not in some new habitat, but in the centre of the range of *H. s. heidelbergensis*; the new traits were not adaptations to some new environment, but at most exaptations enabling the new taxon to expand and invade new environments.

8. Gradualism and punctuation will be distinguishable in the most recent period, because of radio-carbon dating.

This final point is a practical one, not a theoretical model. Still, there are difficulties: $C^{14}$ dating has shed some light on the Neandertal/modern change-over in Europe, but has still not, to everyone's satisfaction at least, succeeded in demonstrating whether this was by evolution or by replacement. On the other hand, U/Th dating techniques in combination with simple stratigraphic methods have been sufficient to shed light on the origin of anatomically modern *H. sapiens*, and K/Ar dating, again in conjunction with traditional methods, has at least enabled the raising of a working hypothesis on the origin of the 3733/*erectus/sapiens* lineage. So all is not lost at the older ages, and in peculiarly vexed and controversial cases even radio-carbon dating has difficulty in providing a resolution.

Nomogenesis? I believe that this might be a suitable term to encompass the view I have presented. Most of Berg's (1926) old criteria are there, and it is, to an un-looked-for extent, 'evolution by law'. Endogenous genetic factors must play an enormous, until lately quite unrecognized, role in human evolution; natural selection has an obvious role to play, but I believe more and more that the evidence allows it mainly a 'fine-tuning' role, and that the course of evolution is determined far more by nomogenesis than by traditional adaptive mechanisms.

# Bibliography

Adams, D. (1982). *Life, the universe and everything*. Pan Books, London.

Alberch, P., Gould, S.J., Oster, G.F. and Wake, D.B. (1979). Size and shape in ontogeny and phylogeny. *Palaeobiology*, **5**, 296–317.

Albrecht, G.H. (1978). The craniofacial morphology of the Sulawesi macaques. *Contributions to Primatology*, No. 13.

Albrecht, G.H. (1980). Latitudinal, taxonomic, sexual and insular determinants of size variation in Pigtail Macaques. *International Journal of Primatology*, **1**, 141–52.

Aldrich-Blake, F.P.G. (1968). A fertile hybrid between two *Cercopithecus* spp. in the Budongo Forest, Uganda. *Folia primatologica*, **9**, 15–21.

Alexeev, V.P. (1986). *The Origin of the Human Race*. Progress Publishers, Moscow.

Anderson, S. and Evensen, M.K. (1978). Randomness in allopatric speciation. *Systematic Zoology*, **27**, 421–30.

Anderson, S. and Koopman, K.F. (1981). Does interspecific competition limit the sizes of ranges of species? *American Museum Novitates*, **2716**, pp. 1–10.

Andrews, P. (1971). *Ramapithecus wickeri* mandible from Fort Ternan, Kenya. *Nature*, **231**, 192–4.

Andrews, P. (1984). An alternative interpretation of the characters used to define *Homo erectus*. *Courier Forschungsinstitut Senckenberg*, **69**, 167–75.

Andrews, P. (1985). Family group systematics and evolution among catarrhine primates. In *Ancestors: the hard evidence* (ed. E. Delson), pp. 14–22. Alan R. Liss, New York.

Andrews, P. and Cronin, J.E. (1982). The relationships of *Sivapithecus* and *Ramapithecus* and the evolution of the orang-utan. *Nature*, **297**, 541–6.

Andrews, P. and Groves, C.P. (1976). Gibbons and brachiation. In *Gibbon and Siamary* (ed. D.M. Rumbaugh), Vol. 4, pp. 167–218. S. Karger, Basel.

Andrews, P. and Simons E. (1977). A new African Miocene gibbon-like genus, *Dendropithecus* (Hominoidea, Primates) with distinctive postcranial adaptations: its significance to the origin of Hylobatidae. *Folia primatologica*, **28**, 161–9.

Andrews, P. and Tekkaya, I. (1980). A revision of the Turkish Miocene hominoid *Sivapithecus meteai*. *Palaeontology*, **23**, 85–95.

Andrews, P., Hamilton, W.R. and Whybrow, P.J. (1978). Dryopithecines from the Miocene of Saudi Arabia. *Nature*, **274**, 249–50.

Andriamiandra, A. (1972). L'appareil genital femelle du Lemur catta Linnaeus 1758: histologie et histo physiologie. *Bulletin de l'Association des Anatomistes*, **1972**, 163–81.

Arambough, C. and Coppens, Y. (1967). Sur la découverte, dans le Pleistocene inférieur de la vallée de l'Omo (Ethiopie), d'une mandibule d'Australopithecien. *Comptes Rendues de l'Académie des Sciences, Paris, D*, **265**, 589–90.

Arambourg, C. and Coppens, Y. (1968). Découverte d'un australopithecien nouveau dans les gisements de l'Omo (Ethiopie). *South African Journal of Science*, February issue pp. 58–9.

Archer, M. and Clayton, G. (ed.) (1984). *Vertebrate zoogeography and evolution in Australia*. Hesperian Press, Carlisle, Western Australia.

Ardrey, R. (1961). *African genesis*. Collins, London.

Armstrong, E. (1985). Allometric considerations of the adult mammalian brain, with special emphasis on primates. In *Size and scaling in primate biology* (ed. W.L. Jungers), pp. 115–46. Plenum, New York.

Arnason, U. (1977). The role of chromosomal rearrangement in mammalian speciation with special reference to Cetacea and Pinnipedia. *Hereditas*, **70**, 113–8.

Aronson, J.L., *et al.* (1971). New geochronologic and palaeomagnetic data for the hominid-bearing Hadar Formation in Ethiopia. *Nature*, **267**, 323–7.

Arredondo, O. and Varona, L.S. (1983). Sobre la validez de *Montaneia anthropomorpha* Ameghino, 1910. *Poeyana*, **255**, 1–21.

Ashton, E.H. (1960). The influence of geographic isolation on the skull of the green monkey (*Cercopithecus aethiops sabaeus*). V. The degree and pattern of differentiation in the cranial dimensions of the St. Kitts Green Monkey. *Proceedings of the Royal Society B*, **151**, 563–83.

Ashton, E.H. and Oxnard, C.E. (1963). The musculature of the primate shoulder. *Transactions of the Zoological Society of London*, **29**, 553–650.

Ashton, E.H., Flinn, R.M., Moore, W.J., Oxnard, C.E. and Spence, T.F. (1981). Further quantitative studies of form and function in the primate pelvis with special reference to *Australopithecus*. *Transactions of the Zoological Society of Lond.*, **36**, 1–98.

Auel, J.M. (1980). *The clan of the cave bear*. Coronet Books, New York.

Ayala, F. (1975). Genetic differentiation during the speciation process. In *Evolutionary biology* (ed. T. Dobzhansky, M.K. Hecht, and W.C. Steere), pp. 1–78. Plenum, New York.

Ayres, J.M. (1985). On a new species of Squirrel Monkey, genus *Saimiri*, from Brazilian Amazonia. *Papeis Avulsos de Zoologia, S. Paulo*, **36**, 147–64.

Baba, M.L., Darga, L.L. and Goodman, M. (1979). Immunodiffusion systematics of the Primates. *Folia primatologica*, **32**, 207–38.

Bailit, H.L., DeWitt, S.J. and Leigh, R.A. (1968). The size and morphology of the Nasioi dentition. *American Journal of Physical Anthropology*, **28**, 271–88.

Baker, J.R. (1974). *Race*. Oxford University Press, London.

Barnicot, N.A. and Hewett-Emmett, D. (1972). Red cell and serum proteins of *Cercocebus, Presbytis, Colobus* and certain other species. *Folia Primatologica*, **17**, 442–7.

Barry, J.C., Behrensmeyer, A.K. and Monaghan, M. (1980). A geological and biostratigraphic framework for Miocene sediments near Khaur village, northern Pakistan. *Postilla*, **183**, 1–19.

Barry, J.C., Jacobs, L.L. and Kelley, J. (1986). An Early Middle Miocene Catarrhine from Pakistan with comments on the dispersal of catarrhines into Eurasia. *Journal of Human Evolution*, **15**, 501–8.

Bartstra, G-J. (1982). The river-laid strata near Trinil, site of *Homo erectus erectus*, Java, Indonesia. *Modern Quaternary Research in SE Asia*, **7**, 97–130.

Bartstra, G-J., Basoeki, and Santosa Azis, B. (1976). Solo valley research 1975. *Modern Quaternary Research in SE Asia*, **2**, 23–36.

Bearder, S.K. and Doyle, G.A. (1974). Ecology of bushbabies, *Galago senegalensis* and *Galago crassicaudatus*, with notes on their behaviour in the field. In *Prosimian biology* (ed. R.D. Martin, G.A. Doyle, and A. Walker), pp. 109-30. Duckworth, London.

Behrensmeyer, A.K. (1975). The taphonomy and paleoecology of Plio-Pleistocene vertebrate assemblages east of Lake Rudolf, Kenya. *Bulletin of the Museum of comparative Zoology Harvard*, **146**, 473-578.

Behrensmeyer, A.K. (1982). Time resolution in fluvial vertebrate assemblages. *Paleobiology*, **8**, 211-27.

Benefit, B.R. and Pickford, M. (1986). Miocene fossil Cercopithecoids from Kenya. *American Journal of Physical Anthropology*, **69**, 441-64.

Beneviste, R.E. and Todaro, G.J. (1982). Gene transfer between Eukaryotes. *Science*, **217**, 1202.

Berg, L. (1969). *Nomogenesis or evolution determined by law* (trans. J.N. Rostovsov). MIT Press, Cambridge, Mass.

Berge, C. (1984). Multivariate analysis of the pelvis for hominids and other extant primates: implications for locomotion and systematics of the different species of australopithecines. *Journal of Human Evolution*, **13**, 555-62.

Berge, C., Orban-Segebarth, R. and Schmid, P. (1984). Obstetrical interpretation of the australopithecine pelvic cavity. *Journal of Human Evolution*, **13**, 573-87.

Bernstein, I.S. (1966). Naturally occurring primate hybrid. *Science*, **154**, 1559-60.

Berrill, N.J. (1955a). *Man's emerging mind*. Dobson Books, London.

Berrill, N.J. (1955b). *The origin of vertebrates*. Oxford University Press.

Berry, R.J. and Jakobson, M.E. (1975). Ecological genetics of an island population of the house mouse. *Journal of Zoology, London*, **175**, 523-40.

Berry, R.J., Jakobson, M.E. and Peters, J. (1978a). The house mouse of the Faroe Islands: a study in microdifferentiation. *Journal of Zoology London*, **185**, 73-92.

Berry, R.J. Peters, J. and Van Aarde, R.J. (1978b). Sub-antarctic house mice: colonization, survival and selection. *Journal of Zoology, London*, **184**, 127-41.

Bickham, J.W. and Baker, R.J. (1979). Canalization model of chromosomal evolution. *Bulletin of the Carnegie Museum Natural Histology*, **13**, 70-84.

Bilsborough, A. (1986). Comment on Skelton *et al.*, Phylogenetic Analysis of Early Hominids. *Current Anthropology*, **27**, 35-6.

Bilsborough, A. and Wood, B.A. (1986). The nature, origin and fate of *Homo erectus*. In *Major topics in primate and human evolution* (ed. B. Wood, L. Martin, and P. Andrews), pp. 295-316. Cambridge University Press.

Binford, L.R. (1981). *Bones: ancient men and modern myths*. Academic Press, New York.

Blumenberg, B. (1983). The evolution of the advanced hominid brain. *Current Anthropology*, **24**, 589-623.

Blumenberg, B. (1984). Allometry and evolution of tertiary hominoids. *Journal of Human Evolution*, **13**, 613-76.

Blumenberg, B. (1985). Biometrical studies upon hominoid teeth: the coefficient of variation, sexual dimorphism and questions of phylogenetic relationship. *Bio-Systems*, **18**, 149-84.

Blumenberg, B. and Lloyd, A.T. (1983). *Australopithecus* and the origin of the genus *Homo*: aspects of biometry and systematics with accompanying catalogue of tooth metric data. *BioSystems*, **16**, 127-67.

Bocquet-Appel, J-P. and Masset, C. (1982). Farewell to paleodemography. *Journal of Human Evolution*, **11**, 321-333.

Boese, G.K. (1973). Social behavior and ecological considerations of West African baboons (*Papio papio*). In *Socioecology and psychology of primates* (ed. R.H. Tuttle), pp. 205–30. Mouton, The Hague.

Bookstein, F.L., Gingerich, P.D. and Kluge, A.J. (1978). Hierarchical linear modelling of the tempo and mode of evolution. *Paleobiology*, **4**, 120–34.

Brace, C.L. (1967). *The stages of human evolution*. Prentice-Hall, Englewood Cliffs, NJ.

Brandon-Jones, D. (1977). The evolution of recent Asian colobines. In *Recent advances in primatology* (eds. D.J. Chivers and K.A. Joysey), pp. 323–5.

Brandon-Jones, D. (1984). Colobus and leaf-monkeys. In *The encyclopaedia of mammals, 1* (ed. D. Macdonald), pp. 398–408. Unwin Animal Library, London.

Bräuer, G. (1981). New evidence on the transitional period between Neanderthal and modern man. *Journal of Human Evolution*, **10**, 467–74.

Bräuer, G. (1984). The 'Afro-European *sapiens*-hypothesis', and hominid evolution in East Asia during the late Middle and Upper Pleistocene. *Courier Forschungsinstitut Senckenberg*, **69**, 145–65.

Brent, L., Rayfield, L.S., Chandler, P., Fiertz, W., Medawar, P.B. and Simpson, E. (1981). Supposed Lamarckian inheritance of immunological tolerance. *Nature*, **290**, 508–12.

Brown, F.H. (1982). Tulu Bor Tuff at Koobi Fora correlated with the Sidi Hakoma Tuff at Hadar. *Nature*, **300**, 631–5.

Brown, F., Harris, J., Leakey, R. and Walker, A. (1985*a*). Early *Homo erectus* skeleton from west Lake Turkana, Kenya. *Nature*, **316**, 788–92.

Brown, F.H., McDougall, I., Davies, T. and Maier, R. (1985*b*). An integrated Plio-Pleistocene chronology for the Turkana Basin. In *Ancestors: the hard evidence* (ed. E. Delson), pp. 82–90, Alan R. Liss, New York.

Brown, W.L. Jr. (1957). Centrifugal speciation. *Quarterly Review of Biblogy*, **32**, 247–77.

Brown, W.L. and Wilson, E.O. (1954). The case against the trinomen. *Systematic Zoology*, **3**, 174–6.

Brown, W.L. and Wilson, E.O. (1956). Character displacement. Systematic Zoology, **5**, 49–64.

Burnet, F.M. (1940). *Biological aspects of infectious disease*. Cambridge University Press, Cambridge.

Bush, G.L. (1969). Sympatric host race formation and speciation in frugivorous flies of the genus *Rhagoletis* (Diptera, Tephritidae). *Evolution*, **23**, 237–51.

Bush, G.L., Case, S.M., Wilson, A.C. and Patton, J.L. (1977). Rapid speciation and chromosomal evolution in mammals. *Proceedings of the National Academy of Sciences of the USA*, **74**, 3942–6.

Butler, H. (1964). The reproductive biology of a strepsirhine (*Galago senegalensis senegalensis*). *International Review of General and Experimental Zoology*, **1**, 241–96.

Cachel, S. (1981). Plate tectonics and the problem of anthropoid origins. *Yearbook of Physical Anthropology*, **24**, 139–72.

Cadien, J.D. (1972). Dental variation in man. In *Perspectives on human evolution* (ed. S.L. Washburn and P. Dolhinow), pp. 199–222. Holt, Rinehart, and Winston, New York.

Campbell, B.G. (1965). *The nomenclature of the Hominidae*. Occasional paper No.

22, Royal Anthropological Institute of Great Britain and Ireland.

Campbell, J.H. (1982). Autoevolution. *Perspectives in evolution* (ed. R. Milkman). pp. 190–201. Sinauer Press, New York.

Carney, J., Hill, A., Miller, J.A. and Walker, A. (1971). Late Australopithecine from Baringo district, Kenya. *Nature*, **230**, 509–14.

Carson, H.L. (1975). The genetics of speciation at the diploid level. *American Naturalist*, **109**, 83–92.

Carson, H.L. (1976). Inference of the time of origin of some *Drosophila spp. Nature*, **259**, 395–6.

Cartmill, M. (1972). Arboreal adaptations and the origin of the Order Primates. *The functional and evolutional biology of primates* (ed. R.H. Tuttle). pp. 97–122. Aldine-Atherton, Chicago, Ill.

Cartmill, M. (1974). *Daubentonia, Dactylopsila*, woodpeckers and klinorhynchy. In *Prosimian biology* (ed. R.D. Martin, G.A. Doyle, and A.C. Walker), pp. 655–72. Duckworth, London.

Cartmill, M. (1975). Strepsirhine basicranial structures and the affinities of the Cheirogaleidae. In *Phylogeny of the primates: a multidisciplinary approach* (ed. W.P. Luckett and F.S. Szalay), pp. 313–54. Plenum, New York.

Cartmill, M. (1981). Morphology, function, and evolution of the Anthropoid postorbital septum. In *Evolutionary biology of the New World monkeys and continental drift* (ed. R.L. Ciochon and A.B. Chiarelli), pp. 243–74. Plenum, New York.

Chamberlain, A.T. and Wood, B.A. (1985). a reappraisal of variation in hominid mandibular corpus dimensions. *American Journal of Physical Anthropology*, **66**, 399–405.

Chamberlain, A.T. and Wood, B.A. (1986). Comment on Skelton *et al.*, Phylogenetic analysis of early hominids. *Current Anthropology*, **27**, 36–7.

Charles-Dominique, P. (1971). Eco-ethologie des prosimiens du gabon. *Biologica Gabonica*, **7**, 121–28.

Chasen, F.N. (1940). A handlist of Malaysian mammals. *Bulletin of the Raffles Museum Singapore*, **15**, i–xx, 1–209.

Cheng Guoliang, Li Suling, and Lin Jinlu (1977). Discussion of the age of *Homo erectus yuanmouensis* and the event of early Matuyama. *Scientia Geoligica Sinica*, 1977, 34–43.

Cherry, L.M., Case, S.M. and Wilson, A.C. (1978). Frog perspective on the morphological difference between humans and chimpanzees. *Science*, **200**, 209–11.

Chiarelli, A.B. (1979). Primate chromosome atlas. In *Comparative karyology of primates* (ed. A.B. Chiarelli, A.L. Koen, and G. Ardito). Mouton, The Hague.

Chiu Chung-lang, Ku Yu-min, Chang Yin-yun, and Chang Sen-shui (1973). Peking man fossils and cultural remains newly discovered at Choukoutien. *Vertebrata Palasiatica*, **11**, 109–31.

Chopra, S.R.K. (1978). New fossil evidence on the evolution of Hominoidea in the Sivaliks and its bearing on the problem of the evolution of early man in India. *Journal of Human Evolution*, **7**, 3–9.

Ciochon, R.L., Savage, D.E., Thaw Tint, and Ba Maw (1985). Anthropoid origins in Asia? New discovery of *Amphipithecus* from the Eocene of Burma. *Science*, **229**, 756–9.

Clarke, R.J. (1976). New cranium of *Homo erectus* from Lake Ndutu, Tanzania. *Nature*, **262**, 484–7.

Clarke, R.J. (1985). A new reconstruction of the Florisbad cranium, with notes on the site. *Ancestors: the hard evidence*, (ed. E. Delson), pp. 301–5. Alan R. Liss New York.

Clarke, R.J. (1985a). *Australopithecus* and early *Homo* in southern Africa. In *Ancestors: the hard evidence* (ed. E. Delson), pp. 171–7. Alan R. Liss, New York.

Coimbra-Filho, A.F. and Mittermeier, R.A. (1973). New data on the taxonomy of the Brazilian marmosets of the genus *Callithrix* Erxleben, 1777. *Folia primatologica*, **20**, 241–64.

Conroy, G.C., Jolly, C.J. Cramer, D. and Kalb, J.E. (1978). Newly discovered fossil hominid skull from the Afar Depression, Ethiopia. *Nature*, **307**, 423–8.

Coon, C.S. (1963). *The origin of races*. Jonathan Cape, London.

Coppens, Y. (1966). Le Tchadanthropus. *L'Anthropologie*, **70**, 5–16.

Coppens, Y. (1983a). Systématique, phylogénie, environnement et culture des australopithèques, hypothèses et synthèse. *Bulletin et Memoires de la Société d'Authropologie, Paris*, **10**, 273–84.

Coppens, Y. (1983a). Les plus anciens fossiles d'hominidés. *Pontifical Academy of Science, Scripta Varia*, **50**, 1–9.

Corbet, G.B. (1970). Patterns of subspecific variation. *Symposium of the Zoological Society of London*, **26**, 105–16.

Corbet, G.B. (1978). *The mammals of the Palaearctic Region: a taxonomic review*. British Museum (Natural History), London.

Corruccini, R.S. Ciochon, R.L. (1977). The phenetic position of *Pliopithecus* and its phylogenetic relationship to the Hominoidea. *Systematic Zoology*, **26**, 290–9.

Corruccini, R.S. and McHenry, H.M. (1981). Cladometric analysis of Pliocene hominids. *Journal of Human Evolution*, **9**, 209–21.

Corruccini, R.S. Baba, M., Goodman, M. Ciochon, R.L. and Cronin, J.E. (1980). Non-linear macromolecular evolution and the molecular clock. *Evolution*, **34**, 1216–9.

Courtenay, J., Groves, C. and Andrews, P. (in press). Inter- or intra-island variation? An assessment of the differences between Bornean and Sumatran orang-utans. In *The orang-utan* (ed. J.H. Schwartz). Oxford University Press, Oxford.

Craw, R.C. (1985). Classic problems of southern hemisphere biogeography re-examined. *Zeitschrift für zoologisches Systematik und Evolutions Forschung*, **35**, 1–10.

Craw, R.C. and Weston, P. (1984). Panbiogeography: a progressive research program? *Systematic Zoology*, **33**, 1–13.

Croizat, L. (1962). *Space, time, form: the biological synthesis*. Published by the author in Caracas. (Not currently available.)

Cronin, J.E. and Meikle, W.E. (1979). The phyletic position of *Theropithecus*: congruence among molecular, morphological, and palaeontological evidence. *Systematic Zoology*, **28**, 259–69.

Cronin, J.E. and Sarich, V.M. (1975). Molecular systematics of the New World monkeys. *Journal of Human Evolution*, **4**, 357–75.

Cronin, J.E. and Sarich, V.M. (1976). Molecular evidence for dual origin of mangabeys among Old World monkeys. *Nature*, **260**, 700–2.

Crowson, R.A. (1970). *Classification and biology.* Heinemann, London.

Curtis, G.H. (1981). Man's immediate forerunners: establishing a relevant time scale in anthropological and archaeological research. *Philosophical Transactions Royal Society of London*, **B292**, 7-20.

Cutler, R.G. (1975). Evolution of human longevity and the genetic complexity governing aging rate. *Proceedings of the Natural Academy of Science of the USA*, **72**, 4664-8.

Dagosto, M. and Musser, G.G. (1986). The status of *Tarsius pumilus. American Journal of Physical Anthropology*, **69**, 192.

Dandelot, P. (1959). Note sur la classification des Cercopithèques du groupe *aethiops. Mammalia*, **23**, 357-68.

Dandelot, P. (1971). Order Primates. In *The mammals of Africa: an identification manual* (ed. J. Meester and H.W. Setzer), part 3. Smithsonian Institution Press, Washington, D.C.

Dao Van Tien (1983). On the north Indochinese gibbons (*Hylobates concolor*) in North Vietnam. *Journal of Human Evolution*, **12**, 367-72.

Dart, R.A. (1953). The predatory transition from ape to man. *International Anthropological and Linguistic Review*, **1**, 201-18.

Day, M.H. (1971). The Omo human skeletal remains. In *The origin of* Homo sapiens, (ed. F. Bordes), pp. 31-5. UNESCO, Paris.

Day, M.H. (1982). The *Homo erectus* pelvis. Punctuation or gradualism? In *Prétirage, ler Congres International de Paleontologie Humaine* (ed. H. de Lumley), pp. 411-21. Union Internationale des Sociétés Préhistoriques et Protohistoriques. Nice.

Day, M.H. (1984). The postcranial remains of *Homo erectus* from Africa, Asia and possibly Europe. *Courier Forschungsinstitut Senckenberg*, **69**, 113-21.

Day, M.H. and Molleson, T.H. (1973). The Trinil femora. In *Human evolution* (ed. M.H. Day), pp. 127-54. Francis, London.

Day, M.H. and Stringer, C.B. (1982). A reconsideration of the Omo Kibish remains and the erectus–sapiens transition. In *L'Homo erectus et la Place de l'Homme de Tautavel parmi les Hominidés fossiles* (ed. H. de Lumley), Vol. 2, pp. 814-46. Louis-Jean Scientific & Literary Publications, Nice.

Day, M.H. and Wickens, E.H. (1980). Laetoli Pliocene hominid footprints and bipedalism. *Nature*, **286**, 385-7.

Day, M.H., Leakey, M.D., and Olson, T.R. (1978). On the status of *Australopithecus afarensis. Science*, **207**, 1102-3.

Day, M.H., Leakey, M.D., and Magori, C. (1980). A new hominid fossil skull (LH18) from the Ngaloba Beds, Laetoli, northern Tanzania. *Nature*, **284**, 55-6.

Dean, M.C. (1985). The eruption pattern of the permanent incisors and first permanent molars in *Australopithecus* (*Paranthropus*) *robustus. American Journal of Physical Anthropology*, **67**, 251-7.

Dean, M.C. and Wood, B.A. (1982). Basicranial anatomy of Plio-Pleistocene hominids from East Africa. *American Journal of Physical Anthropology*, **59**, 157-74.

De Boer, L.E.M. (1982). Karyological problems in breeding Owl Monkeys. *International Zoo Yearbook*, **22**, 119-24.

De Bonis, L. and Melentis, J. (1984). La position phylétique *d'Ouranopithecus. Courier Forschungsinstitut Senckenberg*, **69**, 13-23.

De Bonis, L. and Melentis, J. (1985). La place du genre Ouranopithecus dans l'évolution des Hominidés. *Comptes Rendues de l'Académie des Sciences, Paris,* II, **300**, 429–32.

De Bonis, L., Johanson, D. Melentis, J., and White, T. (1981). Variations métriques de la denture chez les Hominidés primitifs: comparaison entre *Australopithecus afarensis* et *Ouranopithecus macedoniensis. Comptes Rendues de l'Académie des Sciences, Paris,* II, **292**, 373–6.

Deloison, Y. (1985). Comparative study of calcanei of primates and *Pan-Australopithecus–Homo* relationship. In *Hominid evolution: past, present and future* (ed. P.V. Tobias), pp. 143–7. Alan R. Liss, New York.

Delson, E. (1973). Fossil colobine monkeys of the Circum-Mediterranean region and the evolutionary history of the Cercopithecidae (Primates, Mammalia). Unpublished Ph.D. thesis, Columbia University, New York.

Delson, E. (1975). Evolutionary history of the Cercopithecidae. *Contributions to Primatology,* **5**, 167–217.

Delson, E. (1978). *Prohylobates* (Primates) from the Early Miocene of Libya: a new species and its implications for cercopithecoid origins. *Geobios,* **12**, 725–33.

Delson, E. (1980). Fossil macaques, phyletic relationships and a scenario of deployment. In *The macaques: studies in ecology, behaviour and evolution* (ed. D.G. Lindburg), pp. 10–29. Van Nostrand, New York.

Delson, E. (1984). Cercopithecid biochronology of the African Plio-Pleistocene: correlation among eastern and southern hominid-bearing localities. *Courier Forschungsinstitut Senckenberg,* **69**, 199–218.

Delson, E. (1985a). The earliest *Sivapithecus? Nature,* **318**, 107–8.

Delson, E. (1985b). Catarrhine evolution. In *Ancestors: the hard evidence* (ed. E. Delson), pp. 9–13. Alan R. Liss, New York.

Delson, E. and Rosenberger, A.L. (1980). Phyletic perspectives on platyrrhine origins and anthropoid relationships. In *Evolutionary biology of New World monkeys and continental drift* (ed. R.L. Ciochon and A.B. Chiarelli), pp. 445–58. Plenum, New York.

de Lumley, H. and Sonakia, A. (1985). Première découverte d'un *Homo erectus* sur le continent indien à Hathnora, dans la moyenne vallée de la Narmada. *L'Anthropologie,* **89**, 13–61.

de Lumley, H., de Lumley, M-A., Bada, J.L., and Turekian, K.K. (1977). The dating of pre-Neanderthal remains at Caune de l'Arago, Tautavel, Pyrenées-Orientales, France. *Journal of Human Evolution,* **6**, 223–4.

Déne, H., Goodman, M., Prychodko, W., and Moore, G.W. (1976). Immuno-diffusion systematics of the Primates. III. The Strepsirhini. *Folia primatologica,* **25**, 35–61.

Deol, M.S. (1970). The determination and distribution of coat colour variation in the house mouse. *Symposium of the Zoological Society of London,* **26**, 239–50.

de Queiroz, K. (1985). The ontogenetic method for determining character polarity and its relevance to phylogenetic systematics. *Systematic Zoology,* **34**, 280–99.

De Vos, J. and Sondaar, P.Y. (1982). The importance of the Dubois collection reconsidered. *Modern Quaternary Research in SE Asia,* **7**, 35–63.

Dice, L.R. (1947). Effectiveness of selection by owls of deer mice (*Peromyscus maniculatus*) which contrast in colour with their background. *Contributions of the Laboratory of Vertebrate Biology* **34**, 1–20.

Dolinar-Osole, Z. (1956). Nova Pitekantropoidna Oblika hominida iz severne Afrike. *Arheoloski Vestnik*, **7**, 173–80.

Donehower, L. and Gillespie, D. (1979). Restriction site periodicities in highly repetitive DNA of Primates. *Journal of Molecular Biology*, **134**, 805–34.

Doolittle, W.F. and Sapienza, C. (1980). Selfish genes, the phenotype paradigm, and genome evolution. *Nature*, **284**, 601–3.

Dostal, A. and Zapfe, H. (1986). Zahn schmelzprismen von *Mesopithecus pentelicus* Wagner, 1839, im Vergleich mit rezenten Cercopitheciden (Primates: Cercopithecidae). *Folia Primatologica*, **46**, 235–51.

Dover, G. (1982). Molecular drive: a cohesive mode of species evolution. *Nature*, **299**, 111–7.

Doyle, G.A., Anderson, A., and Bearder, S.K. (1971). Reproduction in the lesser bushbaby (*Galago senegalensis moholi*) under semi-natural conditions. *Folia primatologica*, **14**, 15–22.

Dunbar, R.I.M. (1982). Adaptation, fitness and the evolutionary tautology. In *Current problems in sociobiology* (ed. King's College Sociobiology Group), pp. 9–28. Cambridge University Press.

Dutrillaux, B., Fosse, A.M., and Chauvier, G. (1979). Etude cytogénétique de six espèces ou sous-espèces de mangabeys. *Annales de Génétique*, **22**, 88–92.

Dutrillaux, B., Couturier, J., and Chauvier, G. (1980). Chromosomal evolution of 19 species or subspecies of Cercopithecinae. *Annales Génétique*, **23**, 133–43.

Dutrillaux, B., Viegas-Pequignot, E., and Couturier, J. (1980). Très grande analogie de marquage chromosomique entre le lapin (*Oryctologus cuniculus*) et les primates, dont l'homme. *Annales de Génétique*, **23**, 22–5.

Dutrillaux, B., Couturier, J., and Viegas-Pequignot, E. (1981). Chromosomal evolution in primates. *Chromosomes Today*, **7**, 176–91.

Eaglen, R.H. (1983). Parallelism, parsimony, and the phylogeny of the Lemuridae. *International Journal of Primatology*, **4**, 249–73.

Eaglen, R.H. and Simons, E.L. (1980). Notes on the breeding biology of thick-tailed and silvery galagos in captivity. *Journal of Mammalogy*, **61**, 534–7.

Eck, G.C. and Jablonski, N.G. (1984). A reassessment of the taxonomic status and phyletic relationships of *Papio baringensis* and *Papio quadratirostris*. *American Journal of Physical Anthropology*, **65**, 109–34.

Echhardt, R.B. (1974). The dating of *Gigantopithecus*: a critical reappraisal. *Anthropologischer Anzeiger*, **34**, 129–39.

Eco, U. (1983). *The name of the rose*. Picador, London.

Eisentraut, M. (1973). Die Wirbeltierfauna von Fernando Po und West Kamerun, *Bonn Zoological Monographs*, **3**, 1–428.

Eldredge, N. and Cracraft, J. (1980). *Phylogenetic patterns and the evolutionary process*. Columbia University Press, New York.

Eldredge, N. and Gould, S.J. (1972). Punctuated equilibria: an alternative to phyletic gradualism. In *Models in paleobiology* (ed. T. Schopf), pp. 82–115. Freeman and Cooper, San Francisco.

Eldredge, N. and Tattersall, I. (1975). Evolutionary models, phylogenetic reconstruction, and another look at hominid phylogeny. In *Approaches to primate paleobiology* (ed. F.S. Szalay) *Contributions to Primatology*, Vol. 5, pp. 218–42. S. Karger; Basel.

Ellerman, J.R. and Morrison-Scott, T.C.S. (1951). *Checklist of Palaearctic and*

*Indian mammals*. Trustees of the British Museum (Natural History), London.

Elliot, D.G. (1913). *A review of the Primates* (3 vols.). American Museum of Natural History, New York.

Endler, J.A. (1977). *Geographic variation, subspecies and clines*. Princeton University Press.

Endrödy-Younga, S. (1980). The concept of heteromorph speciation. *Annals of the Transvaal Museum*, **32**, 241-7.

Eudey, A.A. (1981). Morphological and ecological characters in sympatric populations of Macaca in the Dawna Range. In *Primate evolutionary biology* (ed. A.B. Chiarelli and R.S. Corruccini), pp. 44-50. Springer-Verlog, Bolin.

Ewens, W.J. (1969). Mean fitness increase when fitnesses are additive. *Nature*, **221**, 1076.

Falk, D. (1979). Cladistic analysis of New World monkey sulcal patterns: methodological implications for primate brain studies. *Journal of Human Evolution*, **8** 637-45.

Falk, D. (1985). Hadar AL 162-28 endocast as evidence that brain enlargement preceded cortical reorganization in hominid evolution. *Nature*, **313**, 45-7.

Feldesman, M.R. (1982). Morphometric analysis of the distal humerus of some Cenozoic catarrhines: the late divergence hypothesis revisited. *American Journal of Physical Anthropology*, **59**, 73-95.

Ferguson, W.W. (1983). An alternative interpretation of *Australopithecus afarensis* fossil material. *Primates*, **24**, 397-409.

Ferguson, W.W. (1984). Revision of fossil hominid jaws from the Plio/Pleistocene of Hadar, in Ethiopia including a new species of the genus *Homo* (Hominoidea: Homininae). *Primates*, **25**, 519-29.

Ferguson, W.W. (1987). Revision of the subspecies of *Australopithecus africanus* (Primates: Hominidae) including a new subspecies from the Late Pliocene of Ethiopia. *Primates*, **28**, 258-65.

Fisher, R.A. (1930). *The genetical theory of Natural Selection*. Clarendon Press, Oxford.

Fleagle, J.G. (1975). A small gibbon-like Hominoid from the Miocene of Uganda. *Folia primatologica*, **24**, 1-15.

Fleagle, J.G. and Bown, T.M. (1983). New primate fossils from Late Oligocene (Colhuehuapian) localities of Chubut, Province, Argentina. *Folia primatologica*, **41**, 240-66.

Fleagle, J.G. and Kay, R.F. (1983). New interpretations of the phyletic position of Oligocene hominoids. In *New interpretations of ape and human ancestry* (ed. R.L. Ciochon and R.S. Corruccini), pp. 181-210. Plenum, New York.

Fleagle, J.G. and Kay, R.F. (1987). The phyletic position of the Parapithecidae. *Journal of Human Evolution*, **16**, 483-532.

Fleagle, J.G. and Simons, E.L. (1978). *Micropithecus clarki*, a small ape from the Miocene of Uganda. *American Journal of Physical Anthropology*, **49**, 427-40.

Fleagle, J.G., Bown, T.M. Obradovich, J.D., and Simons, E.L. (1986). How old are the Fayum primates? In *Primate evolution* (ed. J.G. Else and P.C. Lee), pp. 3-17. Cambridge University Press.

Fooden, J. (1963). Review of Woolly Monkeys (genus *Lagothrix*). *Journal of Mammology*, **44**, 213-47.

Fooden, J. (1964). Rhesus and crab-eating Macaques: intergradation in Thailand. *Science*, **143**, 363-5.

Fooden, J. (1967*a*). Identification of the Stump-tailed Monkey, *Macaca speciosa* I. Geoffroy, 1826. *Folia Primatologica*, **5**, 153–64.

Fooden, J. (1967*b*).Complementary specialization of male and female reproductive structures in the bear macaque, *Macaca arctoides*. *Nature*, **214**, 939–41.

Fooden, J. (1969). Taxonomy and evolution of the monkeys of Celebes (Primates: Cercopithecidae). *Bibliotheca Primatologica*, **10**, 1–148.

Fooden, J. (1971*a*). Report on Primates collected in western Thailand January–April, 1967. *Fieldiana, Zoology*, **59**, 1–62.

Fooden, J. (1971*b*). Female genitalia and taxonomic relationships of *Macaca assamensis*. *Primates*, **12**, 63–73.

Fooden, J. (1975). Taxonomy and evolution on liontail and pigtail macaques (Primates: Cercopithecidae). *Fieldiana, Zoology*, **67**, 1–169.

Fooden, J. (1976). Provisional classification and key to living species of macaques. *Folia primaologica*, **25**, 225–36.

Fooden, J. (1982). Ecogeographic segregation of macaque species. *Primates*, **23**, 574–9.

Fooden, J., Mahabal, A., and Saha, S.S. (1981). Redefinition of Rhesus Macaque—Bonnet Macaque boundary in peninsular India. *Journal Bombay natural Histology Society*, **78**, 463–74.

Fooden, J., Quan Guoqiang, Wang Zongren, and Wang Yingxiang (1985). The stumptail macaques of China. *American Journal of Primatology*, **8**, 11–30.

Ford, H.D. and Ford, E.B. (1930). Fluctuation in numbers, and its influence on variation, in *Melitaea aurinia*, Rott. (Lepidoptera). *Transactions of the Royal Entomological Society of London*, **76**, 345–51.

Ford, S.M. (1986). Systematics of the New World Monkeys. In *Comparative primate biology* (ed. D.R. Swindler and J. Erwin), Vol. 1, pp. 73–135. Alan R. Liss, New York.

Friday, A.E. (1982). Parsimony, simplicity and what actually happened. *Zoological Journal of the Linnaean Soceity*, **74**, 329–35.

Galbreath, G.J. (1983). Karyotypic evolution in *Aotus*. *American Journal of Primatology*, **4**, 245–51.

Gao Jian (1975). Australopithecine teeth associated with *Gigantopithecus*. *Vertebrata Palasiatica*, **13**, 81–8.

Gauckler, A. and Kraus, M. (1970). Kennzeichnung und Verbreitung von *Myotis brandti* (Eversman, 1845). *Zeitschrift für Säugetierkunde*, **35**, 113–24.

Gautier, J-P. (1985). Quelques characteristiques ecologiques du singe des marais: *Allenopithecus nigroviridis* Lang 1923. *Revue d'Ecologie*, **40**, 331–42.

Gautier-Hion, A. (1971). L'ecologie du talapoin du Gabon. *Terre et Vie*, **25**, 427–90.

Gebo, D.L. (1985). The nature of the Primate grasping foot. *American Journal of Physical Anthropology*, **67**, 269–77.

Gebo, D.L. (1986). Anthropoid origins—the foot evidence. *Journal of Human Evolution*, **15**, 421–30.

Geist, V. (1978). *Life strategies, human evolution, environmental design*. Springer, New York.

Gelvin, B.R. (1980). Morphometric affinities of *Gigantopithecus*. *American Journal of Physical Anthropology*, **53**, 541–68.

Georgiadis, N. (1985). Growth patterns, sexual dimorphism and reproduction in African ruminants. *African Journal of Ecology*, **23**, 75–87.

Gingerich, P.D. (1976). Cranial anatomy and evolution of early Tertiary Plesiadapidae (Mammals, Primates). *University of Michigan Papers in Paleontology*, **15**, 1–41.

Gingerich, P.D. (1977*a*). Patterns of evolution in the mammalian fossil record. In *Patterns of evolution* (ed. A. Hallam), pp. 469–500. Elsevier, Amsterdam.

Gingerich, P.D. (1977*b*). New species of Eocene primates and the phylogeny of European Adapidae. *Folia primatology*, **28**, 60–80.

Gingerich, P.D. (1977*c*). Dental variation in Early Eocene *Teilhardina belgica*, with notes on the anterior dentition of some early Tarsiiformes. *Folia primatologica*, **28**, 144–53.

Gingerich, P.D. (1977*d*). Radiation of Eocene Adapidae in Europe. *Geobios, Memoire Speciale*, **1**, 165–82.

Gingerich, P.D. (1978*a*). The stratophenetic approach to phylogeny reconstruction in vertebrate paleontology. In *Phylogenetic analysis and paleontology* (ed. J. Cracraft and N. Eldredge), pp. 41–77.

Gingerich, P.D. (1978*b*). The Stuttgart collection of Oligocene primates from the Fayum Province of Egypt. *Paläontological Zeitschrift*, **52**, 82–92.

Gingerich, P.D. (1981). Early Cenozoic Omomyidae and the evolutionary History of Tarsiiform primates. *Journal of Human Evolution*, **10**, 345–74.

Gingerich, P.D. (1984). Primate evolution: evidence from the fossil record, comparative morphology, and molecular biology. *Yearbook of Physical Anthropology*, **27**, 57–72.

Gingerich, P.D. and Haskin, R.A. (1981). Dentition of early Eocene *Pelycodus jarrovii* and the generic attribution of species formerly referred to *Pelycodus*. *Contributions from the Museum of Paleontology University of Michigan*, **25**, 327–37.

Gingerich, P.D. and Sahni, A. (1979). *Indraloris* and *Sivaladapis*: Miocene adapid primates from the Siwaliks of India and Pakistan. *Nature*, **279**, 415–6.

Gingerich, P.D. and Schoeninger, M. (1977). The fossil record and primate phylogeny. *Journal of Human Evolution*, **6**, 483–505.

Ginsburg, L. and Mein, P. (1980). *Crouzelia rhodanica*, nouvelle espèce de Primate catarrhininen, et essai sur la position systématique des Pliopithecidae. *Bulletin du Museé national d'Histoire naturelle, Paris (2)*, **4C**, 57–85.

Godfrey, L. and Jacobs, K.H. (1981). Gradual, autocatalytic and punctuational models of hominid brain evolution: a cautionary tale. *Journal of Human Evolution*, **10**, 255–72.

Goldschmidt, R. (1940). *The material basis of evolution*. Yale University Press.

Goodman, M. (1963). Man's place in the phylogeny of the Primates as reflected in serum proteins. In *classification and human evolution* (ed. S.L. Washburn), pp. 204–34. Aldine, Chicago, IL.

Goodman, M., Farris, W.Jr., Moore, W., Prychodko, W., Poulik, E., and Sorenson, M. (1974). Immunodiffusion systematics of the primates II: findings on *Tarsius*, Lorisidae and Tupaiidae. In *Prosimian biology* (ed. R.D. Martin, G.A. Doyle and A.C. Walker), pp. 881–90. Duckworth, London.

Goodman, M., Romero-Herrera, A.E., Dene, H., Czelusniak, J., and Tashian, R.E. (1982). Amino acid sequence evidence on the phylogeny of primates and other Eutherians. In *Macromolecular sequences in systematic and evolutionary biology* (ed. M. Goodman), pp. 115–91. Plenum, New York.

Gorczynski, R.M. and Steele, E.J. (1981). Simultaneous yet independent inheritance of somatically acquired tolerance to two distinct H-2 antigenic haplotype determinants in mice. *Nature*, **289**, 678-81.

Gould, S.J. (1976). Ladders, bushes and human evolution. *Natural History*, **85**(4), 24-31.

Gould, S.J. (1977), *Ontogeny and phylogeny*. Harvard University Press.

Gould, S.J. (1980). Is a new and general theory of evolution emerging? *Paleobiology*, **6**, 119-30.

Gould, S.J. and Lewontin, R.C. (1979). The spandrels of San Marco and the panglossian paradigm: a critique of the adaptationist programme. *Proceedings of the Royal Society of London*, **205**, 581-98.

Gould, S.J. and Vrba, E.S. (1982). Exaptation—a missing term in the science of form. *Paleobiology*, **8**, 4-15.

Grehan, J.R. and Ainsworth, R. (1985). Orthogenesis and evolution. *Systematic Zoology*, **34**, 174-92.

Grant, P.R. (1971). Comment on Simberloff's letter. *American Naturalist*, **105**, 194-7.

Greenfield, L.O. (1980). A late divergence hypothesis. *American Journal of Physical Anthropology*, **52**, 351-65.

Grine, F. (1985). Australopithecine evolution: the deciduous dental evidence. In *Ancestors: the hard evidence* (ed. E. Delson), pp. 153-67. Alan R. Liss, New York.

Groves, C.P. (1970*a*). The forgotten leaf-eaters, and the phylogeny of the Colobinae. In *Old World monkeys* (ed. J.R. Napier and P.H. Napier), pp. 555-89. Academic Press, London.

Groves, C.P. (1970*b*). *Gorillas*. Arthur Barker, London.

Groves, C.P. (1972*a*). Systematics of the genus *Nycticebus*. *Proceedings of the 3rd International Congress of Primatology*, **1**, 44-53.

Groves, C.P. (1972*b*). Systematics and phylogeny of gibbons. *Gibbon and Siamang*, **1**, 1-89.

Groves, C.P. (1974*a*). Taxonomy and phylogeny of prosimians. In *Prosimian biology* (ed. R.D. Martin, G.A. Doyle, and A.C. Walker), pp. 449-73.

Groves, C.P. (1974*b*). New evidence on the evolution of the apes and man. *Vestnik Ustredny usted Geologie*, **49**, 53-6.

Groves, C.P. (1978). Phylogenetic and population systematics of the mangabeys (Primates: Cercopithecidae). *Primates*, **19**, 1-34.

Groves, C.P. (1980). Speciation in *Macaca*: the view from Sulawesi: In *The macaques: studies in ecology, behavior and evolution* (ed. D.G. Lindburg), pp. 84-124. Van Nostrand, New York.

Groves, C.P. (1981). Systematic relationships in the Bovini (Artiodactyla, Bovidae). *Zeitchrift für zoologische Systematik und Evolutionsforschung*, **19**, 264-78.

Groves, C.P. (1982). The systematics of tree kangaroos (*Dendrolagus*; Marsupialia, Macropodidae). *Australian Mammalogy*, **5**, 157-86.

Groves, C.P. (1983). Phylogeny of the living species of Rhinoceros. *Zeitschrift für zoologiche Systematik und Evolutionsforschung*, **21**, 293-313.

Groves, C.P. (1986). Systematics of the Great Apes. In *Comparative Primate biology, 1, Systematics, evolution and anatomy* (eds. J. Erwin and D.R. Swindler), pp. 187-217. Alan R. Liss, New York.

Groves, C.P. and Grubb, P. (1982). The species of Muntjac (genus *Muntiacus*) in

Borneo: unrecognised sympatry in tropical deer. *Zoologische Mededelingen Leiden*, **56**, 203–16.

Groves, C.P. and Mazák, V. (1975). An approach to the taxonomy of the Hominidae: Gracile Villafranchian Hominids of Africa. *Casopis pro Mineralogii Geologii*, **20**, 225–47.

Groves, C.P. and Sabater Pi, J. (1985). From ape's nest to human fix-point. *Man*, NS, **20**, 22–47.

Groves, C.P. and Stott, K.W. Ir. (1979). Systematic relationships of gorillas from Kahuzi, Tshiaberimu and Kayonza. *Folia primatologica*, **32**, 161–79.

Grubb, P. (1973). Distribution, divergence and speciation of the drill and mandrill. *Folia primatologica*, **20**, 161–77.

Grubb, P. (1982). Refuges and dispersal in the speciation of African forest mammals. In *Biological diversification in the tropics* (ed. G.T. Prince), pp. 537–53. Columbia University Press.

Guo S., Zhou S., Meng W., Zhang, P., Sun, S., Hao, X., Liu, S., Zhang, F., Hu, R., and Liu, J. (1980). Fission track dating of Peking Man. *Kexue Tongbao*, **25**, 770–2.

Gu Yumin and Liu Yipu (1983). First discovery of *Dryopithecus* in East China. *Acta Authropologica Sinica*, **2**, 305–14.

Haigh, J. and Maynard Smith, J. (1972). Population size and protein variation in man. *Genetical Research, Cambridge*, **19**, 73–89.

Haimoff, E.H., Gittins, S.P., Whitten, A.J. and Chivers, D.J. (1984). A phylogeny and classification of gibbons based on morphology and ethology. In *The lesser apes* (ed. H. Preuschoft, D.J. Chivers, W.Y. Brockelman, and N. Creel), pp. 614–32. Edinburgh Univesity Press.

Haldane, J.B.S. (1932). *The causes of evolution*. Longmans, London.

Hanak, V. (1966). Zur Systematik und Verbreitung der Gattung *Plecotus* Geoffroy, 1818 (Mammalia, Chiroptera). *Lynx*, NS, **6**, 57–66.

Happold, D.C.D. (1973). The Red Crowned Mangabey, *Cercocebus torquatus*, in western Nigeria. *Folia Primatologica*, **20**, 423–8.

Harmon, R.S., Glazek, J. and Nowak, K. (1980). $^{230}$Th/$^{234}$U dating of travertine from the Bilzingsleben archaeological site. *Nature*, **284**, 132–5.

Harris, J.M. (1985). Age and paleoecology of the Upper Laetoli Beds, Laetoli, Tanzania. In *Ancestors: the hard Evidence* (ed. E. Delson), pp. 7–81. Alan R. Liss, New York.

Harrison, T. (1986a). The phylogenetic relationships of the Oreopithecidae. *American Journal of Physical Anthropology*, **71**, 265–384.

Harrison, T. (1986b). New fossil anthropoids from the middle Miocene of East Africa and their bearing on the origin of the Oreopithecidae. *American Journal of Physical Anthropology*, **71**, 265–84.

Hayman, R.W. (1937). A note on *Galago senegalensis inustus* Schwarz. *Annaels and Magazine of Natural History*, **20**, 149–51.

Heads, M. (1985). On the nature of ancestors. *Systematic Zoology*, **34**, 205–15.

Heberer, G. (1963). Ueber einen neuen archanthropinen Typus aus der Oldoway-Schlucht. *Zeitschrift für Morphologie und Anthropologie*, **53**, 171–7.

Hedrick, P.W. (1982). Genetic hitchhiking: a new factor in evolution? *BioScience*, **32**, 845–53.

Hedrick, P., Jain, S. and Holden, L. (1978). Multilocus systems in evolution. *Evolutionary Biology*, **11**, 101–82.

Heim, J-L. (1976). Les hommes fossiles de La Ferrassie I. *Archives de l'Institut de Paleontologie humaine*, 35, 1-331.

Heltne, P.G. and Kunkel, L.M. (1975). Taxonomic notes on the pelage of *Ateles paniscus paniscus, A. p. chamek* (sensu Kellogg and Goldman, 1944) and *A. fusciceps rufiventris* ( = *A. f. robustus* Kellogg & Goldman, 1944). *Journal of Medical Primatology*, 4, 83-102.

Hemmer, H. (1966). Untersuchungen zur Stammesgeschichte der Pantherkatzen (Pantherinae). I. *Veröffentliche Zoologisches Staatssammluug München*, 11, 1-121.

Hemmer, H. (1967). *Allometrie-Untersuchungen zur Evolution des menschlichen Schadels und seiner Rassentypen*. Fischer, Frankfurt and Main.

Hemmer, H. (1972). Notes sur la position phylétique de l'homme de Petralona. *L'Anthropologie*, 76, 155-62.

Hemmer, H. (1974). Untersuchungen zur Stammesgeschichte der Pantherkatzen (Pantherinae). III. Zur Artgeschichte des Lowen *Panthera (Panthera) leo* (Linnaeus 1758). *Veröffentliche Zoologisches Staatssamlung München*, 17, 167-280.

Hennig, W. (1966). *Phylogenetic systematics* (trans. D.D. Davis and R. Zangerl). University of Illinois Press.

Hershkovitz, P. (1949). Mammals of northern Colombia, preliminary report no. 4: monkeys (Primates), with taxonomic revisions of some forms. *Proceedings of the US National Museum*, 98, 323-427.

Hershkovitz, P. (1963). A systematic and zoogeographic account of the monkeys of the genus *Callicebus* of the Amazonas and Orinoco River basins. *Mammalia*, 27, 1-80.

Hershkovitz, P. (1966). Taxonomic notes on Tamarins, genus *Saguinus* (Callithricidae, Primates), with descriptions of four new forms. *Folia Primatologica*, 4, 381-95.

Hershkovitz, P. (1968). Metachromism or the principle of evolutionary change in mammalian tegumentary colours. *Evolution*, 22, 556-75.

Hershkovitz, P. (1970a). Notes on Tertiary Platyrrhine monkeys and description of a new genus from the late Miocene of Colombia. *Folia Primatologica*, 12, 1-37.

Hershkovitz, P. (1970b). Metachromism like it is. *Evolution*, 24, 644-8.

Hershkovitz, P. (1974). A new genus of Late Oligocene monkey (Cebidae, Platyrrhini) with notes on postorbital closure and platyrrhine evolution. *Folia primatologice*, 21, 1-35.

Hershkovitz, P. (1978). *Living New World monkeys (Platyrrhini) with an introduction to primates*. Chicago University Press.

Hershkovitz, P. (1979). The species of sakis, genus *Pithecia*, with notes on sexual dichromatism. *Folia primatologica*, 31, 1-22.

Hershkovitz, P. (1983). Two new species of Night Monkeys, genus *Aotus*: a preliminary report on *Aotus* taxonomy. *American Journal of Primatology*, 4, 209-43.

Hershkovitz, P. (1984). Taxonomy of Squirrel Monkeys genus *Saimiri*: a preliminary report with description of a hitherto unnamed form. *American Journal Primatology*, 7, 155-210.

Hershkovitz, P. (1985). A preliminary taxonomic review of the South American Bearded Saki Monkeys, genus *Chiropotes*, with the description of a new subspecies. *Fieldiana, Zoology*, NS 27, 1-46.

Heuvelmans, B. (1955). *Sur la piste des betes ignorees*, 2 volumes. Plon, Paris.

Hewett-Emmett, D., Venta, P.J., and Tashian, R.E. (1982). Features of gene structure, organization, and expression that are providing unique insights into molecular evolution and systematics. In *Macromolecular sequences in systematic and evolutionary biology* (ed. M. Goodman), pp. 357–405. Plenum, New York.

Hill, (1985). Early Hominid from Baringo, Kenya. *Nature*, **315**, 222–4.

Hill, W.C.D. (1952). The external and visceral anatomy of the olive colobus monkey (*Procolobus verus*). *Proceedings of the zoological Society of London*, **122**, 127–86.

Hill, W.C.O. (1953). *Primates: comparative anatomay and taxonomy, I. Strepsirhini*. Edinburgh University Press.

Hill, W.C.O. (1957). *Primates: comparative anatomy and taxonomy, III. Pithecoidea*. Edinburgh University Press.

Hill, W.C.O. (1960). *Primates: comparative anatomy and Taxonomy, IV, Cebidae Part A*. Edinburgh University Press.

Hill, W.C.O. (1962). *Primates: comparative anatomy and taxonomy, V. Cebidae, Part B*. Edinburgh University Press.

Ho, M.W. and Saunders, P.T. (1979). Beyond neo-Darwinism—an epigenetic approach to evolution. *Journal of Theoretical Biology*, **78**, 573–91.

Hofman, M.A. (1983). Encephalization in hominids: evidence for the model of punctuationalism. *Brain Behavior and Evolution*, **22**, 102–17.

Hoffstetter, R. (1969). Un primate de l'Oligocène inférieur sud-americain: *Branisella boliviana* gen. et sp. nov. *Compte Rendues de l'Académie des Sciences Paris, D*, **269**, 434–7.

Hoffstetter, R. (1974). Phylogeny, and geographical deployment of the Primates. *Journal of Human Evolution*, **3**, 327–50.

Hoffsetter, R. (1977). Origine et principales dichotomies des Primates Simiiformes ( = Anthropoidea). *Comptes Rendues de l'Académie des Sciences Paris*, **284**, 2095–8.

Hoffstetter, R. (1979). Controversés actuels sur la phylogénie et la classification des Primates. *Bulletin et Mémoires de la Société d'Anthropologie Paris*, **6**, 305–32.

Hoffstetter, R. (1982). Les Primates Simiiformes ( = Anthropoidea) (compréhension, phylogénie, histoire biogéographique). *Annales de Paléontologie (Vertébrés-Invertébrés)*, **68**, 241–90.

Hollihn, K-U. (1971). Das Verhalten von Guerezas (*Colobus guereza* und *Colobus polycomos*), Nasenaffen (*Nasalis larvatus*) und Kleideraffen (*Pygathrix nemaeus*) bei der Nahrungsaufnahme und ihre Haltung. *Zeitschrift für Säugetierkundé*, **36**, 65–95.

Holloway, R.L. (1980). The OH7 (Olduvai Gorge, Tanzania) hominid partial endocast revisited. *American Journal of Physical Anthropology*, **53**, 267–74.

Holloway, R.L. (1981). The endocast of the Omo L388–6 juvenile hominid: gracile or robust *Australopithecus*? *American Journal of Physical Anthropology*, **54**, 109–18.

Holloway, R.L. (1983). Cerebral brain endocast pattern of *Australopithecus afarensis* hominid. *Nature*, **303**, 420–2.

Homewood, K.M. and Rodgers, W.A. (1981). A previously undescribed mangabey from southern Tanzania. *International Journal of Primatology*, **2**, 47–55.

Horwich, R.H. (1983). Species status of the Black Howler Monkey, *Alouatta pigra*, of Belize. *Primates*, **24**, 288–9.

Howell, F.C. (1969). Remains of Hominidae from Pliocene/Pleistocene formations in the Lower Omo Basin, Ethiopia. *Nature*, **223**, 1234-9.

Howell, F.C. (1978). Hominidae. In *Evolution of African mammals* (ed. V.J. Maglio and H.B.S Cooke), pp. 154-248. Harvard University Press.

Howells, W.W. (1973). *Cranial variation in modern man: a study by multivariate analysis.* Peabody Museum Papers, Harvard.

Hrdlicka, A. (1928). Catalogue of human crania in the United States National Museum collections. *Proceedings of the US National Museum 71*, Art. 24, 1-140.

Huang Wanpo, Fang Dushen, and Ye Yongxiang (1982). Preliminary study on the fossil hominid skull and fauna of Hexian, Anhui. *Vertebrate Palasiatica*, **20**, 248-57.

Hublin, J.J. (1985). Human fossils from the North African Middle Pleistocene and the origin of *Homo sapiens*. In *Ancestors: the hard evidence* (ed. E. Delson), pp. 283-9. Alan R. Liss, New York.

Hull, D.B. (1970). Cladism gets sorted out. *Paleobiology*, **6**, 131-6.

Hull, D.B. (1979). A craniometric study of the black and white *Colobus* Illiger, 1811 (Primates, Cercopithecoidea). *American Journal of Physical Anthropology*, **51**, 168-81.

Humphries, C.J. (1983). Problems of phylogenetic reconstruction; methods of phylogenetic reconstruction (Book review). *Systematic Zoology*, **32**, 301-10.

Husson, A.M. (1978). *The mammals of Suriname. Zoological Monographs of the Rijksmuseum van Natuurlijke Historie, no. 2.* E.J. Brill, Leiden.

Huxley, T.H. (1870). On the methods and results of ethnology. *Fortnightly Review*, **1**, 257-77.

Isaac, G.Ll. (1976). The activities of early African hominids: a review of archaeological evidence from the time span two and a half to one million years ago. In *Human origins: Louis Leakey and the East African evidence* (ed. G.Ll. Isaac and E.R. McCawn), pp. 483-514. Press, Menlo Park, Calif.

Isaac, G.Ll. and Curtis, G.H. (1974). Age of early Acheulian industries from the Peninj group, Tanzania. *Nature*, **249**, 624-7.

Ishida, H., Pickford, M., Nakaya, H. and Nakano, Y. (1984). Fossil anthropoids from Nachola and Samburu Hills, Samburu district, Kenya. *African Study Monograph*, supplement **2**, 73-85.

Itihara, M., Kadar, D. and Watanabe, N. (1985). Concluding remarks. In *Quaternary geology of the hominid fossil bearing formations in Java* (ed. N. Watanabe and D. Kadar), pp. 367-78. Geological Research and Development Centre, Bandung.

Jacob, T. (1976). Early populations in the Indonesian region. In *The origin of the Australians* ed. R.L. Kirk and A.G. Thorne), pp. 81-94. Australian Institue of Aboriginal Studies, Canberra.

Jacobs, L.L. (1981). Miocene lorisid primates from the Pakistan Siwaliks. *Nature*, **289**, 585-7.

Jaeger, J-J. (1981). Les hommes fossiles du Pléistocene moyen du Maghreb dans leur cadre géologique, chronologique et paléoécologique. In *Homo erectus: papers in honor of Davidson Black* (ed. B.A. Sigmon and J.S. Cybulski), pp. 159-264. University of Toronto Press.

Jarvis, C. (1982). Evolution of horns in ungulates: ecology and paleoecology. *Biological Reviews*, **57**, 261-318.

Jefferies, R.P.S. (1981). Fossil evidence on the origin of Chordates and Echinoderms. *Origine dei grandi Phyla dei Metazoi: atti del Convegni Lincei*, **49**, 487–565. Accademia Nazionale dei Lincei, Roma.

Jelinek, A.J. (1982). The Tabun Cave and paleolithic man in the Levant. *Science*, **216**, 1369–75.

Johanson, D.C. and Edey, M.A. (1981). *Lucy: the beginnings of humankind*. Granada, London.

Johanson, D.C. White, T.D., and Coppens, Y. (1978). A new species of the genus *Australopithecus* (Primates: Hominidae) from the Pliocene of Eastern Africa. *Kirtlandia*, issue 28, pp. 1–14.

Johnson, C. (1976). *Introduction to natural selection*. University Park Press, Baltimore, Md.

Jollos, V. (1913). Experimentelle Untersuchungen an Infusorien. *Biologische Centralblatt*, 33.

Jolly, C.J. (1966). Introduction to the Cercopithecoidea, with notes on their use as laboratory animals. *Symposium of the Zoological Society of London*, **17**, 427–57.

Jolly, C.J. (1970). The seed-eaters: a new model of hominid differentiation based on a baboon analogy. *Man*, NS, **5**, 5–26.

Jolly, C.J. (1972). The classification and natural history of *Theropithecus (Simopithecus)* (Andrews, 1916), baboons of the African Plio-Pleistocene. *Bulletin of the British Museum (natural History), Geology*, **22**, 1–123.

Jouffroy, F.K. and Günther, M.M. (1985). Interdependence of morphology and behaviour in the locomotion of Galagines. In *Primate morphophysiology locomotor analyses and human bipedalism* (ed. S. Kondo), pp. 201–34. University of Tokyo Press.

Jouffroy, F-K. and Lessertisseur, J. (1978). Etude écomorphologique des proportions des membres des primates et spécialement des prosimiens. *Annales des Sciences naturelles Zoologie Paris (12)*, **20**, 99–128.

Jungers, W.L. and Stern, J.T. (1983). Body proportions, skeletal allometry and locomotion in the Hadar hominids: a reply to Wolpoff. *Journal of Human Evolution*, **12**, 673–84.

Karlin, S. (1975). General two-locus selection models: some objectives, results and interpretations. *Theoretical Papecism Biology*, **7**, 364–98.

Kay, R.F. (1975). Allometry and early hominids. *Science*, **189**, 63.

Kay, R.F. (1977). The evolution of molar occlusion in the Cercopithecidae and early Catarrhines. *American Journal of Physical Anthropology*, **46**, 327–52.

Kay, R.F. (1978). Molar structure and diet in extant Cercopithecidae. In *Development, function and evolution of teeth* (ed. P. Butler and K.A. Joysey). Academic Press, New York.

Kay, R.F. (1982). *Sivapithecus simonsi*, a new species of Miocene Hominoid, with comments on the phylogenetic status of the Ramapithecinae. *International Journal of Primatology*, **3**, 113–73.

Kay, R.F. and Simons, E.L. (1983). Dental formulae and dental eruption patterns in Parapithecidae. *American Journal of Physical Anthropology*, **62**, 363–75.

Kay, R.F., Fleagle, J.G., and Simons, E.L. (1981). A revision of the Oligocene apes of the Fayum province, Egypt. *American Journal of Physical Anthropology*, **55**, 293–322.

Kelley, J. (1986). Species recognition and sexual dimorphism in *Proconsul* and *Rangwapithecus, Journal of Human Evolution*, **15**, 461–95.

Kelley, J. and Pilbeam, D. (1986). The Dryopithecines: taxonomy, comparative anatomy and phylogeny of Miocene large hominoids. In *Comparative primate biology, Vol. 1. Systematics, evolution and anatomy* (ed. D.R. Swindler and J. Erwin), pp. 361–411. Alan R. Liss, New York.

Kellogg, R. and Goldman, E.A. (1944). Review of the Spider Monkeys. *Proceedings of the US National Museum*, **96**, 1–45.

Kennedy, G.E. (1983). Some aspects of femoral morphology in *Homo erectus*. *Journal Human Evolution*, **12**, 587–616.

Kennedy, G.E. (1984). Are the Kow Swamp hominids 'archaic'? *American Journal of Physical Anthropology*, **65**, 163–8.

Kettlewell, H.B.D. (1955). Selection experiments on industrial melanism in the Lepidoptera. *Heredity*, **9**, 323–42.

Kimbel, W.H., White, T.D. and Johanson, D.C. (1985). Craniodental morphology of the hominids from Hadar and Laetoli: evidence of '*Paranthropus*' and *Homo* in the Mid-Pliocene of Eastern Africa? In *Ancestors: the hard evidence* (ed. E. Delson), pp. 120–37. Alan R. Liss, New York.

Kimura, M. (1968). Evolutionary rate at the molecular level. *Nature*, **217**, 624–6.

Kimura, M. (1981). Was globin evolution very rapid in the early stages? A dubious case against the rate constancy hypothesis. *Journal of Molecular Evolution*, **17**, 110–13.

Kingdon, J. (1971). *East African mammals: an atlas of evolution in Africa. 1.* Academic Press, London.

Kingdon, J. (1980). The role of visual signals and face patterns in African forest monkeys (guenons) of the genus *Cercopithecus*. *Transactions of the Zoological Society of London*, **35**, 431–75.

Koenders, L., Rumpler, Y., Ratsiarson, J., and Peyrieras, A. (1985). *Lemur macaco flavifrons* (Gray, 1867): a rediscovered subspecies of primates. *Folia Primatologica*, **44**, 210–15.

Koestler, A. (1971). *The case of the midwife toad*. Hutchinson, London.

Konstant, W. (1985). Spider Monkeys in captivity and in the wild. *Primate Conservation*, **5**, 82–109.

Koop, B.F. Goodman, M., Xu, P., Chan, K., and Slighton, J.L. (1986). Primate η-globin DNA sequences and man's place among the great apes. *Nature*, **319**, 234–8.

Kortlandt, A. (1972). *New perspectives on ape and human evolution*. Stichting voor Psychobiologie, Amsterdam.

Krantz, G.S. (1975). An explanation for the diastema of Javan *erectus* skull IV. In *Palaeoanthropology: morphology and paleoecology* (ed. R.H. Tuttle), pp. 361–72. Mouton, The Hague.

Krassilov, V.A. (1980). Directional evolution: a new hypothesis. *Evolutionary Theory*, **4**, 203–20.

Krumbiegel, I. (1978). Die kurzschwanz-stumpfnase *Simias concolor*, und die übrigen Nasenaffen. *Säugetierkundliche Mitteilungen*, **26**, 59–75.

Kuhn, H-J. (1964). Zur Kenntnis von Bau und Funktion des Magens der Schlankaffen (Colobinae). *Folia Primatologica*, **2**, 193–221.

Kuhn, H-J. (1972). On the perineal organ of male *Procolobus badius*. *Journal of Human Evolution*, **1**, 371–8.

Kuroda, S., Kano, T., and Muhindo, K. (1985). Further information on the new monkey species, *Cercopithecus salongo* Thys van den Audenaerde, 1977. *Primates*, **26**, 325-33.

Kurth, G. (1965). Die (Eu)Homininen. In *Menschliche Abstammungslehre* (ed. G. Heberer), pp. 357-425. G. Fischer, Stuttgart.

Larnach, S.L. and Macintosh, N.W.G. (1974). A comparative study of Solo and Australian Aboriginal crania. In *Grafton Elliot Smith: the man and his work* (ed. A.P. Elkin and N.W.G. Macintosh), pp. 95-102. Sydney University Press.

Lavocat, R. (1974). The interrelationships between the African and South American rodents and their bearing on the problem of the origin of South American monkeys. *Journal of Human Evolution*, **3**, 323-6.

Lawrence, B. and Washburn, S.L. (1936). A new eastern race of *Galago demidovii*. *Occasional Papers of the Boston Society of Natural History*, **8**, 255-66.

Leakey, L.S.B. and Leakey, M.D. (1964). Recent discoveries of fossil hominids in Tanganyika: at Olduvai and near Lake Natron. *Nature*, **202**, 3-5.

Leakey, L.S.B., Tobias, P.V., and Napier, J.R. (1964). A new species of the genus *Homo* from Olduvai Gorge. *Nature*, **202**, 5-9.

Leakey, M., Tobias P.V., Martyn, J.E., and Leakey, R.E.F. (1969). An Acheulian industry with prepared core technique and the discovery of a contemporary hominid mandible at Lake Baringo, Kenya. *Proceedings of the Prehistoric Society*, **25**, 48-76.

Leakey, M.D. and Hay, R.L. (1982). *The chronological positions of the fossil hominids of Tanzania*, pp. 753-65. Centre National de la Recherche Scientifique.

Leakey, M.D., Clarke, R.J., and Leakey, L.S.B. (1971). New hominid skull from Bed I, Olduvai Gorge, Tanzania. *Nature*, **232**, 308-12.

Leakey, M.G. (1982). Extinct large colobines from the Plio-Pleistocene of Africa. *American Journal of Physical Anthropology*, **58**, 153-72.

Leakey, M.G. (1985). Early Miocene cercopithecoids from Buluk, Northern Kenya. *Folia primatologica*, **44**, 1-14.

Leakey, R.E.F. (1974). Further evidence of Lower Pleistocene hominids from East Rudolf, North Kenya, 1973. *Nature*, **248**, 653-6.

Leakey, R.E. and Leakey, M.G. (1986*a*). A new Miocene hominoid from Kenya. *Nature*, **324**, 143-6.

Leakey, R.E. and Leakey, M.G. (1986*b*). A second new Miocene hominoid from Kenya, *Nature*, **324**, 146-8.

Leakey, R.E.F. and Walker, A.C. (1976). *Australopithecus, Homo erectus* and the single species hypothesis. *Nature*, **261**, 572-4.

Leakey, R.E.F. and Walker, A.C. (1985). Further hominids from the Plio-Pleistocene of Koobi Fora, Kenya. *American Journal of Physical Anthropology*, **67**, 135-63.

Lee, A.K. and Cockburn, A. (1985). *Evolutionary ecology of marsupials*. Cambridge University Press.

Le Gros Clark, W.E. (1959). *The antecedents of man*. Edinburgh University Press.

Le Gros Clark, W.E. and Leakey, L.S.B. (1952). The Miocene Hominoidea of East Africa. *Fossil Mammals of Africa (British Museum (Natural History) )*, **1**, 1-117.

Le Pichon, X. (1968). See floor spreading and continental drift. *Journal of Geophysical Research*, **73**, 3661-97.

Leutenegger, W. (1977). Sociobiological correlates of sexual dimorphism in body weight in South African Australopiths. *South African Journal of Science*, **73**, 143-4.

Levinton, J.S. and Simon, C.M. (1980). A critique of the punctuated equilibria model and implications for the detection of speciation in the fossil record. *Systematic Zoology*, **29**, 130-42.

Lewin, R. (1982). Molecular drive: how real, how important? *Science*, **218**, 552-3.

Lewis, O.J. (1981). Functional morphology of the joints of the evolving foot. *Symposium of the Zoological Society of London*, **46**, 169-88.

Lewontin, R.C. (1974). *The genetic basis of evolutionary change*. Columbia University Press.

Li Chuan-kuei, (1978). A Miocene gibbon-like primate from Shihhung, Kiangsu Province. *Vertebrata Palasiatica*, **16**, 185-92.

Li Zhixiang and Lin Zhengyu (1983). Classification and distribution of living primates in Yunnan China. *Zoological Research*, **4**, 111-9.

Liu, T. and Ding, M. (1983). Discussion on the age of "Yuanmou Man". *Acta Anthropologica Sinica*, **2**, 40-8.

Liu Chun, Zhu Xiangyuan, and Ye Sujuan (1977). A palaeomagnetic study on the cave-deposits of Zhoukoudian (Choukoutien), the locality of *Sinanthropus*. *Scientia Geologica Sinica*, 1977, 26-33.

Lorenz, R. (1969). Notes on the care, diet and feeding habits of Goeldi's monkey. *International Zoo Yearbook*, Vol. 9, pp. 150-5.

Lovejoy, C.D. (1974). The gait of australopithecines. *Yearbook of Physical Anthropology*, **17**, 147-61.

Lovejoy, C.O. (1981). The origin of man. *Science*, **211**, 341-50.

Løvtrup, S. (1974). *Epigenetics: a treatise on theoretical biology*. John Wiley, Chichester, Sussex.

Lu Qingwu, Xu Qinghua and Zheng Liang (1981). Preliminary research on the cranium of *Sivapithecus yunnanensis. Vertebrata Palasiatica*, **19**, 106.

Luckett, W.P. (1976). Cladistic relationships among primate higher categories: evidence of the fetal membranes and placenta. *Folia Primatologica*, **25**, 245-76.

Luckett, W.P. and Maier, W. (1986). Developmental evidence for anterior tooth homologies in the aye-aye, *Daubentonia. American Journal of Physical Anthropology,* **69**, 233.

Lucotte, G. (1979*a*). Distances électrophorétiques entre les differentes espèces de singes du groupe des Mangabeys. *Annales de Génétique*, **22**, 85-7.

Lucotte, G. (1979*b*). Génétique des populations, spéciation et taxonomie chez les Babouins: II—similitudes génétiques comparées entre differentes espèces: *Papio papio, P. anubis, P. cynocephalus* et *P. hamadryas* basees sur les données du polymorphisme des enzymes erythrocytaires. *Biochemical Systematics and Ecology*, **7**, 245-51.

Lucotte, G. and Lefebvre, J. (1980). Distances électrophorétiques entre les cinq espèces de babouins du genre *Papio* basées sur les mobilités des protéines et enzymes sériques. *Biochemical Systematics and Ecology*, **8**, 317-22.

Ma Shilai and Wang Yingxiang (1986). The taxonomy and distribution of the gibbons in southern China and its adjacent region—with description of three new subspecies. *Zoological Research*, **7**, 394-410.

Ma Xinghua, Qian Fang, Li Pu and Ju Shiqiang (1978). Palaeomagnetic dating of Lantian Man. *Vertebrata Palasiatica*, **16**, 238-43.

Macbeth, N. (1974). *Darwin retried*. Garnstone Press, London.

McCluskey, J., Olivier, T.J., Freedman, L. and Hunt, E. (1974). Evolutionary divergences between populations of Australian wild rabbits. *Nature*, **249**, 278-9.

Mcdonald, J.N. (1981). *North American bison: their classification and evolution*. University of California Press, Berkeley.

McDougall, I. (1985). K-Ar and $^{40}$Ar/$^{39}$Ar dating of the hominid bearing Plio-Pleistocene sequence at Koobi Fora, Lake Turkana, northern Kenya. *Geological Society of America Bulletin*, **96**, 159-75.

MacFadden, P.L., Brock, A. and Partridge, T.C. (1979). Palaeomagnetism and the age of the Makapansgat hominid site. *Earth and Planetary Science Letters*, **44**, 373-82.

McHenry, H.M. (1975). Fossils and the mosaic nature of human evolution. *Science*, **190**, 425-31.

McHenry, H.M. (1978). Fore- and hindlimb proportions in Plio-Pleistocene hominids. *American Journal of Physical Anthropology*, **49**, 15-22.

McHenry, H.M. (1984). The common ancestor: a study of the postcranium of *Pan paniscus, Australopithecus*, and other hominoids. In *The pygmy chimpanzee* (ed. R.L. Susman), pp. 201-30. Plenum, New York.

McHenry, H.M. (1986). The first bipeds: a comparison of the *A. afarensis* and *A. africanus* postcranium and implications for the evolution of bipedalism. *Journal of Human Evolution*, **15**, 177-91.

McKenna, M.C. (1975). Toward a phylogenetic classification of the Mammalia. In *Phylogeny of the primates, a multidisciplinary approach* (ed. W.P. Luckett and F.S. Szalay), pp. 21-46. Plenum, New York.

MacKinnon, J. and MacKinnon, K. (1980). The behaviour of wild spectral tarsiers. *International Journal of Primatology*, **1**, 361-79.

McNeill, J. (1982). Phylogenetic reconstruction and phenetic taxonomy. *Zoological Journal of the Linnaean Society*, **74**, 337-44.

MacPhee, R.D.E (1981). *Auditory regions of primates and Eutherian insectivores*. Contributions to Primatology, Vol. 18.

MacPhee, R.D.E. and Woods, C.A. (1982). A new fossil Cebine from Hispaniola. *American Journal of Physical Anthropology*, **58**, 419-36.

Machado, A. de Barros, (1969). Mamiferos de Angola ainda nao citados ou pouco conhecidos. *Publicaçóes culturais da Companhia de Diamantes de Angola, Lisboa*, **46**, 93-232.

Maglio, V.J. (1973). Origin and evolution of the Elephantidae. *Transactions of the American Philosophical Society*, NS, **63**(3), 1-149.

Mahé, J. (1976). Craniométrie des Lémuriens: analyse multivariables—phylogénie. *Mémoires du Musée National d'Histoire Naturelle, C*, **32**, 1-342.

Maier, W. (1980). Konstruktionsmorphologische Untersuchungen am Gebiss der rezenten Prosimiae (Primates). *Abhandlungen der Senckenburgische-Naturforschenden Gesellschaft*, **538**, 1-158.

Marshall, J.T. (1981). Taxonomy. In *The mouse in biomedical research, I. History, genetics and wild mice* (ed. H.L. Foster, J.D. Small and J.G. Fox), pp. 17-26. Academic Press, New York.

Marshall, J.T. and Sugardjito, J. (1986). Gibbon systematics. In *Comparative primate biology, Vol. 1. Systematics, evolution and autonomy* (ed. D.R. Swindler and J. Erwin), pp. 137-85. Alan R. Liss, New York.

Martin, L.B. (1983). The relationships of the later Miocene Hominoidea. Unpublished PhD. thesis. University College, London.

Martin, L. and Andrews, P. (1984). The phyletic position of *Graecopithecus freybergi* Koenigswald. *Courier Forschunngs Institut Senchkenberg*, **69**, 25-40.

Martin, R.D. (1968). Towards a new definition of Primates. *Man*, NS, **3**, 377-401.

Martyn, J. (1967). Pleistocene deposits and new fossil localities in Kenya. *Nature*, **215**, 476-9.

Matsu'ura, S. (1982). A chronological framing for the Sangiran hominids—fundamental study by the fluorine dating method. *Bulletin of the National Science Museum Tokyo, D (Anthropology)*, **8**, 1-53.

Matsu'ura, S. (1985). A consideration of the stratigraphic horizons of hominid finds from Sangiran by the fluorine method. In *Quaternary geology of the hominid fossil bearing formations in Java* (ed. N. Watanabe and O. Kadar), pp. 359-66. Special Publication No. 4, Geological Research and Development Centre, Bandung.

Maw, B., Ciochon, R.L., and Savage, D.E. (1979). Late Eocene of Burma yields earliest anthropoid primate, *Pondaungia cotteri. Nature*, **282**, 65-7.

Maynard Smith, J. (1966). Sympatric speciation. *American Naturalist*, **100**, 637-50.

Maynard Smith, J. (1981). Macroevolution. *Nature*, **289**, 13-14.

Mayr, E. (1942). *Systematics and the origin of species*. Columbia University Press, New York.

Mayr, E. (1963). *Animal species and evolution*. Harvard University Press.

Mayr, E., Linsley, E.G., and Usinger, R.L. (1953). *Methods and principles of systematic zoology*. McGraw-Hill, New York.

Meester, J. and Setzer, H.W. (1971). *The mammals of Africa: an identification manual*. Smithsonian Institution Press, Washington, DC.

Mehlman, M.J. (1984). A reassessment of the associations and the age of the Lake Eyasi (Tanzania) hominid fossil crania. *African Archaeological Review*.

Meier, B., Albignac, R., Peyrieras, A., Rumpler, Y., and Wright, P. (1987). A new species of *Hapalemur* (Primates) from South East Madagascar. *Folia primatologica*, **48**, 211-15.

Menzies, J.I. (1970). An eastward extension to the known range of the olive colobus monkey (*Colobus verus* van Beneden). *Journal of the West African Science Association*, **15**, 83-84.

Mittermier, R.A., de Macedo Ruiz, H., and Luscombe, A. (1975). A woolly monkey rediscovered in Peru. *Oryx*, **13**, 41-6.

Moorrees, C.F.A. (1957). *The Aleut dentition*. Harvard University Press.

Morgan, E. (1972). *The descent of woman*. Stein and Day, New York.

Morgan, E. (1982). *The aquatic ape*. Souvenir Press, London.

Moss, M.L., Chase, P.S. and Howes, R.I. Jr. (1967). Comparative odontometry of the permanent post-canine dentition of American whites and negroes. *American Journal of Physical Anthropology*, **27**, 125-47.

Mturi, A.A. (1976). New Hominid from Lake Ndutu, Tanzania. *Nature*, **262**, 484-5.

Musgrave, J.H. (1973). The phalanges of Neanderthal and Upper Palaeolithic hands. In *Human evolution (SSHB, vol. XI)* (ed. M.H. Day), pp. 59-85. Taylor and Francis, London.

Musser, G.G. (1969). Results of the Archbold Expeditions. No. 91. A new genus and species of Murid Rodent from Celebes, with a discussion of its relationships. *American Museum Novitates*, No. 2384.

Musser, G.G. (1982). Results of the Archbold Expeditions. No. 110. *Crunomys* and the small-bodied shrew rats native to the Philippines Islands and Sulawesi (Celebes). *Bulletin of the American Museum of Natural History*, **174**, 1–95.

Musser, G.G. and Dagosto, M. (1987). The identity of *Tarsius pumilus*, a pygmy species endemic to the montane mossy Forests of Central Sulawesi. *American Museum Novitates*, No. 2867, 1–53.

Nagel, U. (1973). A comparison of Anubis baboons, Hamadryas baboons and their hybrids at a species border in Ethiopia. *Folia primatologica*, **19**, 104–65.

Napier, J.R. and Napier, P.H. (1967). *A handbook of living primates*. Academic Press, London.

Napier, J.R. and Napier, P.H. (1985). *The natural history of the primates*. British Museum (Natural History), London.

Napier, P.H. (1981). *Catalogue of Primates in the British Museum (Natural History) and elsewhere in the British Isles. Part II: Family Cercopithecidae, Subfamily Cercopithecinae*. British Museum (Natural History), London.

Napier, P.H. (1985). *Catalogue of Primates in the British Museum (Natural History) and elsewhere in the British Isles. Part III: Family Cercopithecidae, Subfamily Colobinae*. British Museum (Natural History), London.

Nelson, G. (1978). Ontogeny, phylogeny, paleontology and the biogenetic law. *Systematic zoology*, **27**, 324–45.

Nelson, G. and Rosen, D.E. (1981). *Vicariance biogeography: a critique*. Columbia University Press, New York.

Niemitz, C. (1984). Taxonomy and distribution of the genus Tarsius Storr, 1780. In *Biology of tarsiers* (ed. C. Niemitz), pp. 1–16. Springer, Berlin.

Oates, J.F. (1985). The Nigerian Guenon, *Cercopithecus erythrogaster*: ecological, behavioural, systematic and historical observations. *Folia Primatologica*, **45**, 25–43.

Oates, J.F. and Trocco, T.F. (1983). Taxonomy and phylogeny of black and white colobus monkeys. *Folia Primatologica*, **40**, 83–113.

Ohno, S. (1970). *Evolution by gene duplication*. Springer, New York.

Olson, T.R. (1978). Hominid phylogenetics and the existence of *Homo* in Member I of the Swartkrans Formation, South Africa. *Journal Human Evolution*, 7, 159–78.

Olson, T.R. (1980). *Galago crassicaudatus* E. Geoffroy, 1812: proposed use of the plenary powers to suppress the holotype and to designate a neotype. *Bulletin of zoological Nomenclative*, **37**, 176–85.

Olson, T.R. (1981*a*). Systematics and zoogeography of the greater galagos. *American Journal of Physical Anthropology*, **54**, 259.

Olson, T.R. (1981*b*). Basicranial morphology of the extant hominoids and Pliocene hominids: the new material from the Hadar Formation, Ethiopia, and its significance in early human evolution and taxonomy. In *Aspects of human evolution* (ed. C.B. Stringer), pp. 99–128. Taylor and Francis, London.

Olson, T.R. (1985*a*). Taxonomic affinities of the immature hominid crania from Hadar and Taung. *Nature*, **316**, 539–40.

Olson, T.R. (1985*b*). Cranial morphology and systematics of of the Hadar Formation Hominids and '*Australopithecus' africanus*. In *Ancestors: the hard evidence* (ed. E. Delson), pp. 102–19. Alan R. Liss, New York.

Olson, T.R. (1986). Species diversity and zoogeography in the Galagidae. *Abstracts of papers for XIth IPS Congress*, Göttingen.

Orgel, L.E. and Crick, F.H.C. (1980). Selfish DNA: the ultimate parasite. *Nature*, **284**, 604–7.

Osborn, H.F. (1936). *Proboscidea: a monograph of the discovery, evolution, migration and extinction of the Mastodonts and Elephants of the world.* American Museum Press, New York.

Oxnard, C.E. (1975). *Uniqueness and diversity in human evolution: morphometric studies of australopithecines.* Chicago University Press.

Oxnard, C.E. (1981). The uniqueness of *Daubentonia. American Journal of Physical Anthropology*, **54**, 1–21.

Oxnard, C.E. (1983), Sexual dimorphisms in the overall proportions of primates. *American Journal of Primatology*, **4**, 1–22.

Oxnard, C.E. (1984*a*). Dental sexual dimorphisms in humans and great apes, and in ramapithecines from China. *American Journal of Physical Anthropology*, **63**, 201.

Oxnard, C.E. (1984*b*). *The order of man.* Yale University Press.

Panchen, A.L. (1982). The use of parsimony in testing phylogenetic hypotheses. *Zoological Journal of the Linnaean Society*, **74**, 305–28.

Pariente, G. (1970). Rétinographies comparées des Lémuriens malgaches. *Comptes Rendues de l'Académie des Sciences, Paris, D*, **270**, 1404–7.

Partridge, T.C. (1973). Geomorphological dating of cave openings at Makapansgat, Sterkfontein, Swartkrans and Taung. *Nature*, **246**, 75–9.

Partridge, T.C. (1982). The chronological positions of the fossil hominids of southern Africa. *L'Homo erectus et la place de l' Homme de Tautavel parmi les hominidés fossiles*, pp. 617–75. Centre National de la Recherche Scientifique.

Passingham, R.E. (1982). *The human primate.* W.H. Freeman, San Fransisco, Calif.

Paterson, H.E.H. (1981). The continuing search of the unknown and unknowable: a critique of contemporary ideas on speciation. *South African Journal of Science*, **77**, 113–9.

Paterson, H.E.H. (1982). Perspective on speciation by reinforcement. *South African Journal of Science*, **78**, 53–7.

Patterson, B., Behrensmeyer, A.K. and Sill, W.D. (1970). Geology and fauna of a new Pliocene locality in North-western Kenya. *Nature*, **226**, 918–21.

Patterson, C. and Rosen, D.E. (1977). Review of ichthyodectiform and other Mesozoic teleost fishes and the theory and practice of classifying fossils. *Bulletin of the American Museum of national History*, **158**, 81–172.

Patton, J.L. and Smith, M.F. (1981). Molecular evolution in *Thomomys*: Phyletic systematics, paraphyly, and rates of evolution. *Journal of Mammalogy*, **62**, 493–500.

Peng Yanzhang, Zhang Yaoping, Ye Zhizhang and Liu Shuilin (1983). Study on the stomachs of three species of snub-nosed monkeys. *Zoological Research*, **4**, 167–75.

Peng Yanzhang, Ye Zhizhang, Liu Ruilin and Zhang Yaoping (1984). Craniofacial morphology of *Rhinopithecus. Zoological Research*, **5**, 7–22.

Perkins, E.M. (1975). Phylogenetic significance of the skin of New World monkeys (Order Primates, Infraorder Platyrrhini). *American Journal of Physical Anthropology*, **42**, 395–424.

Petry, D. (1982). The pattern of phyletic speciation. *Paleobiology*, **8**, 56–66.

Petter, J-J. (1965). The lemurs of Madagascar. In *Primate behavior* (ed. I. De Vore), pp. 292–319. Holt, Rinehart, and Winston, New York.

Petter, J-J. and Petter-Rousseaux, A. (1960). Remarques sur la systematique du genre *Lepilemur. Mammalia*, **24**, 76–86.

Peter, J-J., Albignac, R. and Rumpler, Y. (1977). *Faune de Madagascar, 44.*

Mammiferes Lemuriens (Primates Prosimiens). Orstom, Centre National de la Recherche Scientifique, Paris.

Petter-Rousseaux, A. and Bourliere, F. (1965). Persistence des phenomènes d'ovogénésis chez l'adult de *Daubentonia madagascariensis*. *Folia Primatologica*, **3**, 241-4.

Petter-Rousseaux, A. and Petter, J-J. (1967). Contribution à la systématique des Cheirogaleinae (lémuriens malgaches). *Allocebus*, gen. nov., pour *Cheirogaleus trichotis* Guenther, 1875. *Mammalia*, **31**, 574-81.

Pickford, M. (1985). A new look at *Kenyapithecus* based on recent discoveries in western Kenya. *Journal of Human Evolution*, **14**, 113-43.

Pilbeam, D. and Gould S.J. (1974). Size and scaling in human evolution. *Science*, **186**, 892-901.

Pilbeam, D. *et al.* (1977). Geology and palaeontology of Neogene strata of Pakistan. *Nature*, **270**, 684-9.

Pilbeam, D.R., Behrensmeyer, A.K., Barry, J.C and Shah, S.I. (1979). Miocene sediments and faunas of Pakistan. *Postilla*, pp. 1-45, issue 179.

Pocock, R.I. (1907). A monographic revision of the monkeys of the genus *Cercopithecus*. *Proceedings of the Zoological Society of London*, 677-746.

Pocock, R.I. (1918). On the external characters of lemurs and of *Tarsius*. *Proceedings of the Zoological Society of London*, 19-53.

Pollock, J.J. and Mullin, R.J. (1987). Vitamin C biosynthesis in prosimians: evidence for the anthropoid affinity of *Tarsius*. *American Journal of Physical Anthropology*, **73**, 65-70.

Polyak, S. (1957). *The vertebrate visual system*. Chicago University Press.

Powell, J.R. (1978). The founder-flush speciation theory: an experimental approach. *Evolution*, **32**, 465-74.

Powers, R. (1962). The disparity between known age and age as estimated by cranial suture closure. *Man*, **62**, 52-4.

Preuss, T.M. (1982). The fact of *Sivapithecus indicus*: description of a new, relatively complete specimen from the Siwaliks of Pakistan. *Folia Primatologica*, **38**, 141-57.

Prouty, L.A., Buchanan, P.D., Pollitzer, W.S. and Mootnick, A.R. (1983). A presumptive new hylobatid subgenus with 38 chromosomes. *Cytogenetics and Cell Genetics*, **35**, 141-2.

Puech, P-F., Cianfarani, F. and Roth, H. (1986). Reconstruction of the maxillary dental arcade of Garusi Hominid I. *Journal of Human Evolution*, **15**, 325-32.

Qian, F. (1980). Magnetostratigraphic study on the cave deposits containing fossil Peking Man at Zhoukoudian. *Kexue Tongbao*, **25**, 359.

Qing Fang (1985). On the age of "Yuanmou Man"—a discussion with Liu Tungsheng *et al*. *Acta Anthropologica Sinica*, **4**, 324-32.

Quinet, G.E. (1966). Sur la formule dentaire de deux primates du Landénien continental belge. *Bulletin de l'Institut royale des Sciences naturelles, Bruxelles*, **42**(38), 1-6.

Rahm, U. (1966). Les mammiferes de la foret equatoriale de l'Est du Congo. *Annalen koninklijk Museum voor Middlen-Afrika, Tervuren (8) Science Zoology*, **149**, 39-121.

Rak, Y. (1983). *The australopithecine face*. Academic Press, New York.

Rak, Y. (1986). The Neanderthalis: a newlook at an old face. *Journal of Human Evolution*, **15**, 151-64.

Rak, Y. and Howell, F.C. (1978). Cranium of a juvenile *Australopithecus boisei* from the Lower Omo Basin, Ethiopia. *American Journal of Physical Anthropology*, **48**, 345–66.

Rasmussen, D.T. (1986). Anthropoid origins: a possible solution to the Adapidae-Omomyidae paradox. *Journal Human Evolution*, **15**, 1–12.

Remane, A. (1965). Die Geschichte der Menschenaffen. In G. Heberer, ed., *Menschliche Abstammungslehre* (ed. G. Heberer), pp. 249–309. G. Fischer Verlag. Stuttgart.

Richard, A.F. (1985). *Primates in nature.* W.H. Freeman, New York.

Rightmire, G.P. (1979). Cranial remains of *Homo erectus* from Beds II and IV, Olduvai Gorge, Tanzania. *American Journal of Physical Anthropology*, **51**, 99–116.

Rightmire, G.P. (1980). Middle Pleistocene hominids from Olduvai Gorge, northern Tanzania. *American Journal of Physical Anthropology*, **53**, 225–41.

Rightmire, G.P. (1984). Comparisons of *Homo erectus* from Africa and Southeast Asia. *Courier Forschungsinstitut Senckenberg*, **69**, 83–98.

Rimoli, R. (1977). Una nueva especie de monos (Cebidae: Saimirinae: ?Saimiri) de la Hispaniola. *Cuadernos del Ciendia. (Santo Domingo)*, **242**, 1–16.

Ripley, S. (1979). Environmental grain, niche diversification, and positional behavior in Neogene primates: an evolutionary hypothesis. In *Environment, behavior and morphology: dynamic interactions in primates* (ed. M.E. Morbeck, H. Preuschaft, and N. Comberg), pp. 37–74. Gustav Fischer, New York.

Robinson, J.T. (1954). The genera and species of the Australopithecinae. *American Journal of Physical Anthropology*, **12**, 181–200.

Robinson, J.T. (1956). *The dentition of the Australopithecinae. Memoirs of the Transvaal Museum*, No. 9.

Robinson, J.T. (1972). *Early hominid posture and locomotion.* Chicago University Press.

Rollinson, J. (1975). Interspecific comparisons of locomotor behaviour and prehension in eight species of African forest monkeys—a functional and evolutionary study. PhD thesis, University of London.

Romero-Herrera, A.E., Lehmann, H., Joysey, Joysey, K.A., and Friday, A.E. (1973). Molecular evolution of myoglobin and the fossil record: a phylogenetic synthesis. *Nature*, **246**, 389–95.

Romero-Herrera, A.E., Lehmann, H., Castillo, O., Joysey, K.A., and Friday, A.E. (1976). Myoglobin of the orangutan as a phylogenetic enigma. *Nature*, **261**, 162–4.

Rose, K.D. and Fleagle, J.G. (1980). The fossil history of nonhuman primates in the Americas. In *Neotropical primatology* (ed. R.A. Mittermeier and A.F. Coimbra-Filho), pp. 111–67. Academia Brasileira de Ciencias, Rio de Janeiro.

Rose, M.D. (1984). A hominine hip bone, KNM-ER 3228, from East Lake Turkana, Kenya. *American Journal of Physical Anthropology*, **63**, 371–78.

Rosenberger, A.L. (1977). *Xenothrix* and ceboid phylogeny. *Journal of Human Evolution*, **6**, 461–81.

Rosenberger, A.L. (1978). New species of Hispaniolan monkey: a comment. *Anuario Cientifica Universidad Central, Este*, **3**, 249–51.

Rosenberger, A.L. (1979). Cranial anatomy and implications of *Dolichocebus*, a late Oligocene ceboid primate. *Nature*, **279**, 416–8.

Rosenberger, A.L. (1980). Gradistic views and adaptive radiation of platyrrhine primates. *Zeitschrift für Morphologie und Anthropologie*, **71**, 157–63.

Rosenberger, A.L. (1983). Aspects of the systematics and evolution of the marmosets. *Annals of the 1st Brazilian Primatological Congress*, pp. 159–80.

Rosenberger, A.L. and Coimbra-Filho, A.F. (1984). Morphology, taxonomic status and affinities of the Lion Tamarins, *Leontopithecus* (Callitrichinae, Cebidae). *Folia primatologica*, **42**, 149–79.

Rosenberger, A.L., Strasser, E., and Delson, E. (1985). Anterior dentition of *Notharctus* and the Adapid–Anthropoid hypothesis. *Folia Primatologica*, **44**, 15–39.

Rosenzweig, K.A. and Zilberman, Y. (1967). Dental morphology of Jews from Yemen and Cochin. *American Journal of Physical Anthropology*, **26**, 15–22.

Rumpler, Y. (1974). Contribution de la Cytogénétique à une nouvelle classification des Lémuriens malgaches. In *Prosimian biology* (ed. R.D. Martin, G.A. Doyle, and A. Walker), pp. 865–70. Duckworth, London.

Rumpler, Y. and Albignac, R. (1975). Intraspecific chromosome variability in a lemur from the North of Madagascar: *Lepilemur septentrionalis*, species nova. *American Journal of Physical Anthropology*, **42**, 425–30.

Rumpler, Y. and Dutrillaux, B. (1978). Chromosomal evolution in Malagasy lemurs. III. Chromosome banding studies in the genus *Hapalemur* and the species *Lemur catta*. *Cytogenetics and Cell Genetics*, **21**, 201–11.

Santa Lucca, A.P. (1980). *The Ngandong fossil hominids: a comparative study of a Far Eastern Homo erectus group*. Yale University Publication in Anthropology No. 78.

Sarich, V.M. (1967). Quantitative immunological study of evolution of primate albumins. Unpublished PhD thesis, University of California at Berkeley.

Sarich, V.M. and Cronin J.E. (1976). Molecular systematics of the primates. In *Molecular anthropology* (ed. M. Goodman and R. Tashian), pp. 141–70. Plenum, New York.

Sartono, S. (1976*a*) *On the Javanese Pleistocene hominids: a reappraisal*. Abstracts Union International des Societies Prélistoriques et Protohistoriques, Nice.

Sartono, S. (1976*b*). Genesis of the Solo terraces. *Modern Quaternary Research in SE Asia*, **2**, 1–21.

Sartono, S. (1979). The stratigraphy of the Sambungmacan site, Central Java. *Modern Quaternary Research in SE Asia*, **5**, 83–8.

Sartono, S. (1982). Sagittal cresting in *Meganthropus palaeojavanicus* von Koenigswald. *Modern Quaternary Research in SE Asia*, **7**, 201–10.

Sartono, S. (1985). Datings of Pleistocene man of Java. *Modern Quaternary Research SE Asia*, **9**, 115–25.

Sartono, S., Semah, F., Astadiredja, K.A.S., Sukendarmono, M., and Djubiantono, T. (1981). The age of *Homo modjokertensis*. *Modern Quaternary Research SE Asia*, **6**, 91–102.

Schmid, C.W. and Jelinek, W.R. (1982). The Alu family of dispersed repetitive sequences. *Science*, **216**, 1065–70.

Schmid, P. (1981). Comparison of Eocene nonadapids and Tarsius. In *Primate evolutionary biology* (ed. A.B. Chiarelli and R.S. Corruccini), pp. 6–3. Springer-Velag, Berlin.

Schmid, P. (1982). Die systematische Revision der europaischen Microchoeridae Lydekker, 1887 (Omomyiformes, Primates). Inaugural dissertation, Universität Zurich.

Schmid, P. (1983). Eine Rekonstruktion des Skelettes von A.L. 288-1 (Hadar) und deren Konsequenzen. *Folia Primatologica*, **40**, 283-306.

Schultz, A.H. (1970). The comparative uniformity of the Cercopithecoidea. In *The Old World monkeys* (ed. J.R. Napier and P.H. Napier), pp. 39-51. Academic Press, London.

Schwartz, J.H. (1974). Observations on the dentition of the Indriidae. *American Journal of Physical Anthropology*, **41**, 107-14.

Schwartz, J.H. (1984). The evolutionary relationships of man and orang-utans. *Nature*, **308**, 501-5.

Schwartz, J.H. (1986). Primate systematics and a classification of the Order. In *Comparative primate biology* (ed. D.R. Swindler and J. Erwin), Vol. 1, pp. 1-41. Alan R. Liss, New York.

Schwartz, J.H. and Krishtalka, L. (1977). Revision of Picrodontidae (Primates, Plesiadapiformes): dental homologies and relationships. *Annals of the Carnegie Museum*, **46**, 55-70.

Schwartz, J.H. and Tattersall, I. (1979). The phylogenetic relationships of Adapidae (Primates, Lemuriformes). *Anthropological Papers of the American Museum of Natural History*, **55**, 271-83.

Schwartz, J.H. and Tattersall, I. (1982). A note on the status of *'Adapis' priscus* Stehlin, 1916. *American Journal of Primatology*, **3**, 295-8.

Schwartz, J.H. and Tattersall, I. (1982b). Relationships of *Microadapis sciureus* (Stehlin, 1916), and two new primate genera from the Eocene of Switzerland. *Folia Primatologica*, **39**, 178-86.

Schwartz, J.H. and Tattersall, I. (1983). A review of the European primate genus *Anchomomys* and some allied forms. *Anthropological Papers of the American Museum of Natural History*, **57**, 344-52.

Schwartz, J.H. and Tattersall, I. (1985). Evolutionary relationships of living lemurs and lorises (Mammalia, Primates) and their potential affinities with European Eocene Adapidae. *Anthropological Papers of the American Museum natural History*, **60**, 1-100.

Schwartz, J.H., Tattersall, I., and Eldredge, N. (1978). Phylogeny and classification of primates revisited. *Yearbook of Physical Anthropology*, **21**, 95-133.

Schwarz, E. (1929). On the local races and distribution of the Black and White Colobus Monkeys. *Proceedings of the Zoological Society London*, 588-98.

Schwarz, E. (1931a). On the African long-tailed lemurs or Galagos. *Annals and Magazine of natural History*, **7**, 41-66.

Schwarz, E. (1931b). A revision of the genera and species of Madagascan Lemuridae. *Proceedings of the Zoological Society of London*, **1931**, 399-428.

Schwarz, E. (1932). Der Vertreter der Diana-Meerkatze in Zentral-Afrika. *Revue de Zoologie et Botanique Africaine*, **21**, 251-4.

Schwidetzky, I. (1974). *Grundlagen der Rassensystematik*. Bibliographisches Institut-Wissenschaftsverlag, Mannheim.

Scott, G.B.D. (1980). The primate caecum and appendix vermiformis: a comparative study. *Journal of Anatomy*, **131**, 549-63.

Searle, A.G. (1968). *Comparative genetics of coat colour in mammals*. Logos Press, London.

Semah, F. (1982). Pliocene and Pleistocene geomagnetic reversals recorded in the

Geomolong and Sangiran domes (Central Java). *Modern Quaternary Research in SE Asia*, **7**, 151–64.

Senut, B. (1981). *L'humérus et ses articulations chez les Hominidés plio-pléistocenes.* Cahiers de Palaeanthropologie.

Senut, B. (1986). Long bones of the primate upper limb: monomorphic or dimorphic? *Human Evolution*, **1**, 7–22.

Senut, B. and Tardieu, C. (1985). Functional aspects of Plio-Pleistocene hominid limb bones: implications for taxonomy and phylogeny. In *Ancestors: the hard evidence* (ed. E. Delson), pp. 193–201. Alan R. Liss, New York.

Setoguchi, T. (1985). *Kóndous laventicus*, a new ceboid primate from the Miocene of La Venta, Colombia, South America. *Folia primatologica*, **44**, 96–101.

Setoguchi, T. and Rosenberger, A. (1985). Miocene marmosets: first fossil evidence. *International Journal of Primatology*, **6**, 615–25.

Shea, B.T. (1983). Paedomorphosis and neoteny in the pygmy chimpanzee. *Science*, **222**, 521–2.

Shotake, T. (1981). Population genetical study of natural hybridization between *Papio anubis* and *Papio hamadryas*. *Primates*, **22**, 285–308.

Sibley, C.G. and Ahlquist, J.E. (1984). The phylogeny of the hominoid primates, as indicated by DNA–DNA hybridization. *Journal of Molecular Evolution*, **20**, 2–15.

Siesser, W.G. and Orchiston, D.W. (1978). Micropalaeontological reassessment of the age of Pithecanthropus mandible C from Sangiran, Indonesia. *Modern Quantenary Research in SE Asia*, **4**, 25–30.

Sigmon, B.A. (1982). Comparative morphology of the locomotor skeleton of *Homo erectus* and the other fossil hominids, with special reference to the Tautavel innominate and femora. *L'Homo erectus et la place de l'homme de Tautavel parmi les hominides fossils (ler Congrés international de Paléontologie humaine, Nice, Oct. 1982).* Preprints, pp. 422–46.

Simons, E.L. (1962). Two new primate species from the African Oligocene. *Postilla (Yale)*, **64**, 1–12.

Simons, E.L. (1969). A Miocene monkey (*Prohylobates*) from northern Egypt. *Nature*, **223**, 687–9.

Simons, E.L. (1972). *Primate evolution*. Macmillan, New York.

Simons, E.L. (1986). *Parapithecus grangeri* of the African Oligocene: an archaic catarrhine without lower incisors. *Journal of Human Evolution*, **15**, 205–13.

Simons, E.L. and Bown, T.M. (1985). *Afrotarsius chatrathi*, first tarsiiform primate (?Tarsiidae) from Africa. *Nature*, **313**, 475–7.

Simons, E.L. and Kay, R.F. (1983). *Qatrania*, new basal anthropoid primate from the Fayum, Oligocene of Egypt. *Nature*, **304**, 624–6.

Simons, E.L. and Pilbeam, D.R. (1965). A preliminary revision of the Dryopithecinae, *Folia primatologica*, **3**, 81–152.

Simons, E.L., Bown, T.M., and Rasmussen, D.T. (1986). Discovery of two additional prosimian primate families (Omomyidae, Lorisidae) in the African Oligocene. *Journal of Human Evolution*, **15**, 431–7.

Simons, E.L., Rasmussen, D.T., and Mullin, R.J. (1987). A new species of *Propliopithecus* from the Fayum, Egypt. *American Journal of Physical Anthropology*, **73**, 139–47.

Simpson, G.G. (1944). *Tempo and mode in evolution*. Columbia University Press, New York.

Simpson, G.G. (1945). The principles of classification and a classification of mammals. *Bulletin of the American Museum of Natural History*, **85**, 1–350.

Simpson, G.G. (1951). *Horses*. Oxford University Press, New York.

Simpson, G.G., Roe, A. and Lewontin, R.C. (1960). *Quantitative zoology*. Harcourt, Brace and World, New York.

Singer, M.F. (1982). Highly repeated sequences in mammalian genomes. *International Review of Cytology*, **76**, 67–112.

Skelton, R.R., McHenry, H.M. and Drawhorn, G.M. (1986). Phylogenetic analysis of early hominids. *Current Anthropology*, **27**, 21–43.

Slome, D. (1929). The osteology of a Bushman tribe. *Annals of the South African Museum*, **24**, 33–60.

Smith, R.J. (1984). Allometric scaling in comparative biology: problems of concept and method. *American Journal of Physiology*, **246**, (Regulatory Integrative Comparative Physiology 15), R152–R160.

Smith, RJ. (1985). The present as a key to the past: body weight of Miocene hominoids as a test of allometric methods for paleontological inference. In *Size and scaling in primate biology*, (ed. W.L. Jungers), pp. 437–48. Plenum, New York.

Sokal, R.R. and Sneath, P.H.A (1963). *Numerical taxonomy*. W.H. Freeman, San Francisco, Calif.

Sonakia, A. (1985). Early *Homo* from Narmada Valley, India. In *Ancestors: the hard evidence* (ed. E. Delson), pp. 334–9. Alan R. Liss, New York.

Sondaar, P.Y. (1984). Faunal evolution and the mammalian biostratigraphy of Java. *Courier Forschungsinstitut Senckenberg*, **69**, 219–35.

Stanley, S.M. (1979). *Macroevolution: pattern and process*. W.H. Freeman, San Francisco, Calif.

Stearns, S.C. (1977). The evolution of life history traits. *Annual Review of Ecology and Systematics*, **8**, 145–71.

Steele, E.J. (1979). *Somatic selection and adaptive evolution*. Williams–Wallace Productions International, Toronto.

Steele, E.J., Gorczynski, R.M. and Pollard, J.W. (1984). The somatic selection of acquired characters. In *Evolutionary perspectives in the 1980s* (ed. J.W. Pollard), pp. 217–37. John Wiley, Chichester, Sussex.

Stern, J.T. Jr. and Susman, R.L. (1983). The locomotor anatomy of *Australopithecus afarensis*. *American Journal of Physical Anthropology*, **60**, 279–317.

Steudel, K.M. (1985). Allometric perspectives on fossil catarrhine morphology. In *Size and scaling in primate biology* (ed. W.L. Jungers). pp. 449–75. Plenum, London.

Stevens, P.F. (1983). Report of third annual Willi Hennig Society meeting. *Systematic Zoology*, **32**, 285–91.

Stewart, T.D. (1972). The evolution of man in Asia as seen in the lower jaw. *Proceedings, VIII International Congress of Anthropological and Ethnological Sciences*, **1**, 263–6.

Straus, W.L. Jr. (1963). The classification of Oreopithecus. In *Classification and human evolution* (ed. S.L. Washburn), pp. 146–77. Aldine, Chicago, IL.

Stringer, C.B. (1980). The phylogenetic position of the Petralona cranium. *Anthropos*, **7**, 81–95.

Stringer, C. (1981). The dating of European Middle Pleistocene hominids and the existence of *Homo erectus* in Europe. *Anthropologie*, **19**, 3–14.

Stringer, C.B. (1982). Towards a solution to the Neanderthal problem. *Journal of Human Evolution*, **11**, 431–8.

Stringer, C.B. (1984). The definition of *Homo erectus* and the existence of the species in Africa and Europe. *Courier Forschungsinstitut Senckenberg*, **69**, 131–43.

Stringer, C.B. (1985). Middle Pleistocene hominid variability and the origin of the Late Pleistocene humans. In *Ancestors: the hard evidence* (ed. E. Delson), pp. 289–95. Alan R. Liss, New York.

Stringer, C.B. (1986). The credibility of *Homo habilis*. In *Major topics in primate and human evolution* (ed. B. Wood, L. Martin, and P. Andrews), pp. 266–94. Cambridge University Press.

Stringer, C.B. and Burleigh, R. (1981). The Neanderthal problem and the prospects for direct dating of Neanderthal remains. *Bulletin of the British Museum (National History)*, **35**, 225–41.

Stringer, C.B. and Trinkaus, E. (1981). The Shanidar Neanderthal crania. In *Aspects of human evolution* (ed. C.B. Stringer), pp. 129–65. Taylor and Francis, London.

Struhsaker, T.T. (1970). Phylogenetic implications of some vocalizations of *Cercopithecus* monkeys. In *Old World monkeys* (ed. V.R. Napier and P.H. Napier), pp. 365–444. Academic Press, New York.

Struhsaker, T.T. (1981). Vocalizations, phylogeny and palaeogeography of red colobus monkeys (*Colobus badius*). *African Journal of Ecology*, **19**, 265–83.

Sudre, J. (1975). Un prosimien du Paleogene ancien du Sahara nord-occidental: *Azibius trerki* n. gen. n. sp. *Comptes Rendues de l'Academie des Sciences, Paris, D*, **280**, 1539–42.

Sugawaru, R. (1982). Sociological comparison between two wild groups of anubis–hamadryas hybrid baboons. *African Study Monographs*, **2**, 73–131.

Susman, R.L. and Creel, N. (1979). Functional and morphological affinities of the subadult hand (O.H.7). from Olduvai Gorge. *American Journal of Physical Anthropology*, **63**, 224–25.

Susman, R.L. and Stern, J.T. Jr. (1982). Functional morphology of *Homo habilis*. *Science*, **217**, 931–4.

Szalay, F.S. (1968a). The Picrodontidae, a family of early primates. *American Museum Novitates*, 2329, 55 pp.

Szalay, F.S. (1968b). The beginnings of primates. *Evolution*, **22**, 19–36.

Szalay, F.S. (1973). New Palaeocene primates and a diagnosis of the new suborder Paromomyiformes. *Folia primatologier*, **19**, 73–87.

Szalay, F.S. (1981). Functional analysis and the practice of the phylogenetic method as reflected by some mammalian studies. *American Zoologist*, **21**, 37–45.

Szalay, F.S. and Decker, R.L. (1974). Origins, evolution and function of the tarsus in Late Cretaceous Eutheria and Paleocene primates. In *Primate locomotion* (ed. F.A. Jenkins, Jr.), pp. 223–50. Academic Press, New York.

Szalay, F.S. and Delson, E. (1979). *Evolutionary history of the primates*. Academic Press, New York.

Szalay, F.S. and Katz, C.C. (1973). Phylogeny of lemurs, galagos and lorises. *Folia primatologica*, **19**, 88–103.

Szalay, F.S. and Li, C.-K. (1986). Middle Paleocene Euprimate from Southern China and the distribution of Primates on the Paleogene. *Journal of Human Evolution*, **15**, 387–97.

Szalay, F.S., Li, C-K. and Wang, B-Y. (1989). Middle Paleocene omomyid primate from Anhui Province, China: *Decoredon anhuiensis* (Xu, 1976), new combination Szalay and Li, and the significance of *Petrolemur*. *American Journal of Physical Anthropology*, **69**, 269.

Tamarin, R.H. and Krebs, C.J. (1969). *Microtus* population biology. II. Genetic changes at the transferrin locus in fluctuating populations of two vole species. *Evolution*, **23**, 183–211.

Tanner, N. and Zihlman, A. (1976). Women in evolution, part 1: innovation and selection in human origins. *Signs*, **1**, 585–608.

Tardieu, C. (1981). Morpho-functional analysis of the articular surfaces of the knee-joint in primates. In *Primate evolutionary biology* (ed. A.B. Chiarelli and R.S. Corruccini), pp. 68–80. Springer Verlag, Berlin.

Tardieu, C. (1983). *Analyse morpho-fonctionnelle de l'articulation du genou chez les primates et les Hominidés fossiles*. Cahiers de Palaeanthropologie. Centre National de la Recherche Scientifique, Paris.

Tattersall, I. (1982*a*). *The primates of Madagascar*. Columbia University Press, New York.

Tattersall, I. (1982*b*). Two misconceptions of phylogeny and classification. *American Journal of Physical Anthropology*, **57**, 13.

Tattersall, I. (1986). Notes on the distribution and taxonomic status of some subspecies of *Propithecus* in Madagascar. *Folia Primatologica*, **46**, 51–63.

Tattersall, I. and Schwartz, J.H. (1974). Craniodental morphology and the systematics of the Malagasy lemurs (Primates, Prosimii). *Anthropological Papers of the American Museum of natural History*, **52**, 139–92.

Tattersall, I. and Schwartz, J.H. (1983*a*). A reappraisal of the European Eocene primate genus *Periconodon*. *Palaeontology*, **26**, 227–30.

Tattersall, I. and Schwartz, J.H. (1983*b*). A revision of the European Eocene primate genus *Protoadapis* and some allied forms. *American Museum Novitates*, issue 2762, pp. 1–16.

Tattersall, I. and Schwartz, J.H. (1984). A new genus and species of adapid primate from the middle Eocene of Alsace, and a new genus for '*Adapis*' *ruetimeyeri* Stehlin, 1912. *Folia primatologica*, **41**, 231–9.

Taylor, J., Freedman, L., Olivier, T.J., and McCluskey, J. (1977). Morphometric distances between Australian wild rabbit populations. *Australian Journal of Zoology*, **25**, 721–32.

Temin, H.M. (1976). The DNA provirus hypothesis. *Science*, **192**, 1075–80.

Temin, H.M. and Engels, W. (1984). Movable genetic elements and evolution. In *Evolutionary theory* (ed. J.W. Pollard). John Wiley, Chichester.

Templeton, A.R. (1979). The unit of selection *Drosophila mercatorum*. II. Genetic revolution and the origin of coadapted genomes in parthenogenetic strains. *Genetics*, **92**, 1265–82.

Templeton, A.R. (1980*a*). Modes of speciation and inferences based on genetic distances. *Evolution*, **34**, 19–29.

Templeton, A.R. (1980*b*). Macroevolution. *Evolution*, **34**, 1224–7.

Templeton, A.R. (1980*c*). The theory of speciation via the founder principle. *Genetic*, **94**, 1011–38.

Thaeler, C.S. (1968). An analysis of three hybrid populations of pocket gophers (genus *Thomomys*). *Evolution*, **22**, 543–55.

Thenius, E. (1981). Bemerkungen zur taxomischen und stammesgeschichtlichen Position der Gibbons (Hylobatidae, Primates). *Zeitschrift Für Säugetierkunde*, **46**, 232–41.

Thomas, H. (1984). Les Bovidae (Artiodactyla: Mammalia) du Miocène du Sous-Continent indien, de la Peninsule arabique et de l'Afrique: biostratigraphie, biogéographie et écologie. *Palaeogeography, Palaeoclimatology, Palaeoecology*, **45**, 251–99.

Thomas, H. and Verma, S.N. (1979). Découverte d'un Primate adapiforme (Sivaladapinae subfam. nov.) dans le Miocene moyen des Siwaliks de la région de Ramnagar (Jammu et Cachemire, Inde). *Comptes Rendues de l'Academie Sciences Paris, D*, **289**, 833–6.

Thorington, R.W. (1985). The taxonomy and distribution of Squirrel Monkeys (Saimiri). In *Handbook of Squirrel Monkey research* (ed. L.A. Rosenblum and C.L. Coe), pp. 1–33. Plenum, New York.

Thorndike, E.E. (1968). A microscopic study of the marmoset claw and nail. *American Journal of Physical Anthropology*, **28**, 247–53.

Thorne, A.G. (1981). The centre and the edge: the significance of Australasian hominids to African palaeoanthropology. In *Proceedings of the 8th Panafrican Congress of Prehistory and Quaternary Studies* (ed. R.E. Leakey and B.A. Ogot), pp. 180–1. The International Louis Leakey Memorial Institute for African Prehistory, Nairobi.

Thorne, A.G. and Wolpoff, M.H. (1981). Regional continuity in Australasian Pleistocene hominid evolution. *American Journal of Physical Anthropology*, **55**, 337–49.

Thys van den Audenaerde, D.F.E. (1977). Description of a monkey-skin from East-Central Zaire as a probably new monkey-species. *Review de Zoologie africaine*, **91**, 1000–10.

Tobias, P.V. (1964). The Olduvai Bed I hominine with special reference to its cranial capacity. *Nature*, **202**, 3–4.

Tobias, P.V. (1967*a*). *Olduvai Gorge, 2. The cranium and maxillary dentition of Australopithecus (Zinjanthropus) boisei*. Cambridge University Press.

Tobias, P.V. (1967*b*). Pleistocene deposits and new fossil localities in Kenya. *Nature*, **215**, 479–80.

Tobias, P.V. (1968). Middle and early Upper Pleistocene members of the genus *Homo* in Africa. In *Evolution und Hominisation* (ed. G. Karth), pp. 176–94. G. Fischer, Stuttgart.

Tobias, P.V. (1971). *The brain in hominid evolution*. Columbia University Press.

Tobias, P.V. (1979). A. survey and synthesis of the African hominids of the Late Tertiary and Early Quaternary periods. In *Current argument on early man* (ed. L-K. Konigsson), pp. 86–113. Pergamon Press, Oxford.

Tobias, P.V. (1980). '*Australopithecus afarensis*' and *A. africanus*: critique and an alternative hypothesis. *Palaeontologia Africane*, **23**, 1–17.

Trinkaus, E. (1983). *The Shanidar Neandertals*. Academic Press, London.

Trinkaus, E. (1984). Does KNM-ERJ418A establish *Homo erectus* at 2.0 myr BP? *American Journal of Physical Authropology*, **64**, 137–9.

Valladas, H., Reyss, J.L., Joron, J.L., Valladas, G., Bar-Yosef, O., and Vandermeersch, B. (1988). Thermoluminescence dating of Mousterian "Proto-Cro-

Magnon'' remains from Israel and the origin of modern man. *Nature*, **331**, 614-16.

Vallois, H.V. (1960). The social life of early man: the evidence of the skeletons. In *Social life of early man* (ed. S.L. Washburn), pp. 214-35. Viking Fund Publications, New York.

Vančata, V. (1986). Comment on Skelton *et al.*, Phylogenetic analysis of early hominids. *Current Anthropology*, **27**, 37-8.

Vandebroek, G. (1969). *Evolution des Vertébrés de leur origine à l'homme*. Masson & Cie, Paris.

Van Hooff, J.A.R.A.M. (1967). The facial displays of the Catarrhine monkeys and apes. In *Primate ethology* (ed. D. Morris), pp. 7-68. Aldine, Chicago, IL.

Van Valen, L. (1973). A new evolutionary law. *Evolutionary Theory*, **1**, 1-30.

Verheyen, W. (1962). Contribution à la craniologie comparée des Primates: les genres *Colobus* Illiger 1811 et *Cercopithecus* Linné 1758. *Annalen Koninklijk Museum voor Middenafrika, Tervuren, Ser. 8°, Sciences Zoologique*, **105**, i-ix, 1-255.

Vincent, F. (1969). Contribution à l'étude des prosimiens africains: le Galago de Demidoff, réproduction (biologie, anatomie, physiologie) et comportement. Unpublished D. Sci. Nat. thesis, University of Paris.

Vlcek, E. (1978). A new discovery of *Homo erectus* in Central Europe. *Journal of Human Evolution*, **7**, 239-51.

Vogel, C. (1966). The phylogenetical evaluation of some characters and some morphological trends in the evolution of the skull in catarrhine primates. In *Taxonomy and phylogeny of Old World primates with references to the origin of Man* (ed. B. Chiarelli), pp. 21-55. Rosenberg and Sellier, Turin.

von Hagen, H-O (1978). Zur Verwandtschaft der 'Echten Makis' (Prosimii, Gattung *Lemur*). *Zoolologische Veitrage*, **24**, 91-122.

von Koenigswald, G.H.R. (1952). *Gigantopithecus blacki* von Koenigswald, a giant fossil hominoid from the Pleistocene of South China. *Anthropological Papers of the American Museum Natural History*, **43**, 291-326.

von Koenigswald, G.H.R. (1957*a*). Remarks on Gigantopithecus and other hominoid remains from southern China. *Proceedings, Koninklijk Nederlandsche Akademie von Wetenschap*, **B60**, 153-9.

von Koenigswald, G.H.R. (1957*b*), *Hemanthropus* n.G. not *Hemianthropus*. *Proceedings Koninklijk Nederlandsche Akademie von Wetenschap*, **B60**, 416.

von Koenigswald, W. (1979). Ein Lemurenrest aus dem eozanen Olschiefer der Grube Messel bei Darmstadt. *Paläontologiche Zeitschrift*, **53**, 63-76.

Vbra, E.S. (1975). Some evidence of chronology and palaeocology of Sterkfontein, Swartkrans and Kromdraai from the fossil Bovidae. *Nature*, **254**, 301-4.

Vbra, E.S. (1980). Evolution, species and fossils: how does life evolve? *South African Journal of Science*, **76**, 61-84.

Vrba, E.S. (1981). The Kromdraai australopithecine site revisited in 1980: recent investigations and results. *Annals of the Transvaal Museum*, **33**, 17-60.

Vuillaume-Randriamanantena, M., Godfrey, L.R. and Sutherland, M.R. (1985). Revision of *Hapalemur (Prohapalemur) gallieni* (Standing 1905). *Folia Primatologica*, **45**, 89-116.

Waddington, C.H. (1957). *The strategy of the genes*. Allen and Unwin, London.

Wahrman, J., Goitein, R. and Nevo, E. (1969). Mole Rat Spalax: evolutionary significance of chromosome variation. *Science*, **164**, 82-4.

Walker, A.C. (1974). A review of the Miocene Lorisidae of East Africa. In *Prosimian biology* (ed. R.D. Martin, G.D. Doyle, and A.C. Walker), pp. 435–47. Duckworth, London.

Walker, A. (1975). Remains attributable to *Australopithecus* in the East Rudolf succession. In *Stratigraphy, evolution and palaeoecology in the Lake Rudolf Basin* (eds. Y. Coppers, F.C. Howell, G.Ll. Isaac and R.E. Leakey), pp. 484–9. Chicago University Press.

Walker, A. (1981*a*). Diet and teeth. *Philosophical Transactions of the Royal Society of London, B*, **292**, 57–64.

Walker, A. (1981*b*). The Koobi Fora hominids and their bearing on the origins of the genus *Homo*. In *Homo erectus: papers in honour of Davidson Black* (ed. B.A. Sigman and J.S. Cybulski), pp. 193–215. University of Toronto Press.

Walker, A. and Andrews, P. (1973). Reconstruction of the dental arcades of *Ramapithecus wickeri*. *Nature*, **244**, 313–14.

Walker, A. and Leakey R.E.F. (1978). The hominids of East Turkana. *Scientific American*, **239**(2), 44–56.

Walker, A., Zimmermann, M.R. and Leakey, R.E.F. (1982). A possible case of hypervitaminosis A in *Homo erectus*. *Nature*, **296**, 248–50.

Walker, A., Falk, D., Smith, R. and Pickford, M. (1983). The skull of *Proconsul africanus*: reconstruction and cranial capacity. *Nature*, **305**, 525–7.

Walker, A., Leakey, R.E., Harris J.M. and Brown, F.H. (1986). 2.5-Myr *Australopithecus boisei* from west of Lake Turkana, Kenya. *Nature*, **322**, 517–22.

Walls, G. (1939). Origin of the Vertebrate Eye. *Archiv für Ophthalmologie*, **22**, 452–84.

Walter, R.C. and Aronson, J.L. (1982). Revisions of K/Ar ages for the Hadar hominid site, Ethiopia. *Nature*, **296**, 122–7.

Ward, S.C. and Pilbeam, D.R. (1983). Maxillofacial morphology of Miocene hominoids from Africa and Indo-Pakistan. In *New interpretations of ape and human ancestry* (ed. R.L. Ciochon and R.S. Corruccini), 211–38. Plenum, New York.

Ward, S.C., Kimbel, W.H. and Pilbeam, D. (1983). Subnasal alveolar marphology and the systematic position of *Sivapithecus*. *American Journal of Physical Anthropology* **161**, 157–71.

Washburn, S.L. and Lancaster, C.S. (1968). The evolution of hunting. In *Man the Hunter* (eds. R.B. Lee and I. DeVore), pp. 293–303. Aldine, Chicago, Ill.

Wasser, S.K. (1985). Current conservation status of the Mwanihana Rain Forest, Uzungwa Mountains, Sanje, Tanzania. *Primate Conservation*, **6**, 34.

Watanabe, N. and Kadar, D. (1985). *Quaternary geology of the Hominid fossil bearing formations in Java*. Geological Research and Development Centre, Special Publication, No. 4, Bandung.

Weidenreich, F. (1939). Six lectures on *Sinanthropus* and related problems. *Bulletin of the Geological Society of China*, **19**, 1–110.

Weidenreich, F. (1943). The skull of *Sinanthropus pekinensis*: a comparative study on a primitive hominid skull. *Palaeontologia Sinica* (NS 10), **27**, 1–484.

Weidenreich, F. (1951). Morphology of Solo Man. *Anthropological Papers of the American Museum natural History*, **43**, 205–90.

Weiner, J.S. (1958). The pattern of evolutionary development of the genus *Homo*. *South Africa Journal of Medical Science*, **23**, 111–20.

Weinert, H. (1938). Der erste afrikanische Affenmensch, *Africanthropus njarasensis*. *Der Biologie*, **7**, 125.

Weiss, M.L. and Goodman M. (1972). Frequency and maintenance of genetic variability in natural populations of *Macaca fascicularis*. *Journal of Human Evolution*, **1**, 41-8.

Weitzel, V. (1983). A preliminary analysis of the dental and cranial morphology of *Presbytis* and *Trachypithecus* in relation to diet. Unpublished MA thesis, Australian National University.

Weitzel, V. and Groves, C.P. (1985). The nomenclature and taxonomy of the Colobine Monkeys of Java. *International Journal of Primatology*, **6**, 399-409.

Welker, C. (1981). Zur postnatale Entwicklung und zur fruhen Mutter-Kind-Beziehung der Primaten. *Anthropologisder Anzeiger*, **39**, 261-304.

White, M.J.D. (1978). *Modes of speciation*. W.H. Freeman San Francisco. Calif.

White, M.J.D. (1979). Speciation: is it a real problem? *Scientia*, **114**, 455-68.

White, T.D. (1980). Additional fossil hominids from Laetoli, Tanzania: 1976-1979 specimens. *American Journal of Physical Anthropology*, **53**, 487-504.

White, T.D. (1984). Pliocene hominids from the Middle Awash, Ethiopia. *Courier Forschungsinstitut Senckenberg*, **69**, 57-68.

White, T.D., Johanson, D.C. and Kimbel, W.H. (1981). *Australopithecus africanus*: its phyletic position reconsidered. *South African Journal of Science*, **77**, 445-70.

White, T.D., Moore, R.V. and Suwa, G. (1984). Hadar biostratigraphy and hominid evolution. *Journal of Vertebrate Paleontology*, **4**, 575-83.

Whitten, A.J. and Whitten, J..E.J. (1982). Preliminary observations of the Mentawai macque on Siberut Island, Indonesia. *International Journal of Primatology*, **3**, 445-59.

Whybrow P.J. (1984). Geological and faunal evidence from Arabia for mammal 'migrations' between Asia and Africa during the Miocene. *Courier Forschungsinstitut Senckenberg*, **69**, 189-98.

Whybrow, P.J. and Andrews, P. (1978). Restoration of the holotype of *Proconsul nyanzae*. *Folia primatologica*, **30**, 115-25.

Williamson, P.G. (1981). Palaeontological documentation of speciation in Cenozoic molluscs from Turkana basin. *Nature*, **293**, 437-43.

Williamson, P.G. (1985). Punctuated equilibrium, morphological stasis and the palaeontological documentation of speciation: a reply to Fryer, Greenwood and Peake's critique of the Turkana Basin mollusc sequence. *Biological Journal of the Linnaean Society*, **26**, 307-24.

Willis, J.C. (1924). *Age and area*. Cambridge University Press.

Willis, J.C. (1940). *The course of evolution*. Cambridge University Press.

Wilson, A.C., Carlson, S.S. and White, T.J. (1977). Biochemical evolution. *Annual Review, of Biochemistry*, **46**, 573-639.

Wilson, C.C. and Wilson, W.L. (1976). Behavioral and morphological variation among primate populations in Sumatra. *Yearbook of Physical Anthropology*, **20**, 207-53.

Wilson, J.A. (1966). A new primate from the earliest Oligocene, West Texas: preliminary report. *Folia primatologica*, **4**, 227-48.

Wolff, R.G. (1984). New specimens of the primate *Branisella boliviana* from the Early Oligocene of Salla, Bolivia. *Journal Vertebrate Paleontology*, **4**, 570-4.

Wolin, L.R. (1974). What can the eye tell us about behaviour and evolution? Or: The aye-ayes have it, but what is it? In *Prosimian biology* (eds. R.D. Martin, G.A. Doyle, and A.C Walker), pp. 489–87. Duckworth, London.

Wolpoff, M.H. (1980*a*). Cranial remains of Middle Pleistocene European hominids. *Journal of Human Evolution*, **9**, 339–58.

Wolpoff, M.H. (1980*b*). *Paleoanthropology*. Alfred A. Knopf, New York.

Wolpoff, M.H. and Brace, C.L. (1975). Allometry and early hominids. *Science*, **189**, 61–3.

Wolpoff, M.H., Wu Xinzhi, and Thorne, A.G. (1984). Modern *Homo sapiens* origins: a general theory of hominid evolution involving the fossil evidence from East Asia. In *Origin of modern humans* (ed. F. Smith and F. Spencer), pp. 411–83. Alan R. Liss, New York.

Woo Ju-kang ( = Wu Rukang) (1962). The mandibles and dentition of *Gigantopithecus*. *Palaeontologica Sinica*, NS, *D, No. 11*, **146**, 1–62. (Chinese), 65–94 (English).

Woo Ju-kang, (1964). A newly discovered mandible of the Sinanthropus type—*Sinanthropus lantianensis*. *Scientia Sinica*, **13**, 801–11.

Woo Ju-kang (1966). The hominid skull of Lantian, Shensi. *Vertebrata Palasiatica*, **10**, 1–16.

Woo Ju-kang and Pang Ru-ce (1959). Fossil human skull of early Paleoanthropic stage found at Mapa, Shaoquan, Kwantung Province. *Vertebrata Palasiatica*, **3**, 176–82.

Wood, B.A. (1978). Classification and phylogeny of East African hominids. In *Recent advances in primatology, Vol. 3* (ed. D.J. Chivers and K.A. Joysey), pp. 351–72. Academic Press, London.

Wood, B.A. (1984). The origin of *Homo erectus*. *Courier Forschungsinstitut Senckenberg*, **69**, 99–111.

Wood, B.A. (1985). Early *Homo* in Kenya, and its systematic relationships. In *Ancestors: the hard evidence* (ed. E. Delson) pp. 206–14. Alan R. Liss, New York.

Wood, B.A. and Chamberlain, A.T. (1986). *Australopithecus*: grade or clade? In *Major topics in primate and human evolution* (ed. B. Wood, L. Martin, and P. Andrews), pp. 220–48. Cambridge University Press.

Woollard, H.H. (1925). The anatomy of *Tarsius spectrum*. *Proceedings of the Zoological Society of London*, 1071–84.

Wright, S. (1968). *Evolution and the genetics of populations*. Chicago University Press.

Wu Rukang (1984). The crania of *Ramapithecus* and *Sivapithecus* from Lufeng, China. *Courier Forschungsinstitut Senckenberg*, **69**, 41–8.

Wu Rukang (1987). A revision of the classification of the Lufeng Great Apes. *Acta Anthropologica Sinica*, **6**, 265–71.

Wu Rukang and Oxnard, C.E. (1983). *Ramapithecus and Sivapithecus* from China: some implications for higher primate evolution. *American Journal of Primatology*, **5**, 303–44.

Wu Rukang and Pan Yuerong, (1984). A late Miocene gibbon-like primate from Lufeng, Yunnan Province. *Acta Anthropologica Sinica*, **3**, 185–94.

Wu Xinzhi (1981). A well-preserved cranium of an archaic type of early *Homo sapiens* from Dali, China. *Scientia Sinica*, **24**, 530–9.

Xu Qingha (1984). Climate during the Hexian man's time. *Acta Anthropologica Sinica*, **3**, 383–91.

Xu Qingha and Lu Qingwu (1979). The mandible of *Ramapithecus* and *Sivapithecus* from Lufeng, Yunnan. *Vertebrata Palasiatica*, **17**, 12–13.

Yablokov, A.V. (1966). *Variability of mammals*. Amerind, New Delhi.

Yunis, J.J. and Prakash, O. (1982). The origin of man: a chromosomal pictorial legacy. *Science*, **215**, 1525–30.

Zihlman, A.L. and Hunter, W.S. (1972). A biomechanical interpretation of the pelvis of Australopithecus. *Folia primatologica*, **18**, 1–19.

Zihlman, A.L. and Cramer, D.L (1977). Skeletal differences between Pygmy (*Pan paniscus*) and common Chimpanzees (*Pan troglodytes*). *Folia Primatologica*, **29**, 86–94.

Zingeser, M.R. (1973). Dentition of *Brachyteles arachnoides* with reference to alouattine and ateline affinities. *Folia primatologica*, **20**, 351–90.

# Author index

# Subject index